Springer Series in Reliability Engineering

Series editor

Hoang Pham, Piscataway, USA

More information about this series at http://www.springer.com/series/6917

Ajit Kumar Verma · Srividya Ajit
Hari Prasad Muruva

Risk Management of Non-Renewable Energy Systems

Springer

Ajit Kumar Verma
Stord/Haugesund University College
Haugesund
Norway

Srividya Ajit
Stord/Haugesund University College
Haugesund
Norway

Hari Prasad Muruva
Reactor Safety Division
BARC
Mumbai
India

ISSN 1614-7839 ISSN 2196-999X (electronic)
Springer Series in Reliability Engineering
ISBN 978-3-319-16061-0 ISBN 978-3-319-16062-7 (eBook)
DOI 10.1007/978-3-319-16062-7

Library of Congress Control Number: 2015937739

Springer Cham Heidelberg New York Dordrecht London

Printed on acid-free paper

Springer International Publishing AG Switzerland is part of Springer Science+Business Media (www.springer.com)

To Gurus
Sri. Shirdi Sai Baba, Bhagwan Sri Sathya Sai Baba,
Paramhamsa Swami Sathyananda Saraswati and
Swami Gambhirananda and Mata Amrithananda

To Parents
Sri K.P. Verma and Smt. (Late) Sushila Verma
Sri (Late) B.C. Khanapuri and Smt. Vibhavati Khanapuri
Sri. M. Venkateswarlu and Smt. M. Uma Kakshayani
and
Smt. M. Sirisha and Ms. M. Tiana (wife and d/of Hari Prasad Muruva)
Sri. Pradeep K. Verma and his family (b/of Ajit Kumar Verma)
Mr. A. Rune Johansen MD, SIMSEA AS, Haugesund.

Foreword

I am pleased to provide the foreword for this book on *Risk Management of Nonrenewable Energy Systems*. This text illustrates the need for managing the risk of various energy systems by presenting the basic concepts of risk and reliability and subsequent theoretical developments for modeling complex engineering passive safety systems. The chapters are self-explanatory with many demonstrative examples and case studies.

Designing complex engineering systems for achieving high safety levels is a challenging task in most industries. This is particularly true in the case of nuclear power plants. Risk analysis plays a major role in this context as it is essential for both plant safety and for public acceptance. In evaluating the safety of a plant it is important to consider events that are both internal and external to the plant. An important consideration in the treatment of external events is the possibility of disruption of external sources of electricity, cooling water, and other essential supplies including possible prompt operator action following an extreme external event. In such a situation, active systems will not function due to the failure of external power. There is therefore a need to develop systems that can function under this scenario. This situation can be achieved or mitigated with the help of passive systems.

In recent years most advanced nuclear reactors have implemented passive systems, aimed at improved safety and availability, in order to reduce human error and active component malfunctions. Passive systems play an important role in innovative reactor safety systems and can result in substantial economic benefits and simplicity in operation over active systems. The need for a book that addresses these issues with sufficient rigor without compromising the underlining goal of succinct presentation and simplicity has long been felt, and this book aims to achieve this fine balance.

The book starts with an introduction to different sources of energy and explains the need for plant risk analysis. It provides a review of basic probability and statistics essential for understanding risk and reliability concepts. Details on the different levels of risk analysis in nuclear power plants considering both internal and external events are explained using case studies. A separate chapter is devoted to reliability modeling of passive systems illustrating different modeling techniques with their limitations and emphasizing the need for developing new techniques. The book also emphasizes time-dependent reliability analysis and risk management of plants with details on the different tools available for managing risk and its role in decision making. The book is written in a style that should prove attractive to a wide range of researchers and practicing engineers and become a valuable reference and handbook.

This book provides a readable discourse on the fundamentals of plant risk management and the practical application of these concepts.

Roy Billinton
Distinguished Professor Emeritus
University of Saskatchewan

Preface

Risk, which deals with the consequences of failures along with its likelihood, plays an important role in sectors such as chemical and process plants, power plants, nuclear facilities, and aerospace which can impose potential hazards. Risk analysis is essential for a plant from the safety point of view as well as from the public acceptance point of view. In this book the basic concepts required for carrying out the risk analysis of plants and the methods/techniques/tools required for carrying out the same is provided with case studies. Hence, the book is organized into the following chapters.

Chapter 1 introduces different sources of energy and their limitations, power plants, evolution of the nuclear industry, hazards, history of major accidents, and the need for risk analysis of plants.

Chapter 2 provides a detailed review of probability and statistics essential for understanding the risk and reliability analysis concepts and methods that are discussed in the remaining chapters.

Chapter 3 discusses the risk analysis of nuclear power plants, risk analysis methodology, level 1, level 2, level 3 probabilistic safety assessment (PSA), event tree analysis, fault tree methods, treatment of common cause failures, different failure probability models, and parameter estimation using Bayesian methods.

Seismic probabilistic assessment of nuclear power plants is discussed in detail in Chap. 4. This chapter provides various aspects of seismic PSA, probabilistic seismic hazard analysis, fragility analysis of structures, components and systems, accident sequence progression during seismic events, and contribution of seismic events to risk of a plant.

Currently, most advanced reactors use passive systems to improve safety in the plant and to eliminate human interactions, especially during abnormal situations in plants. The basic concepts of passive systems are discussed in Chap. 5. This chapter discusses the need for passive systems reliability, limitations of conventional reliability estimation methods, different methods that exist for reliability estimation of passive systems, and need for mechanistic modeling approach.

Chapter 6 provides details of time-dependent reliability analysis, types of loading, degradation mechanisms that cause reduction of strength of material, techniques available for implementing time-dependent reliability, extreme value theory, stochastic fatigue loading, the concept of out-crossing approach and its limitations, principles of stress corrosion cracking, and its effect on reliability of components with time.

Risk management in nuclear and thermal power plants is discussed in Chap. 7, which includes different tools available for risk management and how it helps in decision making. The basics of response surface methodology, fuzzy set theory, simulation techniques, and stochastic process theory are discussed in the Appendix.

This book is useful for advanced undergraduate and postgraduate students in nuclear engineering, aerospace engineering, power engineering, industrial engineering, reliability and safety engineering, systems engineering, applied probability and statistics, operations research, and earthquake engineering. This book is also suitable for one-semester graduate courses on *Risk Management of Nonrenewable Energy Systems* in all conventional engineering branches like civil, mechanical, chemical, electrical and electronics, as well as computer science. It will also be a valuable reference for practicing engineers, managers, and researchers involved in reliability and safety activities of complex engineering systems.

March 2015

Ajit Kumar Verma
Srividya Ajit
Hari Prasad Muruva

Acknowledgments

We express our sincere thanks to Mr. R.K. Saraf, Dr. G. Rami Reddy, Mr. H.S. Kushwaha, Dr. A.K. Ghosh, Dr. V.V.S. Sanyasi Rao, Mr. K.K. Vaze, Dr. R.K. Singh, Dr. V. Gopika, Mr. Santosh, and Mr. A.J. Gaikwad of Bhabha Atomic Research Centre, Trombay, Mumbai for their valuable suggestions and continuous encouragement throughout our work.

Special thanks to Mr. M. Jaya Prakash, Mrs. M. Lalitha, and Mr. M. Niranjan Kumar for their encouragement and moral support. We are grateful to Prof. Roy Billinton for the 'Foreword' for this book and to Prof. Hoang Pham for his encouragement and support. We would like to thank Profs. Uday Kumar, Rajesh Karki, Navaraj Karki, and Dr. Adithya Thaduri for their moral support. Finally, our thanks are due to our Norwegian friends Ms. Liv Reidun Grimstvedt, Rector, HSH, Mr. A. Rune Johansen, MD, SIMSEA AS, Profs. Torgrim Log, Monika Log, Sanjay Khattri, Tore Markeset, Jayantha P. Liyanage, Per Skjerpe, Jan Frick, Mohsen Assadi, D. Reggie, Øyvin Kyvik, Brit Julbø, Bjarne Vandeskog, Vidar Frette, Chunyan Xie, Runald Meyer, Svein Nesheim, and Indian friends in Norway, namely Dr. Maneesh Singh, Dr. Rupesh Kumar, Dr. Saurabh Kumar, Dr. Rajesh Kumar, Mr. Devender Kumar, Mr. Dhruv Tapasvi, Ms. Shabana Sayed, and their wonderful families.

Mumbai
March 2015

Ajit Kumar Verma
Srividya Ajit
Hari Prasad Muruva

Contents

Chapter 1
Introduction

1.1 General

Imagine a scenario when one day you are traveling on an electric train and suddenly the train stops; after some time the entire train becomes gloomy, you want to call your friend but the cell phone does not function. You think to yourself what's happening? Why does everything stop ***working***? If one analyzes the above scenario the train may stop suddenly due to power failure. There may be backup in the train in the form of batteries to take care of internal lighting and annunciation, but these too get exhausted after some time. If your cell phone is discharged it cannot be turned on. This type of situation may arise in the near future if we fail to save ***energy***.

In the above discussion there are two terms highlighted, i.e., energy and work. In physics energy is defined as the ability to do work and work is defined as the distance through which an object has moved as a result of applied force. These two terms are important in our everyday life. According to the law of conservation of energy, energy can be neither created nor destroyed, but it can be transformed from one form to another. Hence, it is important to know the different sources and forms of energy.

1.2 Sources of Energy

Energy sources can be classified into two groups, namely renewable and non-renewable energy sources [1]. These are explained briefly in the following sections.

A.K. Verma et al., *Risk Management of Non-Renewable Energy Systems*,
Springer Series in Reliability Engineering, DOI 10.1007/978-3-319-16062-7_1

1.2.1 Renewable Energy

Renewable energy is the energy that can be naturally replenished or renewed. It comes from natural resources like sunlight, wind, water, tides, geothermal, biomass; all these sources can be naturally replenished.

Solar energy is the energy created by sunlight [2, 3]. By using solar cells sunlight can be converted into electricity. With the help of solar thermal panels or parabolic mirrors sunlight can be used to heat water/air or produce steam as shown in Fig. 1.1. The solar panels can be placed in the regions of highest solar radiation.

Wind energy [4, 5] is produced when the kinetic energy of the wind converts into mechanical work by the rotation of turbine blades. These turbines cause the rotation of the magnets that generate electricity. A typical wind turbine is shown in Fig. 1.2 which is used to generate electricity.

Hydro energy is produced through rain water which is harvested in a dam or a river [6]. In this case the high pressure water from a certain elevation causes turbine blades to rotate, which in turn rotates the magnets of a generator to produce electricity as shown in Fig. 1.3.

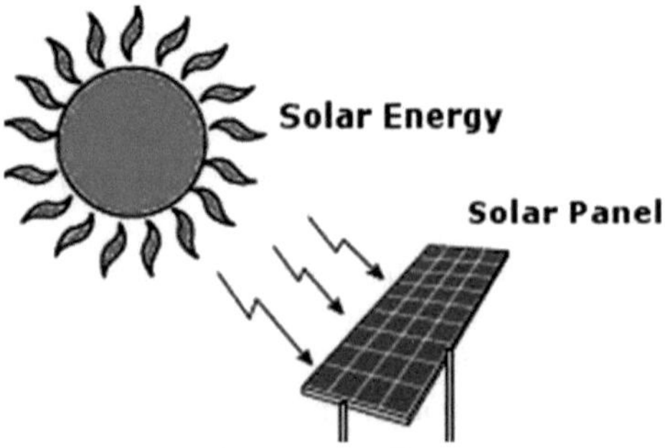

Fig. 1.1 Utilization of solar energy

Fig. 1.2 Wind turbine used for generating electricity

Fig. 1.3 Hydro energy

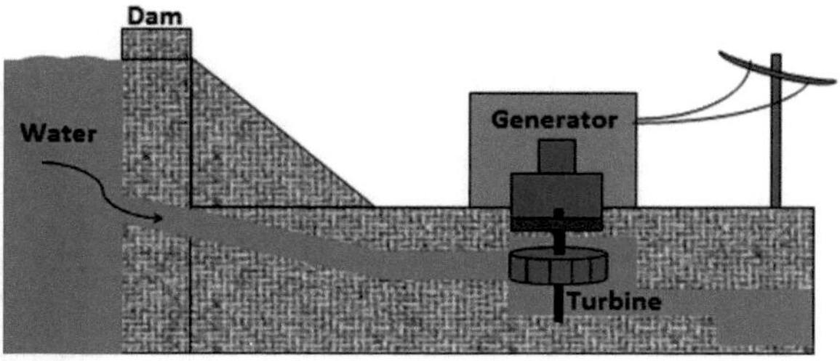

Tidal energy converts the energy of tides into electricity and is also a form of hydro energy [7, 8]. At this point one needs to know the difference between tidal power and wave power. In case of wave power energy is produced due to the surface waves of the ocean. There is always rise and fall of sea level in a day due to the gravitational forces exerted on the Earth by the Sun and Moon and due to the rotation of the Earth. This rise and fall of sea level forms the tides. The rise is called high tide and the fall is called low tide. There are two high tides and two low tides every day. Tidal dams are built near sea shores for generation of electricity. Water flows into the dam during high tide and flows out of during low tide, which results in turning the turbine as shown in Fig. 1.4. However, waves are caused by winds mixing with the surface layer of the ocean. It is not possible to predict the wave patterns and their strength.

Geothermal energy is the energy in the form of heat generated in the Earth's core [9]. The earth is divided into different layers as shown in Fig. 1.5. The earth's core is made up of solid iron surrounding which very hot magma (molten rock) forms the outer core. Slow decay of radioactive particles produce very high temperatures inside the Earth. The layer surrounding the outer core is called the mantle and is made up of magma and rock. The outermost layer of the Earth is called the crust that forms the continents and ocean floors. The heat generated from the mantle is absorbed by the rocks and water in the deep underground. This heated water from the underground can be pumped to the earth surface to produce electricity. This geothermal energy is also a kind of renewable energy source because deep inside the earth heat is produced continuously and water is replenished by rainfall [10].

Biomass energy is produced by burning biomass [11–13]. Biomass is biological material encompassing all living matter in an area. Wood, crops, crop waste, wastes

Fig. 1.4 Tidal energy

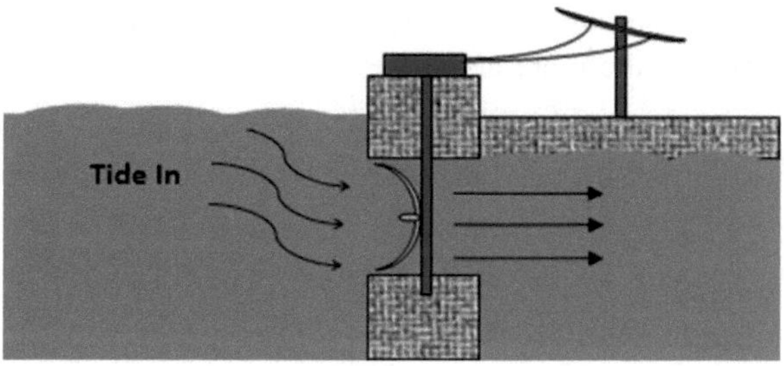

Fig. 1.5 Geothermal energy

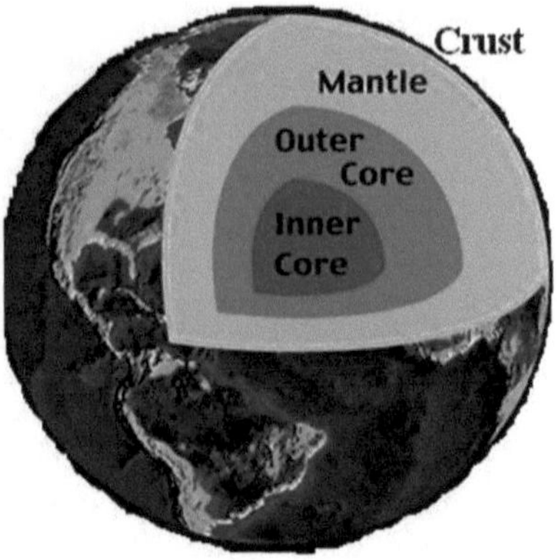

Fig. 1.6 Biomass energy

of plant, mineral, and animal matter comprise much of an area's biomass (Fig. 1.6). Much of this material is considered "garbage" and ends up in landfills. Rubbish buried in the ground naturally yields methane gas during anaerobic decomposition (the decaying of organic matter in the absence of oxygen). As an energy source, biomass can either be used directly, or converted into other energy products such as biofuel. In direct usage, by direct combustion of biomass (plant matter) electricity can be generated with steam turbines. In the indirect case, biomass (plant or animal matter) can be converted into fibers or other industrial chemicals such as biofuels that can be later used as fuel for vehicles. Biomass is renewable, as the production of waste products continues forever and, also, one can always plant and grow more trees, so these are also renewable.

1.2.2 Nonrenewable Energy

The term itself implies that these resources for generating energy cannot be renewed, which means a **nonrenewable energy source** is a natural resource that cannot be generated, grown, or reproduced. Once the resources are consumed it will no more be available for future needs and cannot be sustained in long term. These resources are also considered to be consumed much faster than they are created by nature [14]. Fossil fuels (such as coal, petroleum, and natural gas) and nuclear power (uranium) are examples.

Fossil Fuel Energy: Fossil fuels are formed from the buried remains of plants and animals that lived millions of years ago and are found within the rocks of the Earth's surface. Natural resources such as coal, petroleum (crude oil), and natural gas [15, 16] take thousands of years to form naturally and cannot be replaced as fast as they are being consumed. Eventually, natural resources will become too costly to harvest and humanity will need to find other sources of energy. At present, the main energy sources used by humans are nonrenewable fossil fuels; as a result of continual use since the first internal combustion engine in the seventeenth century, this fuel is still in high demand with conventional infrastructure and transport which are fitted with the combustion engine. The continual use of fossil fuels at the current rate will increase global warming and cause more severe climate changes. Electricity generation through fossil fuels is further explained in the section on fossil fuel power plants.

Nuclear Energy: Nuclear energy is the energy that is stored in the nucleus of an atom. If the stored energy is released it can be utilized to produce electricity. Energy from the atoms can be released by means of either nuclear fission or nuclear fusion. In nuclear fission energy is released when an atom is split into smaller atoms, whereas in the case of nuclear fusion energy is produced when two atoms are fused together. Nuclear fission is used to produce electricity in nuclear power plants. In nuclear power plants the concept of nuclear fission is used to generate heat, later this heat is used to produce steam from water for electricity generation. Electricity generation through nuclear fission is further explained in the section on nuclear power plants.

1.3 Fossil Fuel Power Plants

Power plants in which electricity is generated through burning of coal, fuel oil, or natural gas are called fossil fuel power plants. In general, fossil fuel power plants consist of a steam generator (furnace and steam drum), a turbine, and a generator. In coal-based power plants coal is used as basic fuel to generate heat energy. A typical coal-based power plant is shown in Fig. 1.7.

The initial process in power generation is coal handling. First, the raw coal is transferred from the unloading site to the storage yard by means of belt conveyors.

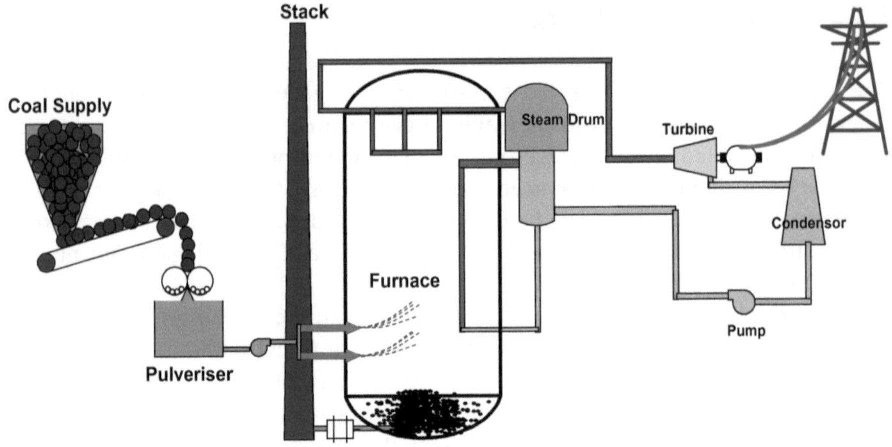

Fig. 1.7 A typical coal-based thermal power plant

Next, this coal is elevated to different levels and supplied to a crusher house where the raw coal is crushed into small pieces suitable for milling. Later, with the help of coal feeders the crushed coal is transported to the pulverizers. The pulverizers are rotating drum grinders or ball mills or any other type of grinders. With the help of pulverisers the crushed coal is made into very fine powder. The pulverized coal is blown into the furnace with the help of hot and cold air mixture through burners located on the walls of the furnace. As soon as the coal enters the furnace it is ignited rapidly which will form a large fireball at the center of the furnace. There is a hopper at the bottom of the furnace to collect the ash that is generated during the ignition process and is always filled with water to quench the ash.

The thermal radiation of the fireball will heat the water that is circulated through the boiler tubes running along the furnace walls. The water as it circulates in the water tubes will absorb heat and convert to steam which is separated from water in a steam drum at the top of the furnace. The saturated steam is fed through steam tubes hanging in the hottest part of the furnace at the exit of the combustion gases from the furnace. The saturated steam is further superheated to a certain temperature before entering the turbine. When steam passes through the turbine it will start rotating, which in turn rotates the electrical generators to produce power. Steam from the exhaust of the turbine gets condensed in the condensers and the condensed water is pumped back to the steam drum through feed water system.

1.4 Nuclear Power Plants

A nuclear power plant (NPP) consists of a reactor core in which heat is generated by controlled nuclear fission [17, 18]. A coolant is circulated with the help of pumps through the reactor core which eventually removes the heat from the core and the heated coolant is used to produce steam in steam generators (SG). The steam that is produced in the SG rotates the turbines when it passes through, which in turn rotates the electrical generators to produce power. Since nuclear fission creates radioactivity, the reactor core is surrounded by a protective shield. This shield absorbs radiation and prevents radioactive material from being released into the environment. In addition, the reactors are further surrounded by a primary and secondary containment structures to protect the radio activity release into the atmosphere in case of an accident and also to protect the reactor against external impacts. A general schematic of a typical NPP is shown in Fig. 1.8.

There are different types of nuclear reactors in different parts of the world. Based on their designs they are categorized under different generations such as generation I, II, III, III$^+$, and IV (Fig. 1.9) [19] and are briefly explained as follows.

Generation-I Reactors

These reactors were built in the early stages of development of nuclear power reactors. They were built mainly to demonstrate the technology as a proof of concept of utilizing nuclear energy for peaceful purposes such as production of electricity. Reactors built during the 1950s and 1960s fall under this category. They were mainly prototype reactors demonstrating the technology. Some of the examples include

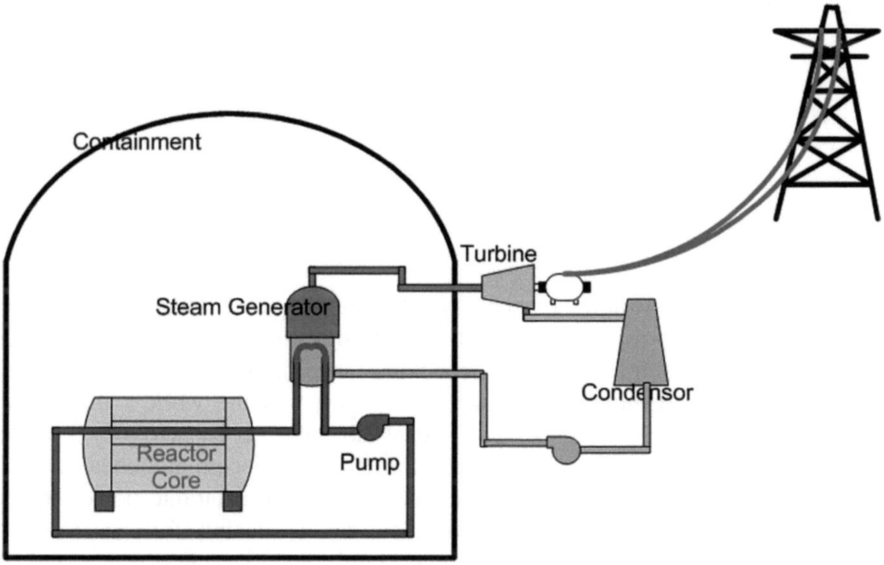

Fig. 1.8 A typical nuclear power plant

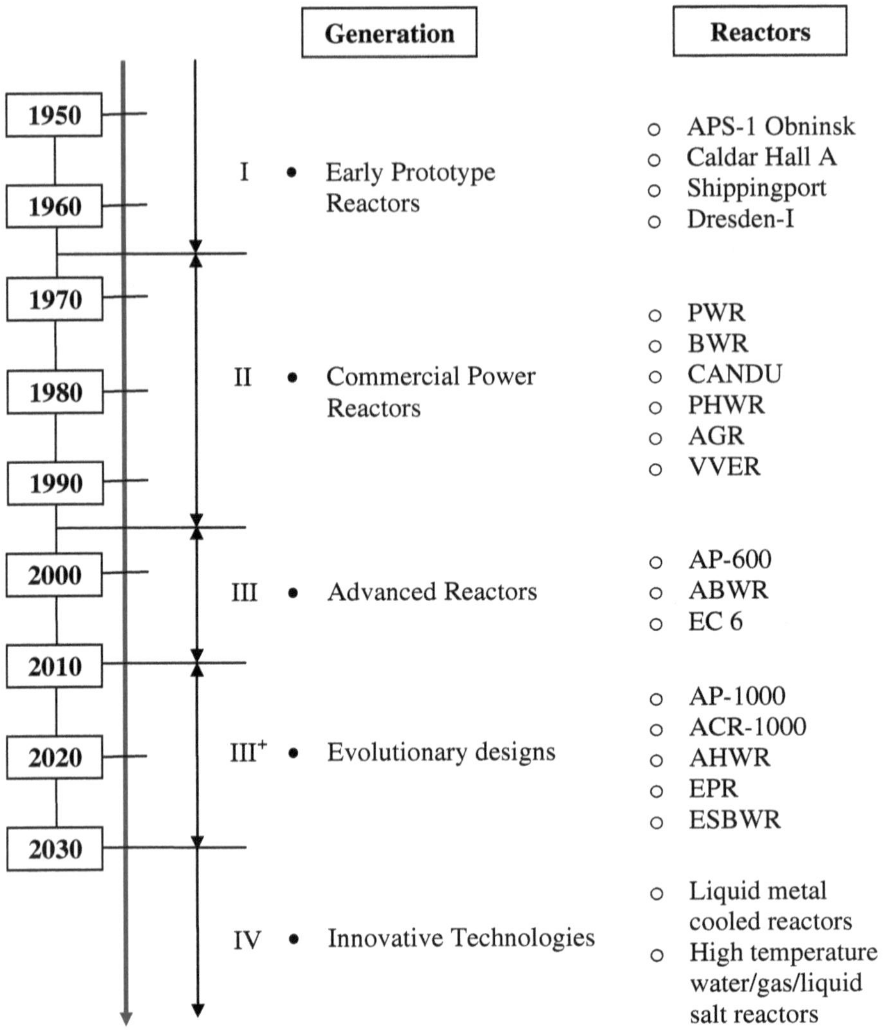

Fig. 1.9 Different generations of nuclear reactors

- APS-1 Obninsk (Atomic Power Station 1 Obninsk, Soviet Union) was the first civilian nuclear power station in the world with a net capacity of 5 MWe; it started its operation in 1954 and decommissioned in 2002.
- Caldar Hall A (Cumberland, United Kingdom) was the world's first power station to generate electricity on an industrial scale (net capacity of 60 MWe) from nuclear energy. The reactor started in 1956 and decommissioned in 2003.
- Shippingport (Pennsylvania, United States of America) atomic power station was a full scale atomic electric power plant devoted mainly for peaceful uses. The reactor started in 1957 and operated until 1982. The reactor was built to

serve as a prototype for commercial electrical power generation. After several years of operation in 1977 it was converted to a pressurized light water breeder reactor (PLWBR).
- Dresden-I (Illinois, United States) NPP operated between 1960 and 1978.

Generation II Reactors
The reactors under this category are of commercial type and are designed to be economical. They are designed with better safety features with an operational lifetime of 40 years. Typical Gen II reactors include

- Pressurized Water Reactors (PWR)
- CANada Deuterium Uranium Reactors (CANDU)
- Boiling Water Reactors (BWR)
- Pressurised Heavy Water Reactors (PHWR)
- Advanced Gas cooled Reactors (AGR)
- Voddo-Codyanoi Energetichesky Reactors (VVER) (Water Water Power Reactor).

These reactors began operation in the late 1960s and mostly use traditional active safety features with few passive features. Most of the world's commercial power plants are of PWR and BWR type.
Generation III Reactors
Gen III reactors are of Gen II type with evolutionary design improvements in the areas of fuel technology, safety systems (mostly implementing passive features), thermal efficiency, etc. These reactors are aimed at longer operational life, typically 60 years of plant operation due to their better reactor technology. Some of the Gen III reactors are listed below.

- AP-600 was one of the first Gen III reactors designed by Westinghouse. It is 600 MWe advanced PWR.
- Advanced Boiling Water Reactor (ABWR).
- Enhanced CANDU 6 (EC6) reactor is 700 MWe heavy-water moderated and heavy-water cooled pressure tube reactor which is the evolution of the proven CANDU 6 design.

Today there are only four Gen III reactors of ABWR type in operation.
Generation III$^+$ Reactors
These reactors are evolutionary designs of Gen III reactors with improved designs and safety. Some of the Gen III$^+$ reactors are

- AP1000 reactor (improvement over AP600)
- Advanced CANDU reactor (ACR-1000)
- Advanced Heavy Water Reactor (AHWR)
- European Pressurized Reactor (EPR)
- Economic Simplified Boiling Water Reactor (ESBWR) based on ABWR
- VVER-1200/392M reactor of AES-2006 type.

Generation IV Reactors

Gen IV reactors are innovative technologies focusing on other alternatives of nuclear power, but still in research mode. Some of the reactors in this category are

- Liquid metal cooled reactors
- High temperature water/gas/liquid salt-based pebble bed thermal reactors.

1.5 Definition of Risk

The term risk is defined as likelihood (frequency or probability) of occurrence of any undesirable event that will have negative effect on human lives and the consequences from such an event [20, 21]. This can be explained with a simple example; consider two instances, in the first instance a person is walking on a busy road where vehicle movement is high and in the second instance the same person is walking on a road where there is no vehicle movement. In the first case it is probable that the person can be hit by a vehicle (motor cycle, car etc.) and he can be injured or he may lose his life. Whereas in the second instance since there is no vehicle movement the person getting hit by a vehicle is ruled out and the same is with injury or fatality. Hence, the person walking on a busy road is on high risk compared to the latter case.

Here, the risk is the person may be injured or lose his life if he is hit by a vehicle. There are two possibilities: first, the person is hit by a vehicle (accident), the other is having an accident the state of the person, whether he is injured or has lost his life. In statistical terms the risk of a person walking on a busy road can be defined as the likelihood that the person is hit by a vehicle and the consequences in terms of injury or fatality. Consequences can be of different types, for example injuries, fatalities, radiation dose absorbed, fatalities due to radiation, damage to property, etc. Now we examine the risk due to natural hazards as well as man-made hazards. These are discussed in the following sections.

1.6 Risk from Various Hazards

Hazard can be defined as any source of potential damage. These sources can be either natural or man-made. A natural event that causes harm to humans is called a natural hazard. Similarly, if the source of hazard is man-made and it causes threat to humans it is called a man-made hazard. Natural Hazards are the result of naturally occurring processes that have been operating since the formation of the Earth. These are mostly geological processes. Following is a list of natural hazards that are noticeable:

- Earthquakes
- Volcanic eruptions

- Tsunami
- Landslides
- Floods
- Droughts
- Hurricanes
- Tornadoes
- Asteroid impacts.

If these processes do not cause damage to humans, then they are simply natural events. However, if they affect humans negatively, then they are termed as natural hazards. Now coming to man-made hazards, these are generated by humans and affect them negatively. Some of the man-made hazards that are noteworthy are as follows:

- Nuclear
- Biological
- Chemical
- Fire
- Travel (road, rail, air, water)
- Epidemic
- Terrorism.

Now what is the risk expected from all these hazards? Consider the risk that is most important is the number of fatalities per year due to some hazard. Table 1.1 gives the comparison of risk from different hazards [22].

Example 1.1 Calculate the risk due to road accidents for the following data: number of accidental deaths occurred in a year 190,000, total population is 100×10^7.

Solution:

$$\begin{aligned} \text{Total population} &= 100 \times 10^7 \\ \text{Societal Risk} &= 190{,}000\text{accidental-deaths}/\text{year} \\ \text{Average Individual Risk} &= (190{,}000 \text{ Deaths}/\text{Year})/100 \times 10^7 \\ &= 1.9\text{E-04 Deaths}/\text{Person-Year} \\ &= 1/5300\text{Deaths}/\text{Person-Year} \end{aligned}$$

Table 1.1 Comparison of death rate of different hazards

S.no	Event	Fatalities/yr
1	Cancer deaths	7.5e-4
2	Road accidents	1.9e-4
3	Earthquakes	2.0e-4
4	Floods	4.0 e-5
5	Fire	5.1e-5

The meaning of the above result is in any given year is that approximately 1 out of every 5,300 people in the entire population will suffer an accidental death.

1.7 History of Accidents

In this section, an overview of major accidents that occurred in the past is discussed. These accidents are compared from the consequence point of view. Here, consequences are either in terms of fatalities or monetary loss.

1.7.1 Three Mile Island Accident

This accident occurred on March 28, 1979 in Pennsylvania, United States [18]. The accident took place in Unit 2 of the two Three Mile Island nuclear reactors. The reactor was a pressurized water reactor. The accident was a partial core meltdown resulting in release of small amounts of radioactivity into the atmosphere. On the International Nuclear Event Scale (INES) this event was classified as level 5 out of 7. It was the worst accident in the history of U.S. commercial nuclear power plants.

The accident started with the failure of primary feed water pumps which feed the water to the SG in the secondary side. Since the SGs were no longer receiving the feed water, emergency shutdown was performed. This actuated the control rods to insert into the core and the chain reaction was stopped. However, after the shutdown some amount of decay heat stays in the core that needs to be removed. Since the SG secondary side does not receive the feed water the decay heat was no longer being removed from the core. Due to the failure of primary feed water pumps auxiliary feed water pumps started automatically. However, the valves in the auxiliary loop were closed for routine maintenance (violation of technical specification), and in effect feed water could not be pumped.

Since the decay heat had no longer been removed, primary system pressure was raised and caused pressure relief valve to open which was attached to the pressurizer. After relieving the pressure the valve was stuck open due to mechanical fault and did not close back. Due to this the coolant in the primary loop escaped from the system and it was a typical case of loss of coolant accident. The operators could not realize the problem and they even turned down emergency core cooling system to actuate. Finally, this led to a partial core meltdown. The prime cause of the accident was a violation of technical specifications and operator errors.

1.7.2 Chernobyl Accident

This accident occurred on April 26, 1986 in Ukraine, Soviet Union of Russia [18]. The accident took place in reactor 4 of the Chernobyl plant. The reactor was a boiling water pressure tube type reactor (RBMK). The accident was a core disruption resulting in the release of large amounts of radioactivity into the atmosphere. This event was classified as level 7 (maximum classification) on INES levels. It has been considered the worst nuclear accident in history. The accident was a result of a power excursion during turbine test functioning in the case of a power grid failure.

In the Chernobyl reactor, diesel generators (DG) were provided as a backup system in case of power grid failure, which operate the coolant pumps to take care of decay heat removal from the core. However, DGs start operating after 15 s of power failure and it takes 60–75 s to attain full speed. During this time it is important to run the coolant pumps to remove the decay heat from the core. In order to fill the 1 min gap it was decided to run these pumps with turbine power. The test was unsuccessfully carried out during previous attempts. The test was planned to carry out before going for reactor shutdown for maintenance activities. The test had to be conducted at a lower power level not less than 700–800 MW thermal power.

During the experiment the operators initially could bring down the power to a stated level. However, due to formation of poisoning in the core the reactor power started to decrease continuously without the operation of control rods. In order to maintain power the operators withdrew the control rods. This led to instability in the core and called for emergency shutdown of the reactor. As the control rods entered the core a power spike had occurred, the core got heated up, causing some of the fuel rods to fracture, the control rods got jammed and could not shutdown the reactor. Due to high steam generation in the core the reactor lid was lifted from the top of the core due to high steam pressure. This was when the first explosion occurred. After a few seconds second explosion took place, which compromised the reactor containment vessel and dispersed the core. As the superheated graphite moderator came into contact with the air it caught fire. By this time the entire core was exposed to the atmosphere and large amounts of radioactivity was released into the environment.

It was reported that 50 % of the area of Ukraine was in some way contaminated. Over 200,000 people had to be evacuated and resettled. The total costs including cleanup, resettlement, and compensation to victims was estimated to be roughly $200 billion [17].

1.7.3 Fukushima Accident

The accident occurred on March 11, 2011 in Fukushima, Japan. The accident took place in reactors 1, 2, and 3 of 6 reactors of Fukushima Daiichi nuclear power plants. The reactors were boiling water reactors. The accident was a core meltdown resulting in release of radioactivity into the atmosphere. This event was classified as level 7 on INES levels. The accident was a result of station blackout caused by seismic and tsunami.

A 9.0 magnitude earthquake under sea occurred in the northwestern Pacific Ocean at a depth of 32 km with its epicenter approximately 72 km from Oshika Peninsula of Tohoku of Japan on March 11, 2011. This was considered to be the most powerful earthquake ever to hit Japan. The earthquake triggered powerful tsunami waves that created wave heights of up to 14 m in Fukushima area and travelled up to 10 km inland in Sendai area.

There were 6 reactors in Fukushima Daiichi area. Among these, reactor 4 was defueled and reactors 5 and 6 were under cold shutdown for maintenance. A 9.0 magnitude earthquake struck the plants, which resulted in automatic shutting down of the other 3 reactors and ceased the chain reaction. The earthquake resulted in failure of power grid and emergency DG had started supplying power to the emergency cooling pumps which circulate coolant for removing the decay heat from the core. After 50 min of operation tsunami waves of height 14 m had arrived and flooded the low level areas in the plant and caused DGs to fail, which resulted in halting emergency core cooling.

As decay heat was not removed from the core the water levels in the core dropped below the top of the hot fuel and steam began reacting with the fuel cladding producing large amounts of hydrogen. As the pressure in the primary containment increased to higher levels it was vented to the atmosphere. After an hour of venting a huge hydrogen explosion occurred in the upper area of the reactor building. As a result of failure of decay heat from the core the entire fuel had melted in Unit 1 and fallen to the bottom of the reactor pressure vessel. However, in case of Unit 2 and 3 it was only a partial core melt and in Unit 4 there was only a hydrogen explosion in the reactor building.

If one observes the above accidents the causes behind these accidents can be due to any one of the following errors such as design errors, manufacturing errors, operating errors and human errors. If one can eliminate these types of errors the accidents can be avoided. But, there are certain situations under which one cannot eliminate the accidents such as naturally occurring disasters namely, earthquakes, floods, storm etc. All the systems or structures cannot be designed for all the environmental conditions or loadings. For example a very strong earthquake can damage structures which are not designed for that level of earthquake. However, precautions can be taken to avoid severe consequences. The analysis which deals with the likelihood occurrences of undesirable events and the consequences due to that event is called risk analysis. This is further explained in the following sections.

1.8 Need for Risk Assessment

As discussed earlier the term risk has two components, first, likelihood of occurrence of an undesirable event and, second, the consequences from that event. Risk from a plant depends on both these components. The likelihood of an event can be higher or lower, but in general the plant design should be such that the events having higher consequences should have lower frequency of occurrence. Similarly, events with higher frequency of occurrence should have lower consequences. With the help of risk analysis one can identify the different undesirable events in the plant and the consequences coming from those events when they occur. Appropriate measures can be taken to reduce the frequency of occurrence of those events or reducing the consequences. This is graphically represented in frequency–consequences (F–C) curve as shown in Fig. 1.10.

Risk has a different role from the public acceptance point of view. Whenever one wants to establish a power plant there is always concern about the safety and risk of the plant. Irrespective of the usage of the plant there will always be agitation from the public with respect to the plant risk. In such situations one is always interested in the quantification of the risk instead of qualitative explanation. It should be proven that the risk from a given plant is much lesser than the risk due to natural hazards such as lightning, earthquakes, flooding, etc. Hence, even for the public acceptance of a plant, risk analysis is very much needed. With respect to system safety, risk analysis will help in improving the design of the system and can identify the weak links in the system, etc.

From the above discussion it is clear that risk assessment of a plant is very important. The next question that arises is how to assess the risk of a plant. In general there are different steps in risk assessment as described below [21, 23, 24].

- Qualitative risk assessment
  - Hazard identification
  - Event identification
  - Identification of mitigating systems
  - Event progression

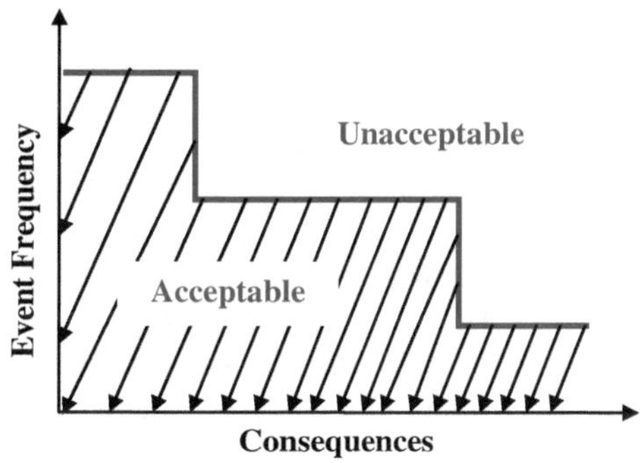

Fig. 1.10 Frequency–consequence curve

- Quantitative risk assessment
  - Likelihood estimation of events
  - Failure probability estimation of mitigating systems
  - Consequence analysis
  - Risk estimation.

As described above, risk assessment can be broadly categorized into qualitative risk assessment and quantitative risk assessment. In qualitative risk assessment only the qualitative aspects of risk are discussed, whereas in quantitative risk assessment risk is quantified by combining both likelihood of occurrence of events and the consequences of these events. There are different techniques and tools available for carrying out the risk assessment of a plant. These are described in detail in the subsequent chapters.

1.9 Organization of the Book

From the above it is obvious that risk analysis is essential from safety as well as public acceptance points of view. In this book the basic concepts required for carrying out risk analysis of plants and the methods/techniques/tools required for carrying out the same is provided with case studies. The book is organized into the following chapters.

This chapter introduces different sources of energy and their limitations, power plants, evolution of nuclear industry, different hazards, history of major accidents, and need for risk analysis of plants.

Chapter 2 provides a detailed review of probability and statistics essential for understanding the risk and reliability analysis concepts and methods that are discussed in the remaining chapters.

Chapter 3 discusses the risk analysis of nuclear power plants, risk analysis methodology, level 1, level 2, level 3 probabilistic safety assessment (PSA), event tree analysis, fault tree methods, treatment of common cause failures, different failure probability models, and parameter estimation using Bayesian methods.

Seismic probabilistic assessment of nuclear power plants is discussed in detail in Chap. 4. This chapter provides different aspects of seismic PSA, probabilistic seismic hazard analysis, fragility analysis of structures, components and systems, accident sequence progression during seismic events, and contribution of seismic events to risk of a plant.

Nowadays most of the advanced reactors use passive systems to improve the safety of the plant and to eliminate human interactions, especially during abnormal situations in the plants. The basic concepts of passive systems are discussed in Chap. 5. This chapter discusses the need for passive systems reliability, limitations of conventional reliability estimation methods, different methods existing for reliability estimation of passive systems, and need for mechanistic modeling approach.

Chapter 6 provides details of time-dependent reliability analysis, different types of loading, different degradation mechanisms which cause reduction of strength of a material, different techniques available for implementing time-dependent reliability, extreme value theory, stochastic fatigue loading, the concept of out-crossing approach and its limitations, principles of stress corrosion cracking, and its effect on reliability of components with time.

Risk management of nuclear and thermal power plants are discussed in Chap. 7, which includes different tools available for risk management and how it helps in decision making. The basics of response surface methodology, fuzzy set theory, simulation techniques, and stochastic process theory are discussed in the Appendix.

References

1. Shepard ML, Chaddock JB, Cocks FH, Harman CM (1976) Introduction to energy technology. Ann Arbor Science Publishers Inc, Ann Arbor
2. Garg HP, Prakash J (2000) Solar energy: fundamentals and applications, 1st revised edn. Tata-McGraw Hill, New Delhi
3. Oswald WJ, Golueke CG (1960) Biological transformation of solar energy. Advan Appl Microbiol 2:223–262
4. Erich H (2000) Wind turbines: fundamentals, technologies, application and economics, 1st edn. Springer, New York
5. Gipe P (2009) Wind energy basics, 2nd edn. Chelsea Green Publishing Company, USA
6. Wagner H-J, Mathur J (2011) Introduction to hydro energy systems. 1st edn. Springer, Berlin
7. Dalton FK (1961) Tidal electric power generation. J Roy Astron Soc Can 55(1):22–33; 55 (2):57–72
8. Charlier RH, Finkl CW (2009) Ocean energy: tide and tidal power, 1st edn. Springer, Berlin
9. Glassley WE (2010) Geothermal energy: renewable energy and environment, Kindle edn. CRC Press, Taylor & Francis Group, London, Boca Raton
10. http://lgdata.s3-website-us-east-1.amazonaws.com/docs/1242/187273/GeothermalI.pdf
11. Aiba S, Humphrey AE, Millis NF (1973) Biochemical engineering, 2nd edn. Academic Press, New York
12. Klass DL (1998) Biomass for renewable energy, fuels and chemicals, 1st edn. Academic Press, USA
13. Hollaender AK, Monty J, Pearlstein RM, Schmidt-Bleek F, Snyder WT, Volkin E (1972) An inquiry into biological energy conversion. University of Tennessee, Knoxville
14. http://www.authorstream.com/Presentation/markevans-1485126-non-renewable-resources/
15. Brame JSS, King JG (1955) Fuel—solid, liquid and gaseous, 5th edn. Edward Arnold (Publishers) Ltd, London
16. Gupta OP (1997) Elements of fuels, furnaces and refractories. Khanna Publishers, India
17. Glasstone S, Sesonske A (1998) Nuclear reactor engineering: reactor design basics, vol 1, 4th edn. CBS Publishers, New Delhi
18. Glasstone S, Sesonske A (1998) Nuclear reactor engineering: reactor systems engineering, vol 2, 4th edn. CBS Publishers, New Delhi
19. Goldberg SM, Rosner R (2011) Nuclear reactors: generation to generation. American Academy of Arts and Sciences, Cambridge
20. Modarres M (1985) Statistical uncertainty analysis in reactor risk estimation. Nucl Eng Des 85:385–399

21. Vose D (2000) Risk analysis-a quantitative guide. Wiley, New York
22. http://www.worldlifeexpectancy.com/world-health-rankings
23. IAEA Safety Series No. 50-P-4 (1992) Procedure for conducting probabilistic safety assessment of nuclear power plants (level 1). International Atomic Energy Agency, Vienna
24. Kaplan S, Garrick BJ (1981) On the quantitative definition of risk. Risk Anal 1:11–37

Chapter 2
Basics of Reliability

2.1 Introduction

In today's competitive world reliability has become a major concern in almost all the industries. Reputation of a company is very closely associated with the reliability of its products. The more reliable a product is, the more likely the company is to have a favorable reputation. The reliability of a product also affects the customer satisfaction. An unreliable product will affect customer satisfaction severely. When it comes to the customer satisfaction quality of the product also matters. At this point one should understand the difference between quality and reliability. Quality of a product is the degree of conformance to applicable specifications and it is dependent on the manufacturing process. Whereas reliability is maintaining the quality of a product over some period of time.

Nowadays, it has become very common to mention the warranty periods for the products by the companies. This has got both advantage and disadvantage. For an example, a person goes to a shop for purchasing a mobile phone. Assume that there are three companies offering the same product for the same price but with different warranty periods. It is obvious that the person may choose the product for which warranty period is more. Some people may go by the company's reputation. Whichever the company has got more reputation people may choose those products irrespective of their cost and warranty periods. If the person is going by the warranty period, then whichever the company specifies more warranty period it will capture the market. This may be the advantage, on the contrary if the product quality is poorer and the product fails to perform its function within the warranty period, the replacement and repair costs will negatively affect company's profits. Hence, it is very much important to maintain the quality of a product for longer times.

A.K. Verma et al., *Risk Management of Non-Renewable Energy Systems*,
Springer Series in Reliability Engineering, DOI 10.1007/978-3-319-16062-7_2

In these discussions one should note that the time factor is mentioned only for certain duration. Because, no component in this world last for infinity time. Even the human beings have their time limitation on this earth. When the human beings have got the time limitation, the products produced by them will certainly have the time limitation. In certain cases it may be much longer but not infinity time. With this discussion one can define the reliability as "the probability that a component or unit performs its intended function for a given period of time under specified environmental conditions." This definition is governed by four factors namely:

1. Probability
2. Intended function
3. Time duration
4. Environmental conditions.

One can always ask a question that why reliability is a probabilistic concept? Reliability is associated with probability because one cannot tell when a component is going to fail. Consider a case of centrifugal pump that is put into the operation, if the pump is of high quality and reliable it can function for longer period, even then one cannot judge when this component will cease from functioning. This can fail immediately after put into the operation or may fail within an hour or in a day or in a year, etc. Here the time to failure of a component is not certain, hence, reliability should be dealt with probability concepts. In the above definition the second term is given as intended function because every component is designed for performing certain task.

Coming to the time duration, as discussed earlier no component will last for infinity time or forever. Hence, there is always some mission time associated with each and every component. This mission time can vary from component to component and also from industry to industry. Consider the case of satellites and their launching vehicles. The mission time of the components used in the space launching vehicles will be until it keeps the satellites in their respective orbital. However, satellite mission time may vary from one launching to another depending on their purpose.

One more factor that is important is environmental conditions, because not all the components will perform under all the environmental conditions. Each and every component has got its own limitations. As an example some components will be designed for performing at very low temperatures and some components may be designed for higher temperatures. Consider a simple case of thermometer which is used for measuring the temperature of a human body. This can well measure the temperature of human body in its specified limits. If the same thermometer is used for measuring the temperature of hot water it will certainly fail because it is not designed for those temperatures, in certain cases that bulb of the thermometer will rupture. Here the question comes can it be possible to design a component working under all environmental conditions. It may be possible in certain cases but it will cost more from manufacturing side.

Finally, is there any component that exists in this world that can deliver its intended function under all environmental conditions for infinity time or forever?

The answer is 'No.' If at all any component exists that is manufactured and has not failed so far will certainly fail at some point of time in the future but that time may be in some cases very longer. The specific causes of failures of components and equipments in a system can be many. Some are known and others are unknown due to the complexity of the system and its environment. A few of them are listed below.

(a) Poor design (component or system),
(b) Wrong manufacturing techniques,
(c) Lack of total knowledge and experience,
(d) Complexity of equipment,
(e) Poor maintenance policies, and
(f) Human errors, etc.

As was discussed earlier the basis for reliability is probability theory, hence, it is very much important to learn the concepts of probability and is explained in the following sections.

2.2 Probability Theory

Probability theory is used where it is not possible to predict the exact outcome of any experiment. The following terms are generally used in the probability theory. These terms are just defined for the sake of completeness of the theory, but for detailed study the reader is requested to refer some statistics books [1–3].

2.2.1 Random Experiment

An experiment that can result in to different outcomes, even though it is repeated in the same manner every time, is called a **random experiment**.

2.2.2 Sample Space

The set of all possible outcomes of a random experiment is called the **sample space** of the experiment. The sample space is denoted as S.

- A sample space is **discrete** if it consists of a finite or countable infinite set of outcomes.
- A sample space is **continuous** if it contains an interval (either finite or infinite) of real numbers.

2.2.3 Event

An **event** is a subset of the sample space of a random experiment.

- The **union** of two events is the event that consists of all outcomes that are contained in either of the two events. Consider the two events as E_1 and E_2 then one can denote the union as $E1 \cup E2$.
- The **intersection** of two events is the event that consists of all outcomes that are contained in both of the two events. One can denote the intersection as $E1 \cap E2$.
- The **complement** of an event in a sample space is the set of outcomes in the sample space that are not in the event. The complement of the event E can be represented as E^1.
- Two events, denoted as $E1$ and $E2$, such that $E1 \cap E2 = \phi$ are said to be **mutually exclusive**. These terms are illustrated in Fig. 2.1.

2.2.4 Probability

Probability is used to quantify the likelihood, or chance, that an outcome of a random experiment will occur.

- Whenever a sample space consists of N possible outcomes that are equally likely, the probability of each outcome is $1/N$.
- For a discrete sample space, the *probability of an event E*, denoted as $P(E)$, equals the sum of the probabilities of the outcomes in E.

Example 2.1 A random experiment can result in one of the outcomes $\{a, b, c, d\}$ with probabilities 0.1, 0.3, 0.5, and 0.1, respectively. Let A denote the event $\{a, b\}$, B the event $\{b, c, d\}$, and C the event $\{d\}$. Then find $P(A)$, $P(B)$, and $P(C)$.

Solution:

$$P(A) = 0.1 + 0.3 = 0.4$$
$$P(B) = 0.3 + 0.5 + 0.1 = 0.9$$
$$P(C) = 0.1$$

2.2.5 Axioms of Probability

Probability is a number that is assigned to each member of a collection of events from a random experiment that satisfies the following properties:

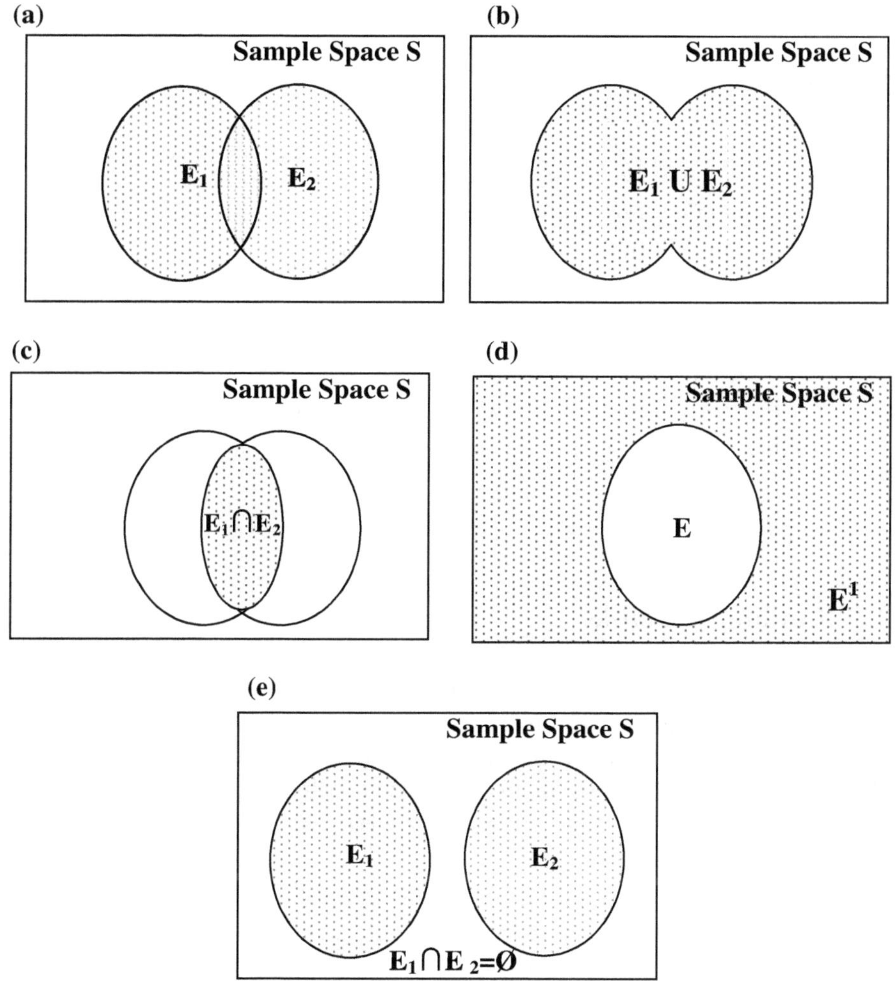

Fig. 2.1 Illustration of **a** sample space and events. **b** Union of two events. **c** Intersection of two events. **d** Complement of an event. **e** Mutually exclusive events

If S is the sample space and E is any event in a random experiment,

(1) $P(S) = 1$

(2) $0 \leq P(E) \leq 1$

(3) For two events $E1$and $E2$ with $E1 \cap E2 = \Phi[\mathrm{P}(\Phi) = 0]$

$$P(E1 \text{ U } E2) = P(E1) + P(E2) \tag{2.1}$$

Example 2.2 Consider Example 2.1. Find *P(S)*, $P(A\cap C)$, $P(A\cup C)$. Where *S* is the sample space.

Solution:

$$
\begin{aligned}
S &= \{a, b, c, d\} \\
P(S) &= P(a) + P(b) + P(c) + P(d) \\
P(S) &= 0.1 + 0.3 + 0.5 + 0.1 = 1.0 \\
A &= \{a, b\} \Rightarrow P(A) = P(a) + P(b) = 0.1 + 0.3 = 0.4 \\
C &= \{d\} \quad \Rightarrow P(C) = P(d) = 0.1 \\
A \cap C &= \Phi \\
P(A \cap C) &= 0 \\
P(A \cup C) &= P(A) + P(C) - P(A \cap C) = P(A) + P(C) \\
P(A \cup C) &= 0.4 + 0.1 = 0.5
\end{aligned}
$$

2.3 Random Variable

A variable that associates a number with the outcome of a random experiment is referred to as a **random variable**. A random variable is a function that assigns a real number to each outcome in the sample space of a random experiment. A random variable is denoted by an uppercase letter such as *X*. After an experiment is conducted, the measured value of the random variable is denoted by a lowercase letter such as x = 60 MPa.

- A **discrete** random variable is a random variable with a finite (or countably infinite) range. Examples of **discrete** random variables: number of scratches on a surface, proportion of defective parts among 1000 tested. Consider a random variable *N* which represents number of cracks on a pipe. This can be treated as a discrete random variable whose sample space can take finite range such as 1, 2, 3, 4, etc. This is shown in Fig. 2.2.
- A **continuous** random variable is a random variable with an interval (either finite or infinite) of real numbers for its range. Examples of **continuous** random variables: Strength, length, pressure, temperature, time, voltage, weight. Consider a random variable *S*, which represents tensile strength of a material. This can be treated as a continuous random variable and can take any value on a real number range (e.g., 300–500 MPa). This is graphically represented as shown in Fig. 2.3.

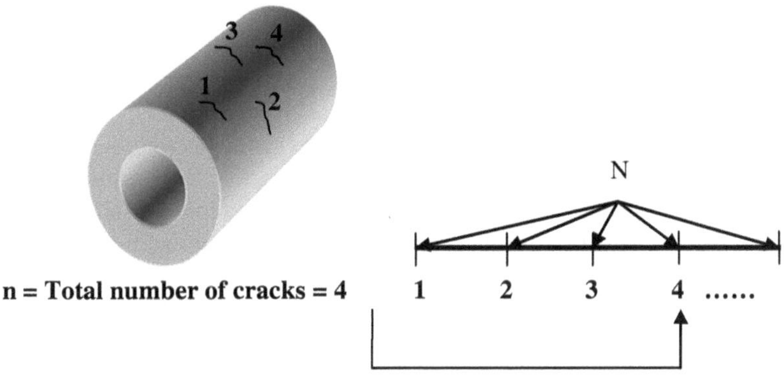

Fig. 2.2 Graphical representation of a discrete random variable

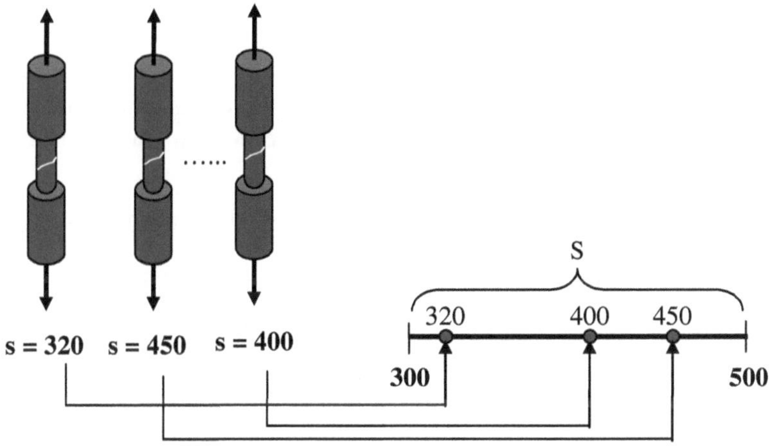

Fig. 2.3 Graphical representation of a continuous random variable

2.3.1 *Discrete Random Variable*

2.3.1.1 Probability Distribution

The **probability distribution** of a random variable *X* is a description of the probabilities associated with the possible values of *X*. For a discrete random variable, the distribution is often specified by just a list of the possible values along with the probability of each. In some cases, it is convenient to express the probability in terms of a formula.

2.3.1.2 Probability Mass Function

Consider a cantilever beam as shown in Fig. 2.4. The beam is subjected to the external loading in which the load is placed at discrete points. The load on the remaining points where load is not mentioned is zero (neglecting the self weight of the beam). Now the loading can be described by a function that specifies the mass at each of the discrete points. Similarly, for a discrete random variable *X*, its distribution can be described by a function that specifies the probability at each of the possible discrete values for *X*.

For a discrete random variable *X* with possible values $x_1, x_2, \ldots x_n$, a **probability mass function** is a function such that

$$\begin{aligned} &(1) \quad f(x_i) \geq 0 \\ &(2) \quad \sum_{i=1}^{n} f(x_i) = 1 \\ &(3) \quad f(x_i) = P(X = x_i) \end{aligned} \qquad (2.2)$$

Here $P(X = x)$ represents the probability that the discrete random variable *X* takes the value *x*. Probability mass function of a discrete random variable is shown in Fig. 2.5.

Example 2.3 *X* is a discrete random variable having the probabilities as given in Table 2.1 for different values of *x*, calculate (a) $P(X > -2)$ (b) $P(-1 \leq X \leq 1)$ (c) $P(X \leq -1 \text{ or } X = 2)$. Plot the probability mass function (Fig. 2.6).

Solution:

(a)	$P(X > -2)$	$= P(X = -1) + P(X = 0) + P(X = 1) + P(X = 2)$	7/8
(b)	$P(-1 \leq X \leq 1)$	$= P(X = -1) + P(X = 0) + P(X = 1)$	3/4
(c)	$P(X \leq -1 \text{ or } X = 2)$	$= P(X = -2) + P(X = -1) + P(X = 2)$	1/2

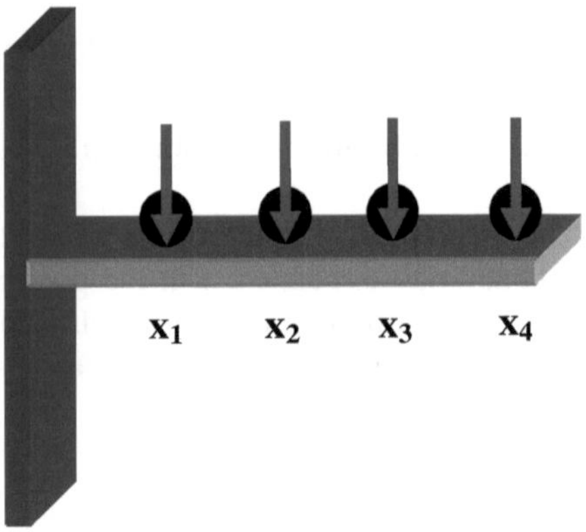

Fig. 2.4 Loadings at discrete points on a long thin beam

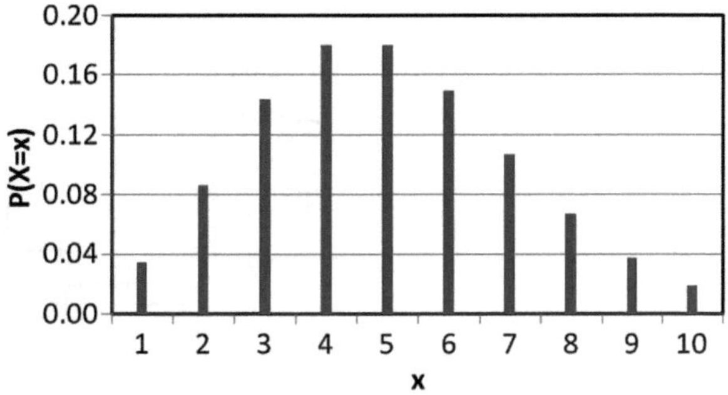

Fig. 2.5 Probability mass function of a discrete random variable

Table 2.1 Probability data of random variable X

x	−2	−1	0	1	2
$f(x)$	1/8	2/8	2/8	2/8	1/8

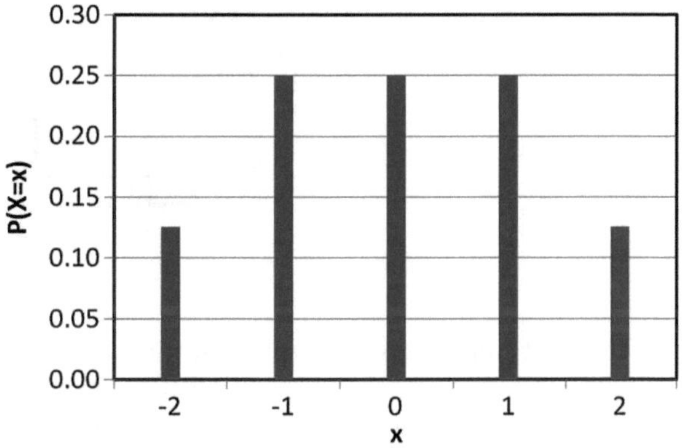

Fig. 2.6 Probability mass function of x

2.3.1.3 Cumulative Distribution Function

The cumulative distributive function (CDF) of variable X, denoted as $F(x)$, can be given as

$$F(x) = P(X \leq x) = \sum_{x_i \leq x} f(x_i) \tag{2.3}$$

For a discrete random variable X, $F(x)$ satisfies the following properties.

$$\begin{aligned} &(1) \quad F(x) = P(X \leq x) = \sum_{x_i \leq x} f(x_i) \\ &(2) \quad 0 \leq F(x) \leq 1 \\ &(3) \quad \text{if}\, x \leq y, \ \text{then}\, F(x) \leq F(y) \end{aligned} \tag{2.4}$$

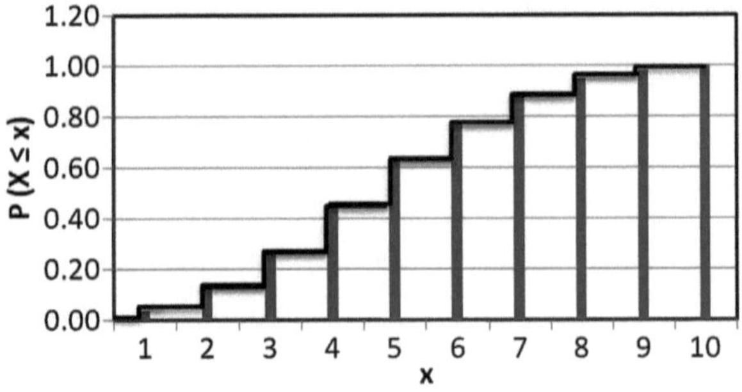

Fig. 2.7 Cumulative distribution function of a discrete random variable

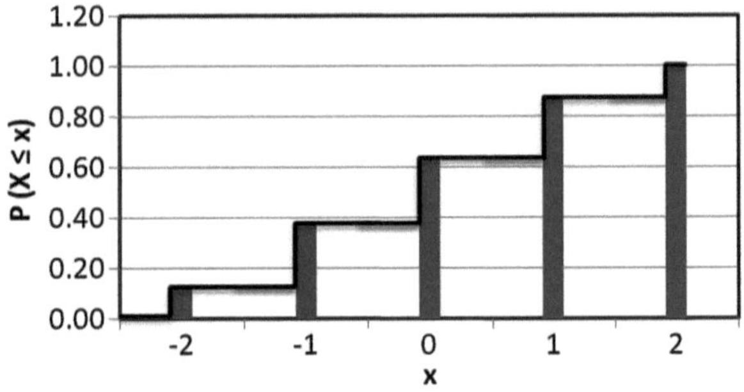

Fig. 2.8 Cumulative distribution function of *x*

The CDF represents the probability that the random variable assumes a value less than or equal to any specified value. Probability distribution function of a discrete random variable is shown in Fig. 2.7.

Example 2.4 Consider Example 2.3, plot cumulative distribution function (Fig. 2.8) and calculate

$$\begin{array}{llll} \text{(a)} & P(X \leq -1) & \text{(b)} & P(X \leq 0) \\ \text{(c)} & P(X \leq 1) & \text{(d)} & P(X \leq 2) \end{array}$$

Solution:

(a)	$P(X \leq -1) = P(X = -2) + P(X = -1)$	3/8
(b)	$P(X \leq 0) = P(X = -2) + P(X = -1) + P(X = 0)$	5/8
(c)	$P(X \leq 1) = P(X = -2) + P(X = -1) + P(X = 0) + P(X = 1)$	7/8
(d)	$P(X \leq 2) = P(X = -2) + P(X = -1) + P(X = 0) + P(X = 1) + P(X = 2)$	1

2.3.1.4 Mean

The **mean** or **expected value** of the discrete random variable X, denoted as μ or $E(X)$, can be written as

$$\mu = E(X) = \sum_x xf(x) \tag{2.5}$$

The mean value generally represents the average of all the possible outcome of an experiment.

2.3.1.5 Variance

The variance of X, denoted as σ^2 or $V(X)$, is given as

$$\begin{aligned} \sigma^2 &= V(X) = E(X-\mu)^2 \\ \sigma^2 &= \sum_x (x-\mu)^2 f(x) = \sum_x x^2 f(x) - \mu^2 \end{aligned} \tag{2.6}$$

The standard deviation of X is represented as σ which is equal to the square root of the variance. The variance or standard deviation represents the spread in the data. If the standard deviation is zero means there is no spread in the data, that is irrespective of number of times one conducts the experiment all the time one observes a single outcome. Thus, in this case it is not necessary to apply probability theory, the outcome is deterministic in nature.

Example 2.5 Consider Example 2.3, calculate mean and variance for the given data as shown in Table 2.2

The mean and variance of the given data are obtained as 0 and 12/8, respectively.

Example 2.6 The probability mass function of a discrete random variable X is given as follows, find the following values.

Table 2.2 Calculation of mean and variance

x	$f(x)$	$xf(x)$	$(x-\mu)^2$	$(x-\mu)^2 f(x)$
−2	1/8	−2/8	4	4/8
−1	2/8	−2/8	1	2/8
0	2/8	0	0	0
1	2/8	2/8	1	2/8
2	1/8	2/8	4	4/8
$\mu = \Sigma xf(x) = 0$			$\sigma^2 = \Sigma(x-\mu)^2 f(x) = 12/8$	

$$
\begin{array}{ll}
f(x) = \frac{2x+1}{22} & x = 0, 1, 3, 5 \\
\text{(i)} \quad P(X = 3) & \text{(ii)} \quad P(X \leq 1) \\
\text{(iii)} \quad P(1 \leq X < 4) & \text{(iv)} \quad P(X > -5) \\
\text{(v)} \quad \text{Mean}(\mu) & \text{(vi)} \quad \text{Variance}(\sigma^2)
\end{array}
$$

Solution:

$$
P(X = 0) = \frac{2 \times 0 + 1}{22} = \frac{1}{22} \quad P(X = 1) = \frac{2 \times 1 + 1}{22} = \frac{3}{22}
$$

$$
P(X = 3) = \frac{2 \times 3 + 1}{22} = \frac{7}{22} \quad P(X = 5) = \frac{2 \times 5 + 1}{22} = \frac{11}{22}
$$

$$
\begin{array}{ll}
\text{(i)} \quad P(X = 3) & = \frac{7}{22} \\
\text{(ii)} \quad P(X \leq 1) & = P(X = 0) + P(X = 1) = \frac{1}{22} + \frac{3}{22} = \frac{4}{22} \\
\text{(iii)} \quad P(1 \leq X < 4) & = P(X = 1) + P(X = 3) = \frac{3}{22} + \frac{7}{22} = \frac{10}{22} \\
\text{(iv)} \quad P(X > -5) & = P(X = 0) + P(X = 1) + P(X = 3) + P(X = 5) \\
& = \frac{1}{22} + \frac{3}{22} + \frac{7}{22} + \frac{11}{22} = \frac{22}{22} = 1 \\
\text{(v)} \quad \text{Mean}(\mu) & = \sum xf(x) \\
& = 0f(0) + 1f(1) + 3f(3) + 5f(5) \\
& = 0P(X = 0) + 1P(X = 1) + 3P(X = 3) + 5P(X = 5) \\
\mu & = 0 \times \frac{1}{22} + 1 \times \frac{3}{22} + 3 \times \frac{7}{22} + 5 \times \frac{11}{22} = \frac{79}{22} = 3.59 \\
\text{(vi)} \quad \text{Variance}(\sigma^2) & = \sum (x - \mu)^2 f(x) \\
\sigma^2 & = (0 - 3.59)^2 f(0) + (1 - 3.59)^2 f(1) + (3 - 3.59)^2 f(3) + (5 - 3.59)^2 f(5) \\
\sigma^2 & = 12.89 \times \frac{1}{22} + 6.71 \times \frac{3}{22} + 0.35 \times \frac{7}{22} + 1.99 \times \frac{11}{22} = \frac{57.36}{22} = 2.61
\end{array}
$$

Some generally used discrete distributions and their corresponding parameters are briefly explained in the following subsections:

2.3.1.6 Uniform Distribution

- Each outcome of an experiment has equal probability of occurrence.
- Consider a discrete random variable X follows uniform distribution. Then each of the outcome in its range, say, x_1, x_2, …, x_n, has equal probability of occurrence. Then the probability mass function can be written as

$$
f(x_i) = P(X = x_i) = \frac{1}{n} \tag{2.7}
$$

- Let 1, 2, 3, 4, …., 10 are the outcomes of a random experiment, then the probability of occurrence of each outcome will be

$$
f(x_i) = P(X = x_i) = \frac{1}{10} = 0.1
$$

- The probability mass function of random variable is shown in Fig. 2.9.

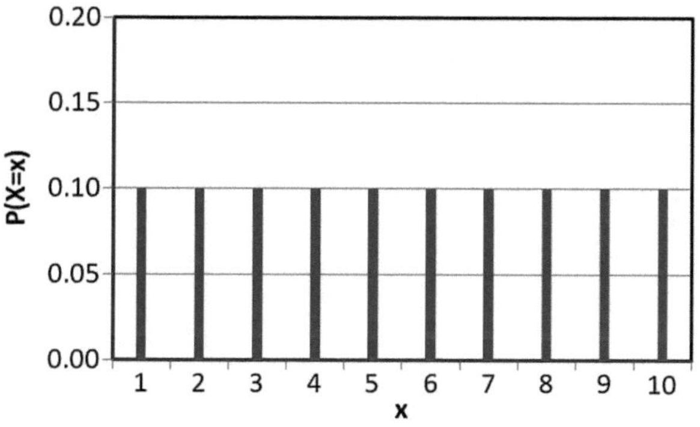

Fig. 2.9 Probability mass function of a discrete uniform random variable

- If X is a discrete uniform random variable on the consecutive integers a, $a + 1$, $a + 2$, …., b for $a \leq b$. Then

$$\begin{aligned} \text{Mean} &= \frac{b+a}{2} \\ \text{Variance} &= \frac{(b-a+1)^2-1}{12} \end{aligned} \tag{2.8}$$

2.3.1.7 Binomial Distribution

A random experiment consists of n Bernoulli trials such that

- The trials are considered to be independent
- There are only two outcomes of each trial either "success" or "failure"
- The probability of a success in each trial, denoted as p, is considered as constant
- If a random variable X follows binomial distribution then probability mass function of X can be given as

$$f(x) = \binom{n}{x} p^x (1-p)^{n-x} \qquad x = 0, 1, \ldots, n \tag{2.9}$$

- where n is the total number of trials, x is the number of success, p is the probability of success, $(1 - p)$ is the probability of failure. For the case of $n = 10$, $p = 0.7$ the probability mass function is shown in Fig. 2.10.
- The mean and variance of binomial distribution can be given as follows:

$$\begin{aligned} \text{Mean} &= np \\ \text{Variance} &= np(1-p) \end{aligned} \tag{2.10}$$

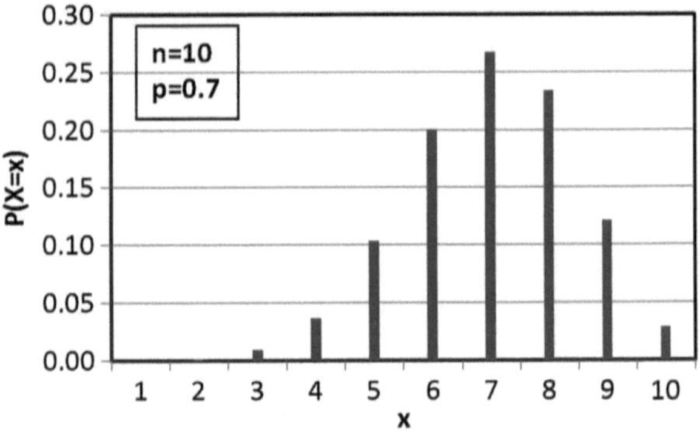

Fig. 2.10 Probability mass function of binomial random variable

2.3.1.8 Poisson Distribution

- Consider a random variable X follows Poisson distribution then probability mass function of X can be given as

$$f(x) = \frac{e^{-\lambda t}(\lambda t)^x}{x!} \qquad x = 0, 1, 2, \ldots \tag{2.11}$$

- where x is the total number of occurrences of an event in a given interval of time t. λ is the rate of occurrence of events.
- Following are the characteristics of Poisson distribution:
  - the probability of occurrence of more than one event in a small interval of time is zero,
  - The rate of occurrence of events is constant
  - The probability of occurrence of one event is independent of other event
- The mean and variance of Poisson distribution are same and are given as

$$\begin{aligned} \text{Mean} &= \lambda t \\ \text{Variance} &= \lambda t \end{aligned} \tag{2.12}$$

- The probability mass function is shown in Fig. 2.11.

2.3.2 *Continuous Random Variable*

2.3.2.1 Probability Density Function

Consider a cantilever beam as shown in Fig. 2.12. The beam is subjected to the external loading (Kg/m) in which the load is uniformly distributed along the length

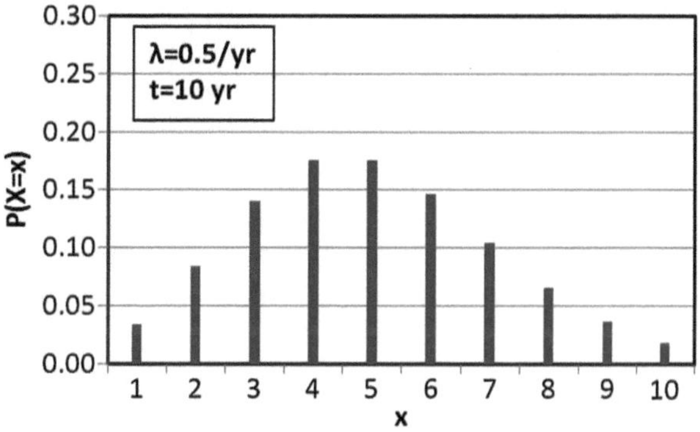

Fig. 2.11 Probability mass function of Poisson distribution

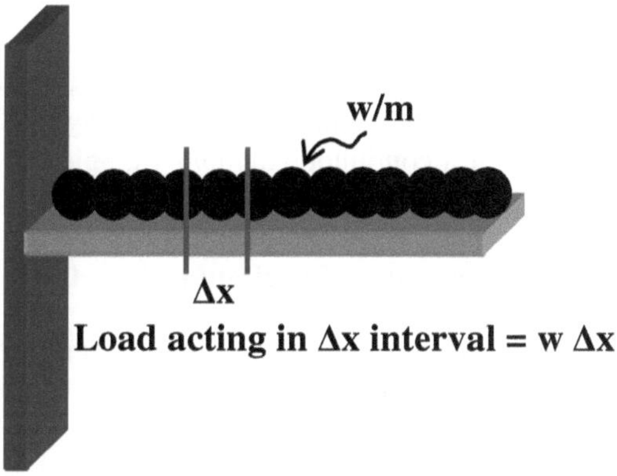

Fig. 2.12 Density function of loading on a long thin beam

of the beam. From this it is not possible to find the amount of loading that is acting at a point. But if one considers the small section of the beam (of length Δx), then one can find easily the total loading acting in that section. It is simply the summation of all the loadings in that interval. Now the loading can be described by a function that specifies the density, i.e., load per unit length of the beam. Similarly, for a continuous random variable X, its distribution can be described by a function that specifies the probability per unit interval of x, i.e., probability density function. For a continuous random variable X, a probability density function (PDF) is a function such that

$$\begin{aligned}
&(1)\quad f(x_i) \geq 0\\
&(2)\quad \int_{-\infty}^{\infty} f(x)\mathrm{d}x = 1\\
&(3)\quad P(a \leq X \leq b) = \int_a^b f(x)\,\mathrm{d}x = \text{area under } f(x) \text{ from } a \text{ to } b\\
&\qquad \text{for any } a \text{ and } b
\end{aligned} \tag{2.13}$$

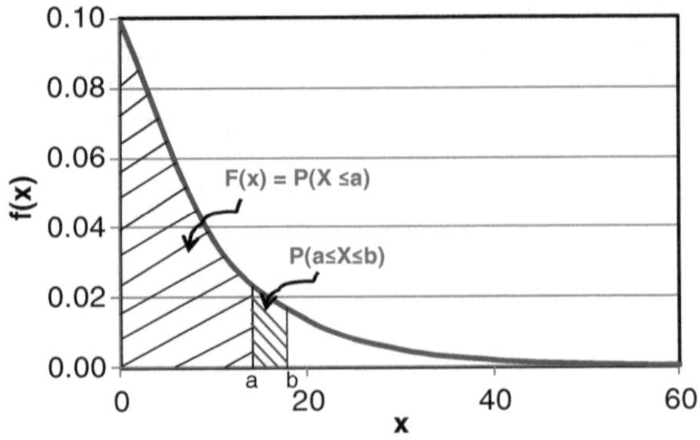

Fig. 2.13 Probability density function of a continuous random variable

A probability density function provides a simple description of the probabilities associated with a random variable. Figure 2.13 shows the probability density function of a continuous random variable.

2.3.2.2 Cumulative Distribution Function

The cumulative distributive function (CDF) of a continuous random variable X can be written as follows:

$$F(x) = P(X \leq x) = \int_{-\infty}^{x} f(x)\mathrm{d}x \quad \text{for } -\infty \leq x \leq \infty \tag{2.14}$$

As shown in Fig. 2.13 CDF is the area under the PDF curve in the interval between lower limit of the random variable to the specified x value. The probability density function of a continuous random variable can be determined from the cumulative distribution function by differentiation as follows

$$f(x) = \frac{\mathrm{d}F(x)}{\mathrm{d}x} \tag{2.15}$$

The CDF of a continuous random variable is shown in Fig. 2.14.

2.3.2.3 Mean

Suppose X is a continuous random variable with probability density function $f(x)$. The mean or expected value of X can be denoted as $E(X)$ and mathematically can be represented as

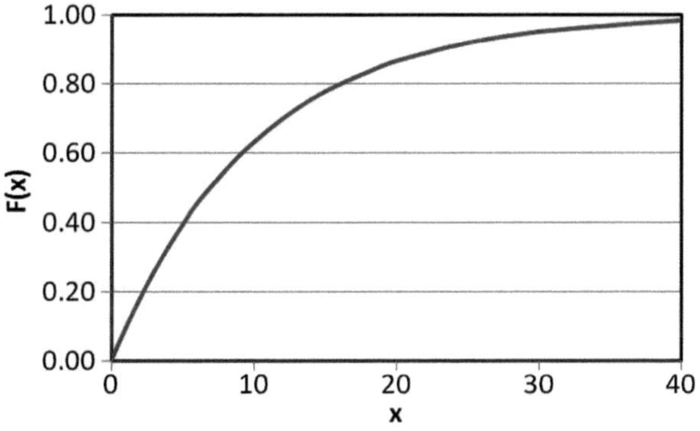

Fig. 2.14 Cumulative distribution function of a continuous random variable

$$\mu = E(X) = \int_{-\infty}^{\infty} xf(x)\mathrm{d}x \tag{2.16}$$

2.3.2.4 Variance

The variance of X can be denoted as σ^2 or $V(X)$ and is given as follows:

$$\begin{aligned}\sigma^2 &= V(X) = E(X-\mu)^2 \\ &= \int_{-\infty}^{\infty} (x-\mu)^2 f(x)\mathrm{d}x\end{aligned}$$

$$\sigma^2 = \int_{-\infty}^{\infty} x^2 f(x)\mathrm{d}x - \mu^2 \tag{2.17}$$

The standard deviation of X can be denoted as σ and is equivalent to square root of variance.

Example 2.7 Consider X as a continuous random variable whose probability density function is given as $f(x) = 1/18\, x^2$ for $-3 \le X \le 3$. Determine the following probabilities.

(a) $P(X > 0)$ (b) $P(X > 1)$ (c) $P(-2 \le X \le 2)$
(d) $P(X < -2)$ (e) Determine x suchthat $P(X > x) = 0.05$
(f) Mean (g) Variance

Solution:

$$f(x) = \tfrac{1}{18}x^2 \qquad -3 \le X \le 3$$

(a) $P(X > 0) = \int_0^3 f(x)\mathrm{d}x = \int_0^3 \tfrac{1}{18}x^2\mathrm{d}x$

$$= \left[\frac{1}{18}\frac{x^3}{3}\right]_0^3 = \frac{1}{2}$$

(b) $P(X > 1) = \int_1^3 f(x)\mathrm{d}x = \int_1^3 \tfrac{1}{18}x^2\mathrm{d}x$

$$= \left[\frac{1}{18}\frac{x^3}{3}\right]_1^3 = \frac{26}{54}$$

(c) $P(-2 \le X \le 2) = \int_{-2}^{2} f(x)\mathrm{d}x = \int_{-2}^{2} \tfrac{1}{18}x^2\mathrm{d}x$

$$= \left[\frac{1}{18}\frac{x^3}{3}\right]_{-2}^{2} = \frac{16}{54}$$

(d) $P(X < -2) = \int_{-3}^{-2} f(x)\mathrm{d}x = \int_{-3}^{-2} \tfrac{1}{18}x^2\mathrm{d}x$

$$= \left[\frac{1}{18}\frac{x^3}{3}\right]_{-3}^{-2} = \frac{19}{54}$$

(e) $P(X > x) = 0.05$

$$\int_x^3 f(x)\mathrm{d}x = 0.05$$

$$\int_x^3 \tfrac{1}{18}x^2\mathrm{d}x = 0.05$$

$$\left[\frac{1}{18}\frac{x^3}{3}\right]_x^3 = 0.05$$

$$\frac{27-x^3}{54} = 0.05$$

$$x^3 = 27 - 54 \times 0.05 = 24.3$$

$$x = (24.3)^{1/3} = 2.8964$$

(f) Mean $= \mu = E(X) = \int_{-\infty}^{\infty} xf(x)\mathrm{d}x$

$$\mu = \int_{-3}^{3} xf(x)\mathrm{d}x = \int_{-3}^{3} x\tfrac{1}{18}x^2\mathrm{d}x$$

$$= \int_{-3}^{3} \tfrac{1}{18}x^3\mathrm{d}x$$

$$\mu = \left[\frac{1}{18}\frac{x^4}{4}\right]_{-3}^{3} = \frac{9}{4} = 2.25$$

(g) Variance $= \sigma^2 = \int_{-\infty}^{\infty} x^2 f(x)\mathrm{d}x - \mu^2$

$$\sigma^2 = \int_{-3}^{3} x^2 f(x)\mathrm{d}x - \mu^2$$

$$= \int_{-3}^{3} x^2\tfrac{1}{18}x^2\mathrm{d}x - \mu^2 = \int_{-3}^{3} \tfrac{1}{18}x^4\mathrm{d}x - \mu^2$$

$$\sigma^2 = \left[\frac{1}{18}\frac{x^5}{5}\right]_{-3}^{3} - \left(\frac{9}{4}\right)^2 = \frac{586}{90} - \frac{81}{16} = 1.4486$$

Important note: Here one should understand the difference between probability mass function (PMF) and PDF. In PMF one can calculate probability that the random variable takes a particular value. Whereas, in case of PDF it is not possible to find the probability that the random variable takes a particular value. Since the continuous random variables can take any value in the real number range it is very difficult to find the probability at a particular value. But one can find the probability value for a given interval. That is one can find what is the probability that the random variable will take a value in between certain range or interval. For example in case of continuous random variable one can calculate $P(a \leq X \leq b)$, which is nothing but the area under the PDF curve in between this range. But it is not possible to calculate $P(X = a)$. In this case it will be simply **Zero**. These two can be mathematically represented as follows:

$$\begin{aligned} P(a \leq X \leq b) &= \int_a^b f(x)\,\mathrm{d}x \\ P(X = a) &= 0 \end{aligned} \tag{2.18}$$

'PDF' simply represents probability per unit interval. The total area under any PDF curve will be always one. Similarly, the summation of all the probabilities in PMF will be one. Some generally used continuous distributions are briefly discussed in the following subsections.

2.3.2.5 Uniform Distribution

- Consider a continuous random variable X follows uniform distribution. Then each of the outcome in its range, say, from a to b, will have equal probability density value. Then the probability density function can be written as

$$f(x_i) = \frac{1}{b-a} \quad a \leq X \leq b \tag{2.19}$$

- The probability density function of X is shown in Fig. 2.15.

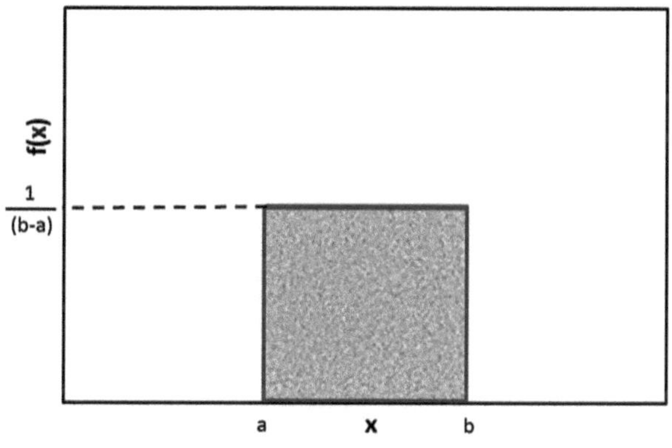

Fig. 2.15 Probability density function of uniform distribution

- The cumulative distribution function of X can be obtained as follows

$$
\begin{aligned}
F_X(x) &= \int_{-\infty}^{x} f(x)\mathrm{d}x \\
&= \int_{a}^{x} f(x)\mathrm{d}x = \int_{a}^{x} \frac{1}{b-a}\mathrm{d}x \\
&= \frac{1}{b-a}\int_{a}^{x} \mathrm{d}x \\
&= \frac{1}{b-a}[x]_a^x \\
F_X(x) &= \frac{x-a}{b-a}
\end{aligned}
\tag{2.20}
$$

- The mean and variance of X can be calculated as follows:

$$
\begin{aligned}
\text{Mean} = \mu = E(X) &= \int_{-\infty}^{\infty} xf(x)\mathrm{d}x \\
\mu &= \int_{a}^{b} xf(x)\mathrm{d}x = \int_{a}^{b} x\frac{1}{b-a}\mathrm{d}x \\
&= \frac{1}{b-a}\int_{a}^{b} x\mathrm{d}x \\
&= \frac{1}{b-a}\left[\frac{x^2}{2}\right]_a^b \\
\mu &= \frac{b^2-a^2}{2(b-a)} \\
\mu &= \frac{(b+a)(b-a)}{2(b-a)} \\
\mu &= \frac{b+a}{2}
\end{aligned}
\tag{2.21}
$$

$$\text{Variance} = \sigma^2 = \int_{-\infty}^{\infty} x^2 f(x)\mathrm{d}x - \mu^2$$

$$\sigma^2 = \int_a^b x^2 f(x)\mathrm{d}x - \mu^2$$

$$= \int_a^b x^2 \frac{1}{b-a}\mathrm{d}x - \mu^2 = \frac{1}{b-a}\int_a^b x^2\mathrm{d}x - \mu^2$$

$$= \frac{1}{b-a}\left[\frac{x^3}{3}\right]_a^b - \left(\frac{b+a}{2}\right)^2 = \frac{b^3-a^3}{3(b-a)} - \left(\frac{b+a}{2}\right)^2$$

$$\sigma^2 = \frac{(b-a)(b^2+ab+a^2)}{3(b-a)} - \left(\frac{b+a}{2}\right)^2$$

$$\sigma^2 = \frac{b^2+ab+a^2}{3} - \frac{b^2+2ab+a^2}{4}$$

$$= \frac{b^2-2ab+a^2}{12}$$

$$\sigma^2 = \frac{(b-a)^2}{12} \tag{2.22}$$

2.3.2.6 Exponential Distribution

- Consider a continuous random variable X follows exponential distribution. Then the probability density function of X can be written as

$$f(x) = \lambda e^{-\lambda x} \quad \text{for } 0 \leq x \leq \infty \tag{2.23}$$

  where λ is parameter of the distribution
- The probability density function of X is shown in Fig. 2.16.

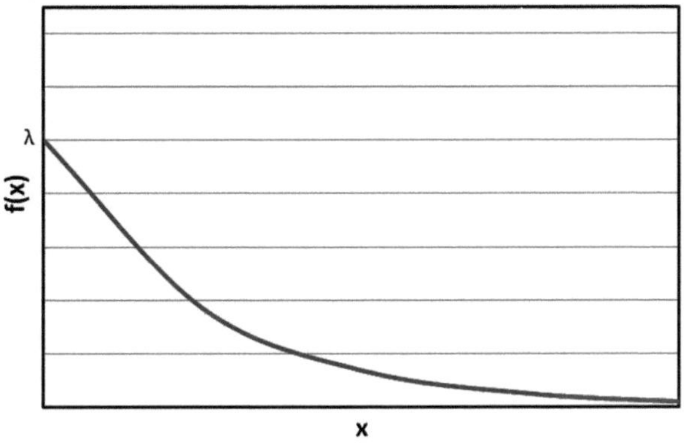

Fig. 2.16 Probability density function of exponential distribution

- The CDF of exponential distribution function can be calculated as

$$\begin{aligned} F_X(x) = P(X \leq x) &= \int_0^x f(x)\mathrm{d}x \quad \text{for } 0 \leq X \leq \infty \\ &= \int_0^x \lambda e^{-\lambda x}\, \mathrm{d}x \\ &= \lambda \int_0^x e^{-\lambda x}\, \mathrm{d}x \\ &= \lambda \left[\frac{e^{-\lambda x}}{-\lambda}\right]_0^x \\ &= -\left[e^{-\lambda x}\right]_0^x \\ &= -\left[e^{-\lambda x} - 1\right] \end{aligned}$$

$$F_X(x) = 1 - e^{-\lambda x} \tag{2.24}$$

- The mean and variance of the exponential distribution are given as

$$\begin{aligned} \text{Mean} = \mu = E(X) &= \int_{-\infty}^{\infty} xf(x)\mathrm{d}x \\ &= \int_0^{\infty} xf(x)\mathrm{d}x = \int_0^{\infty} x\lambda e^{-\lambda x}\mathrm{d}x \\ &= \lambda \int_0^{\infty} xe^{-\lambda x}\mathrm{d}x \\ &= \lambda \left[x\frac{e^{-\lambda x}}{-\lambda}\right]_0^{\infty} - \lambda \int_0^{\infty} \frac{e^{-\lambda x}}{-\lambda}\mathrm{d}x \\ &= \lambda[0] - \left[\frac{e^{-\lambda x}}{-\lambda}\right]_0^{\infty} \end{aligned}$$

$$\text{Mean} = \frac{1}{\lambda} \tag{2.25}$$

$$\text{Variance} = \sigma^2 = \int_{-\infty}^{\infty} x^2 f(x)\mathrm{d}x - \mu^2$$

$$= \int_{0}^{\infty} x^2 f(x)\mathrm{d}x - \mu^2$$

$$= \int_{0}^{\infty} x^2 \lambda e^{-\lambda x}\mathrm{d}x - \mu^2 = \lambda \int_{0}^{\infty} x^2 e^{-\lambda x}\mathrm{d}x - \mu^2$$

$$= \frac{2}{\lambda^2} - \left(\frac{1}{\lambda}\right)^2$$

$$\text{Variance} = \sigma^2 = \frac{1}{\lambda^2}$$

$$\sigma = \frac{1}{\lambda} \tag{2.26}$$

2.3.2.7 Normal Distribution

- Consider a continuous random variable X follows normal distribution. Then the probability density function of X can be written as

$$f(x) = \frac{1}{\sigma\sqrt{2\pi}} e^{-\frac{1}{2}\left(\frac{x-\mu}{\sigma}\right)^2} \qquad -\infty \leq X \leq \infty \tag{2.27}$$

  where μ, σ are the parameters of the distribution
- The probability density function of X is shown in Fig. 2.17.

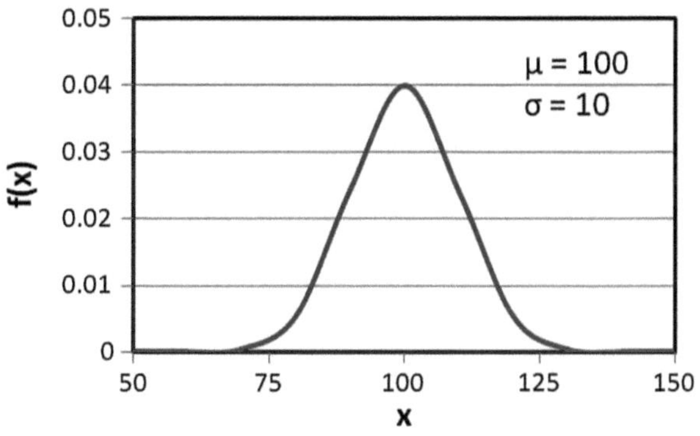

Fig. 2.17 Probability density function of normal distribution

- The mean and variance of the normal distribution are given as

$$\text{Mean} = \mu$$
$$\text{Variance} = \sigma^2$$
$$\text{Standard deviation} = \sigma$$

- Here mean (μ) represents the location parameter and standard deviation (σ) represents the scaling parameter. For a given μ and σ values the PDF of X is shown in Fig. 2.17. The distribution is symmetric about the mean value.
- As the mean value changes the PDF curve also moves either to the left or right as shown in Fig. 2.18. That means the location of the curve is shifting without changing the shape of the curve.
- As the standard deviation changes the scaling of the curve changes without changing the mean value. Here scaling means height and width of the curve changes.
- For the same value of mean, when the standard deviation increases the width of the PDF curve increases and height of the curve decreases which represents there is more spread in the data. Similarly, when the standard deviation decreases width of the curve reduces and height of the curve increases which represents there is less spread in the data. This can be seen in Fig. 2.19.

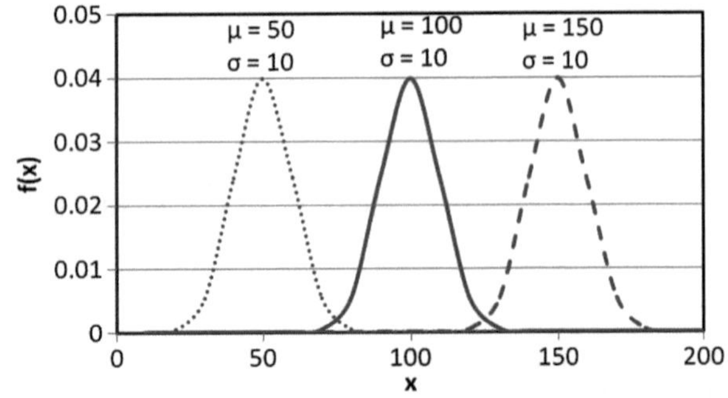

Fig. 2.18 Shifting of PDF curve with change in mean value

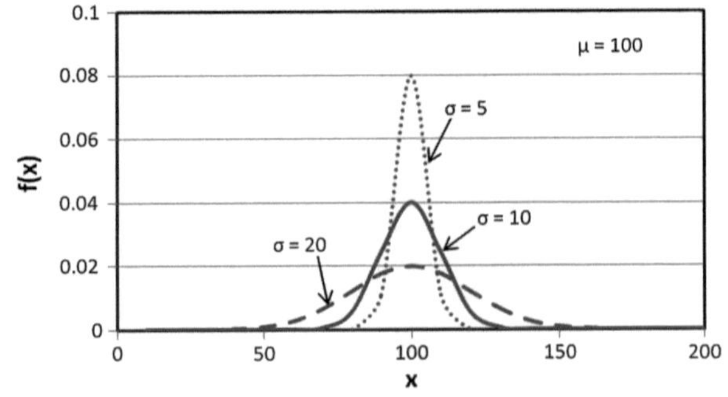

Fig. 2.19 Scaling of PDF curve with change in standard deviation value

- As the standard deviation approaches zero, that is, there is no spread in the data, then the PDF curve looks like simply a straight line at the mean value. That means irrespective of the number of experiments the outcome is same, that is, its mean value.
- The cumulative distribution function of normal random variable can be derived as follows:

$$F_X(x) = P(X \leq x) = \int_{-\infty}^{x} f(x)\mathrm{d}x \qquad \text{for } -\infty \leq X \leq \infty$$

$$= \int_{-\infty}^{x} \frac{1}{\sigma\sqrt{2\pi}} e^{-\frac{1}{2}\left(\frac{x-\mu}{\sigma}\right)^2} \mathrm{d}x$$

$$= \frac{1}{\sigma\sqrt{2\pi}} \int_{-\infty}^{x} e^{-\frac{1}{2}\left(\frac{x-\mu}{\sigma}\right)^2} \mathrm{d}x$$

$$\text{Let } z = \frac{x-\mu}{\sigma} \Rightarrow \frac{\mathrm{d}z}{\mathrm{d}x} = \frac{1}{\sigma}$$

$$F_X(x) = \frac{1}{\sigma\sqrt{2\pi}} \int_{-\infty}^{z} e^{-\frac{z^2}{2}} \sigma\, \mathrm{d}z$$

$$= \frac{1}{\sqrt{2\pi}} \int_{-\infty}^{z} e^{-\frac{z^2}{2}} \mathrm{d}z$$

$$F_X(x) = \Phi(z) = \Phi\left(\frac{x-\mu}{\sigma}\right) = P(Z \leq z) = F_Z(z) \tag{2.28}$$

- The cumulative distribution function of X is shown in Fig. 2.20.

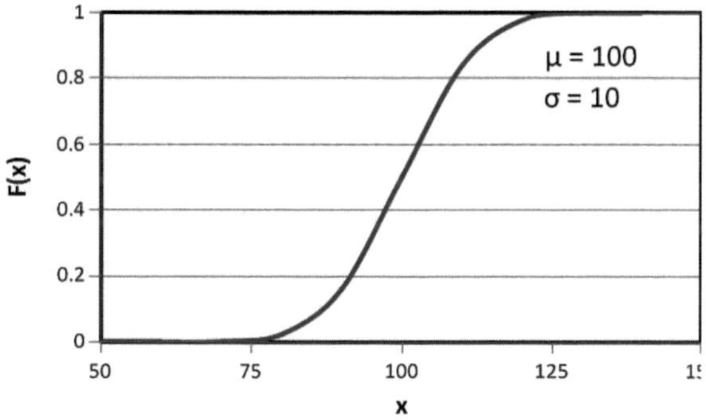

Fig. 2.20 Cumulative distribution function of normal distribution

- Here z is called standard normal random variable whose mean and standard deviation can be calculated as

$$
\begin{aligned}
z &= \frac{x-\mu}{\sigma} \\
\mu_z &= E\left(\frac{x-\mu}{\sigma}\right) \\
&= \frac{E(x)-E(\mu)}{\sigma} \\
&= \frac{\mu-\mu}{\sigma} \\
\mu_z &= 0
\end{aligned}
$$

$$
\begin{aligned}
z &= \frac{x-\mu}{\sigma} \\
V_z &= V\left(\frac{x-\mu}{\sigma}\right) \\
&= \frac{1}{\sigma^2} V(x-\mu) \\
&= \frac{1}{\sigma^2}[V(x)-V(\mu)] \\
V_z &= \frac{1}{\sigma^2}\left[\sigma^2-0\right] \\
\sigma_z^2 &= 1 \Rightarrow \sigma_z = 1
\end{aligned}
$$

- $\Phi(z)$ represents the cumulative distribution function value of z. For calculating CDF values there are standard normal tables existed from which one can directly get the CDF value for a given 'z' value.
- The total area under the standard normal density function also equals to one. Depending on the σ limits the area under the PDF curve will change and is tabulated in the following Table 2.3 and is also shown in Fig. 2.21.

Table 2.3 Area under PDF curve for different σ limits

σ limits	Range	Area (%)
1 σ	$(\mu - 1\sigma)$ to $(\mu + 1\sigma)$	67.73
2 σ	$(\mu - 2\sigma)$ to $(\mu + 2\sigma)$	95.00
3 σ	$(\mu - 3\sigma)$ to $(\mu + 3\sigma)$	99.73

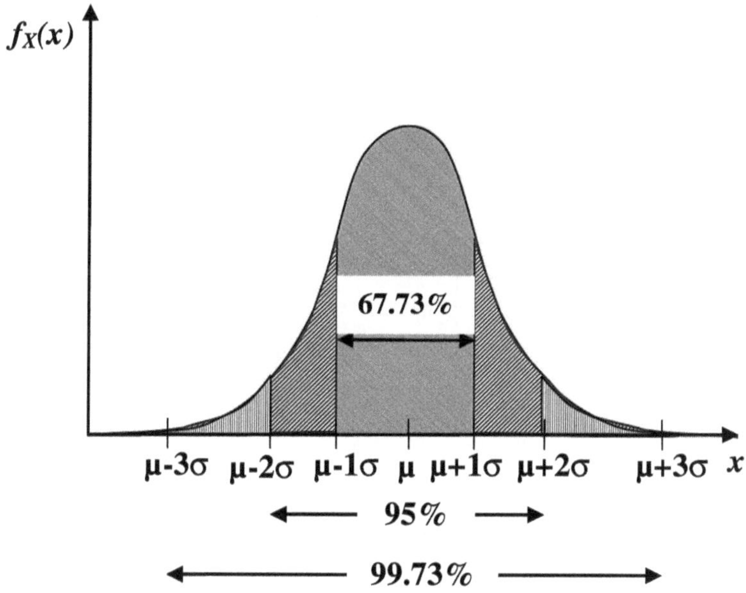

Fig. 2.21 Area under PDF curve for different σ limits

2.3.2.8 Lognormal Distribution

- Consider a continuous random variable Y follows normal distribution and the parameters of the distribution are μ and σ. If Y follows normal distribution then exp(Y) follows lognormal distribution. This is mathematically given as follows:

$$
\begin{aligned}
&Y \quad \text{Normal} \\
&\exp(Y) = e^Y \quad \text{Lognormal} \\
&\text{Let } X = e^Y \Rightarrow Y = \ln(X) \\
&Y \quad \text{Normal} \\
&X \quad \text{Lognormal}
\end{aligned}
$$

- The probability density function of Y can be given as follows:

$$f(y) = \frac{1}{\sigma_y\sqrt{2\pi}} e^{-\frac{1}{2}\left(\frac{y-\mu_y}{\sigma_y}\right)^2} \qquad -\infty \leq Y \leq \infty \tag{2.29}$$

- Now the PDF of X can be obtained by applying transformation rule as follows:

$$
\begin{aligned}
f(x)\mathrm{d}x &= f(y)\mathrm{d}y \\
f(x) &= f(y)\frac{\mathrm{d}y}{\mathrm{d}x} \\
y &= \ln(x) \\
\frac{\mathrm{d}y}{\mathrm{d}x} &= \frac{1}{x} \\
f(x) &= \frac{1}{x}f(y) \\
f(x) &= \frac{1}{\sigma_y\sqrt{2\pi}}\frac{1}{x}e^{-\frac{1}{2}\left(\frac{y-\mu_y}{\sigma_y}\right)^2}
\end{aligned} \tag{2.30}
$$

X limits

$$
\begin{aligned}
y &= \ln(x) \Rightarrow x = e^y \\
y &= -\infty \Rightarrow x = e^{-\infty} = 0 \\
y &= \infty \Rightarrow x = e^{\infty} = \infty
\end{aligned}
$$

By substituting $y = \ln(x)$

$$
f(x) = \frac{1}{\sigma_y\sqrt{2\pi}}\frac{1}{x}e^{-\frac{1}{2}\left(\frac{\ln x-\mu_y}{\sigma_y}\right)^2} \quad 0 \le X \le \infty
$$

- The PDF of lognormal distribution is shown in Fig. 2.22.

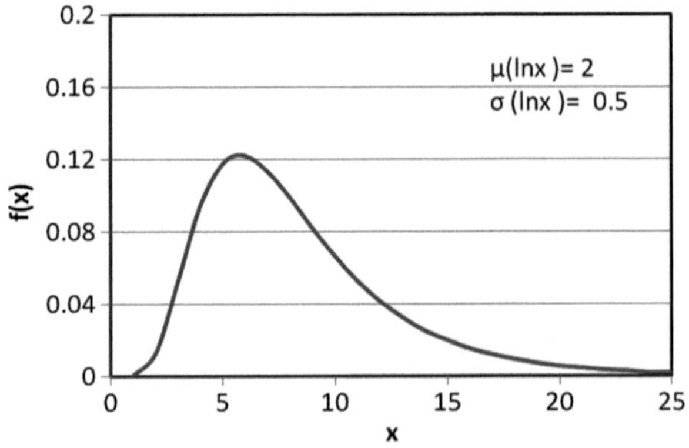

Fig. 2.22 Probability density function of lognormal distribution

- The cumulative distribution of lognormal distribution can be obtained in the similar lines of normal distribution as follows:

$$F_X(x) = P(X \leq x) = \int_0^x f(x)\mathrm{d}x \quad \text{for } 0 \leq X \leq \infty$$

$$= \int_0^x \frac{1}{\sigma_y \sqrt{2\pi x}} e^{-\frac{1}{2}\left(\frac{\ln x - \mu_y}{\sigma_y}\right)^2} \mathrm{d}x$$

$$= \frac{1}{\sigma_y \sqrt{2\pi}} \int_0^x \frac{1}{x} e^{-\frac{1}{2}\left(\frac{\ln x - \mu_y}{\sigma_y}\right)^2} \mathrm{d}x$$

$$\text{Let} \quad z = \frac{\ln x - \mu_y}{\sigma_y} \Rightarrow \frac{\mathrm{d}z}{\mathrm{d}x} = \frac{1}{\sigma_y}\frac{1}{x} \tag{2.31}$$

$$F_X(x) = \frac{1}{\sigma_y \sqrt{2\pi}} \int_{-\infty}^{z} \frac{1}{x} e^{-\frac{z^2}{2}} \sigma_y x \, \mathrm{d}z$$

$$= \frac{1}{\sqrt{2\pi}} \int_{-\infty}^{z} e^{-\frac{z^2}{2}} \mathrm{d}z$$

$$F_X(x) = \Phi(z) = \Phi\left(\frac{\ln x - \mu_y}{\sigma_y}\right) = P(Z \leq z) = F_Z(z)$$

- The CDF of lognormal distribution is shown in Fig. 2.23.
- As it is seen from the probability density function lognormal distribution is defined with the parameters of the normal distribution parameters, i.e., the parameters of the random variable Y (μ_Y, σ_Y).
- Now the question comes what are the parameters of lognormal distribution?

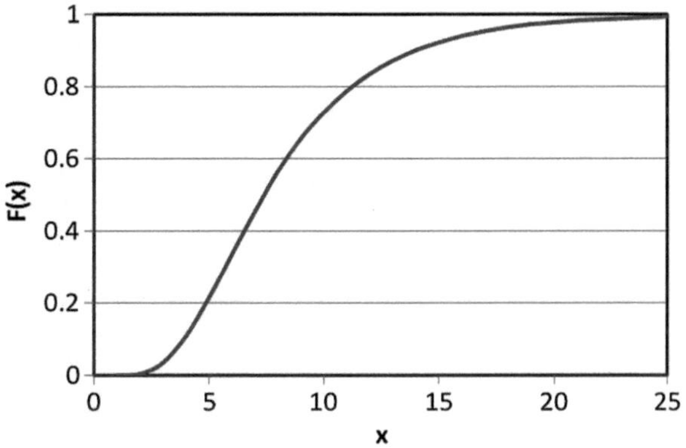

Fig. 2.23 Cumulative distribution function of lognormal distribution

- The lognormal distribution parameters are also similar to normal distribution parameters such as μ_X, σ_X. These parameters can be estimated as follows. We know that

$$X = e^Y$$
$$E(X) = E(e^Y)$$

From the definition of characteristic function

$$E(X) = \int_{-\infty}^{\infty} xf(x)\mathrm{d}x$$

$$E(e^Y) = \int_{0}^{\infty} e^y f(y)\mathrm{d}y$$

$$\text{But} \quad f(y) = \frac{1}{\sigma_y\sqrt{2\pi}} e^{-\frac{1}{2}\left(\frac{y-\mu_y}{\sigma_y}\right)^2}$$

$$E(e^Y) = \int_{-\infty}^{\infty} e^y \frac{1}{\sigma_y\sqrt{2\pi}} e^{-\frac{1}{2}\left(\frac{y-\mu_y}{\sigma_y}\right)^2} \mathrm{d}y$$

$$= \frac{1}{\sigma_y\sqrt{2\pi}} \int_{-\infty}^{\infty} e^y e^{-\frac{1}{2}\left(\frac{y-\mu_y}{\sigma_y}\right)^2} \mathrm{d}y$$

$$\text{Let} \quad z = \frac{y-\mu_y}{\sigma_y} \Rightarrow \frac{\mathrm{d}z}{\mathrm{d}y} = \frac{1}{\sigma_y}$$

$$y = z\sigma_y + \mu_y$$

$$E(e^Y) = \frac{1}{\sigma_y\sqrt{2\pi}} \int_{-\infty}^{\infty} e^{z\sigma_y+\mu_y} e^{-\frac{z^2}{2}} \sigma_y \mathrm{d}z$$

$$= \frac{1}{\sqrt{2\pi}} \int_{-\infty}^{\infty} e^{\mu_y} e^{z\sigma_y-\frac{z^2}{2}} \mathrm{d}z$$

$$E(e^Y) = \frac{1}{\sqrt{2\pi}} e^{\mu_y} \int_{-\infty}^{\infty} e^{z\sigma_y-\frac{z^2}{2}} \mathrm{d}z$$

$$= e^{\mu_y} \frac{1}{\sqrt{2\pi}} \int_{-\infty}^{\infty} e^{z\sigma_y-\frac{z^2}{2}} \mathrm{d}z$$

$$z\sigma_y - \frac{z^2}{2} = \frac{2z\sigma - z^2}{2}$$
$$= -\frac{(z^2 - 2z\sigma)}{2}$$
$$= -\frac{1}{2}\left(z^2 - 2z\sigma + \sigma^2 - \sigma^2\right)$$
$$z\sigma_y - \frac{z^2}{2} = -\frac{1}{2}\left[(z-\sigma)^2 - \sigma^2\right]$$

$$E\left(e^Y\right) = \frac{1}{\sqrt{2\pi}} e^{\mu_y} \int_{-\infty}^{\infty} e^{-\frac{1}{2}\left[(z-\sigma)^2-\sigma^2\right]} \mathrm{d}z$$
$$= e^{\mu_y} \frac{1}{\sqrt{2\pi}} \int_{-\infty}^{\infty} e^{-\frac{(z-\sigma)^2}{2}} e^{\frac{\sigma^2}{2}} \mathrm{d}z$$
$$= e^{\mu+\frac{\sigma^2}{2}} \frac{1}{\sqrt{2\pi}} \int_{-\infty}^{\infty} e^{-\frac{(z-\sigma)^2}{2}} \mathrm{d}z$$

Let $u = z - \sigma$

$$\frac{\mathrm{d}u}{\mathrm{d}z} = 1$$
$$f(u)\mathrm{d}u = f(z)\mathrm{d}z$$
$$f(u) = f(z) = \frac{1}{\sqrt{2\pi}} e^{-\frac{z^2}{2}}$$

$$\int_{-\infty}^{\infty} f(u)\mathrm{d}u = \int_{-\infty}^{\infty} f(z)\mathrm{d}z = 1$$

$$E\left(e^Y\right) = e^{\mu+\frac{\sigma^2}{2}} \frac{1}{\sqrt{2\pi}} \int_{-\infty}^{\infty} e^{-\frac{u^2}{2}} \mathrm{d}u$$

$$E\left(e^Y\right) = e^{\mu+\frac{\sigma^2}{2}} \tag{2.32}$$

- **Standard deviation (σ_X)**

$$\begin{aligned} &\text{we know that} \\ &V(X) = E\left[(X-\mu_X)^2\right] \\ &\qquad = E(X^2) - \mu_X^2 \end{aligned}$$

$$\begin{aligned} \text{But}\quad & X = e^Y \\ V(X) &= E(e^{2Y}) - \mu_X^2 \\ \mu_X &= E(X) = E(e^Y) = e^{\mu+\frac{\sigma^2}{2}} \end{aligned}$$

$$\begin{aligned} V(X) &= E(e^{2Y}) - \left(e^{\mu+\frac{\sigma^2}{2}}\right)^2 \\ &= E(e^{2Y}) - e^{2\left(\mu+\frac{\sigma^2}{2}\right)} \\ &= E(e^{2Y}) - e^{2\mu+\sigma^2} \end{aligned}$$

- As we have $E(e^Y) = e^{\mu+\frac{\sigma^2}{2}}$
- Similarly $E(e^{tY}) = e^{\mu t+\frac{t^2\sigma^2}{2}}$
- If $t = 2$ then $E(e^{2Y}) = e^{2\mu+\frac{4\sigma^2}{2}} = e^{2\left(\mu+\sigma^2\right)}$

$$\begin{aligned} V(e^Y) &= e^{2\left(\mu+\sigma^2\right)} - e^{2\mu+\sigma^2} \\ &= e^{2\mu+\sigma^2}\left[\frac{e^{2\left(\mu+\sigma^2\right)}}{e^{2\mu+\sigma^2}} - 1\right] \\ &= e^{2\mu+\sigma^2}\left[e^{2\left(\mu+\sigma^2\right)-2\mu-\sigma^2} - 1\right] \\ V(e^Y) &= e^{2\mu+\sigma^2}\left[e^{\sigma^2} - 1\right] \end{aligned}$$

$$\begin{aligned} E(e^Y) &= e^{\mu+\frac{\sigma^2}{2}} \\ \mu_X &= e^{\mu_Y+\frac{\sigma_Y^2}{2}} \end{aligned}$$

$$\begin{aligned} V(e^Y) &= e^{2\mu+\sigma^2}\left[e^{\sigma^2} - 1\right] \\ \sigma_X^2 &= e^{2\mu+\sigma^2}\left[e^{\sigma^2} - 1\right] \end{aligned}$$

$$\sigma_X = \sqrt{e^{2\left(\mu+\frac{\sigma^2}{2}\right)}\left[e^{\sigma^2} - 1\right]}$$

$$\sigma_X = e^{\left(\mu+\frac{\sigma^2}{2}\right)}\sqrt{e^{\sigma^2}-1}$$
$$\sigma_X = \mu_X\sqrt{e^{\sigma^2}-1}$$

$$\mu_X = e^{\mu_Y+\frac{\sigma_Y^2}{2}}$$

$$\sigma_X = \mu_X\sqrt{e^{\sigma_Y^2}-1} \tag{2.33}$$

- Similarly, if we know μ_X and σ_X we can find out μ_Y and σ_Y as follows:

$$\sigma_Y^2 = \ln\left[\left(\frac{\sigma_X}{\mu_X}\right)^2 + 1\right]$$
$$\mu_Y = \ln\ \mu_X - \frac{1}{2}\sigma_Y^2 \quad \text{where } Y = \ln\ X \tag{2.34}$$

2.3.2.9 Weibull Distribution

- Consider the random variable X follows Weibull distribution [4]. Then the probability density function of X can be written as

$$f(x) = \frac{\beta}{\alpha}\left(\frac{x}{\alpha}\right)^{\beta-1} e^{\left[-\left(\frac{x}{\alpha}\right)^{\beta}\right]}, \quad \text{for } x > 0 \tag{2.35}$$

- where α is the scale parameter and β is the shape parameter.
- The PDF of Weibull distribution is shown in Fig. 2.24. For different values of β the shape of the PDF curve changes as shown in Fig. 2.24.
- When $\beta = 1$ Weibull distribution will be similar to exponential distribution as shown below

$$f(x) = \frac{\beta}{\alpha}\left(\frac{x}{\alpha}\right)^{\beta-1} e^{\left[-\left(\frac{x}{\alpha}\right)^{\beta}\right]}, \quad \text{for } x > 0$$
$$\text{when} \quad \beta = 1$$
$$f(x) = \frac{1}{\alpha}\left(\frac{x}{\alpha}\right)^{1-1} e^{\left[-\left(\frac{x}{\alpha}\right)^{1}\right]} = \frac{1}{\alpha}e^{\left[-\left(\frac{x}{\alpha}\right)\right]}$$

$$\text{let} \quad \lambda = \frac{1}{\alpha}$$
$$f(x) = \lambda e^{-\lambda x}$$

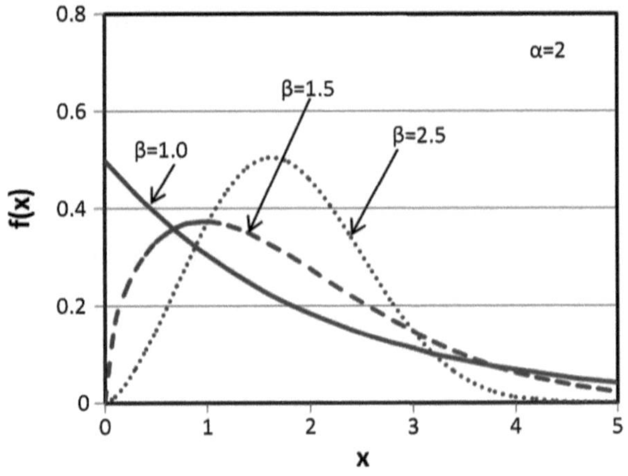

Fig. 2.24 Probability density function of Weibull distribution

- The cumulative distribution can be obtained as

$$F_X(x) = P(X \le x) = \int_0^x f(x)\mathrm{d}x \quad \text{for } 0 \le X \le \infty$$

$$= \int_0^x \frac{\beta}{\alpha}\left(\frac{x}{\alpha}\right)^{\beta-1} e^{\left[-\left(\frac{x}{\alpha}\right)^{\beta}\right]}\,\mathrm{d}x$$

$$= \frac{\beta}{\alpha}\frac{1}{\alpha^{\beta-1}}\int_0^x x^{\beta-1} e^{\left[-\left(\frac{x}{\alpha}\right)^{\beta}\right]}\,\mathrm{d}x$$

$$\text{Let} \quad z = \left(\frac{x}{\alpha}\right)^{\beta} \Rightarrow \frac{\mathrm{d}z}{\mathrm{d}x} = \frac{1}{\alpha^{\beta}}\beta x^{\beta-1}$$

$$\mathrm{d}x = \frac{\alpha^{\beta}}{\beta x^{\beta-1}}\mathrm{d}z$$

$$F_X(x) = \frac{\beta}{\alpha^{\beta}}\int_0^z x^{\beta-1} e^{[-z]}\,\frac{\alpha^{\beta}}{\beta x^{\beta-1}}\mathrm{d}z$$

$$= \int_0^z e^{-z}\mathrm{d}z$$

$$= \left[\frac{e^{-z}}{-1}\right]_0^z = 1 - e^{-z}$$

$$F_X(x) = 1 - e^{-\left(\frac{x}{\alpha}\right)^{\beta}} \tag{2.36}$$

- The CDF of Weibull distribution is shown in Fig. 2.25.

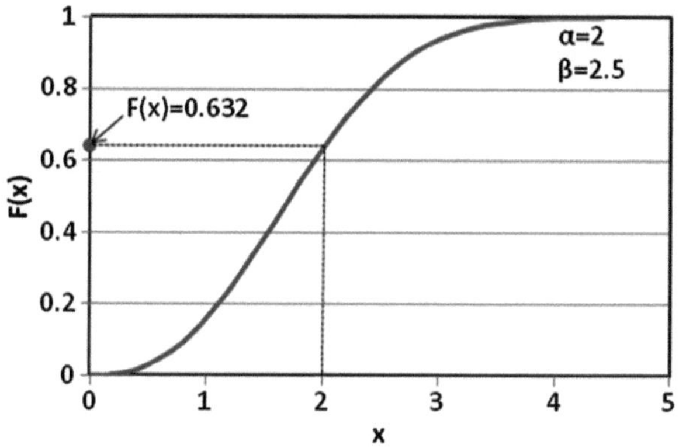

Fig. 2.25 Cumulative distribution function of Weibull distribution

- When $F(x) = 0.632$ then

$$F_X(x) = 1 - e^{-\left(\frac{x}{\alpha}\right)^\beta} = 0.632$$

$$e^{-\left(\frac{x}{\alpha}\right)^\beta} = 1 - 0.632 = 0.368$$

$$e^{\left(\frac{x}{\alpha}\right)^\beta} = \frac{1}{0.368} = 2.7174$$

$$\ln\left(e^{\left(\frac{x}{\alpha}\right)^\beta}\right) = \ln(2.7174) = 1$$

$$\left(\frac{x}{\alpha}\right)^\beta = 1$$

$$\frac{x}{\alpha} = 1^{\frac{1}{\beta}} = 1$$

$$x = \alpha$$

$$\text{when} \quad \text{CDF} = F(x) = 0.632$$

$$\alpha = x$$

- This can be seen in the CDF curve as shown in Fig. 2.25.

2.4 The Reliability Function

In this section the mathematical representation of reliability as a function of time is derived. As was discussed earlier reliability is a probabilistic concept and is also associated with time. To derive the reliability function first one should know what type of random variable that will enter into the analysis. Since reliability is also associated with the functioning of component it is not possible to tell when exactly the component will cease to function or fails. Hence, the times to failure of a component is a random variable and this can take any value in the positive real number range. So in the derivation of reliability function time to failure of a component is considered as a random variable and it is continuous in nature.

Consider the time to failure of a component is a continuous random variable and is represented as 'T.' The probability distribution can be well represented with its probability density function $f_T(t)$. Now the reliability function can be derived by using the definition of the CDF. From the definition of the CDF, the probability of a unit failing by time **t** is given by [5–7]:

$$F(t) = P(T \leq t) = \int_0^t f(t)\mathrm{d}t \quad \text{for } 0 \leq t \leq \infty \tag{2.37}$$

Since this function defines the probability of failure by a certain time, one can consider this as the unreliability function or failure probability function.

This can be explained with a simple example. Consider a component (light bulb) which is kept under testing to study its performance. Assume total number of components put under testing is 100. All the components are identical but they fail independently with each other. That is failure of one component does not affect the failure of the other component. The test is continued until all the bulbs have failed and the time to failure of each bulb is collected. The data can be grouped into intervals. That is, 0–10, 10–20, 20–30 h. In this way from the data one can find out how many components have failed in each interval. Now if one wants to find the fraction of components that is failed in each interval, this can be easily find out by dividing the total number of components failed in that interval with the total number of components. If one wants to find number of components failed till a particular time, this is simply counting all the components failed until that time. Now the fraction of these failed components with the total number of components will give the cumulative distribution function value. That means CDF will tell how many number of components have failed till a particular time in terms of probability.

Now the question comes what is the reliability of the component for a particular time? This is nothing but total number of components that have survived till that time in terms of probability. In other words this will be total number of components that have failed after a particular time in probability scale. This will represent fraction of components that have failed after a particular point of time with the total number of components. In terms of PDF this can be represented as

$$R(t) = P(T > t) = \int_t^\infty f(t)\mathrm{d}t \quad \text{for } 0 \leq t \leq \infty \tag{2.38}$$

The CDF and reliability functions can be graphically represented as shown in Fig. 2.26.

In Fig. 2.26 the area 'Left' to the particular time t represents fraction of total number of components that have failed till that time and is nothing but CDF or failure probability function. The area 'Right' to the time t represents the fraction of

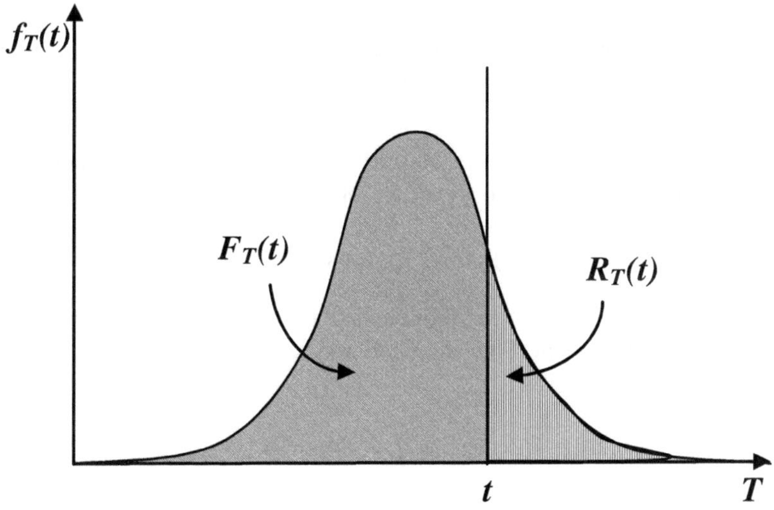

Fig. 2.26 Graphical representation of failure probability and reliability function

total number of components failed after that time and is the reliability function. As we know the total area under the PDF curve is one, then reliability function is complement to the failure probability function or CDF and also they are mutually exclusive events. This can be mathematically represented as

$$\begin{aligned}
&F(t) + R(t) = 1\\
&\int_0^t f(t)\mathrm{d}t + \int_t^\infty f(t)\mathrm{d}t = 1 \quad \text{for } 0 \le t \le \infty\\
&F(t) = 1 - R(t)\\
&R(t) = 1 - F(t)
\end{aligned} \tag{2.39}$$

From the above the PDF can be represented in terms of cumulative distribution function or reliability function as follows:

$$\begin{aligned}
&f(t) = \frac{\mathrm{d}F(t)}{\mathrm{d}t} = -\frac{\mathrm{d}R(t)}{\mathrm{d}t}\\
&\int_0^\infty f(t)\mathrm{d}t = 1\\
&0 \le F(t) \le 1\\
&0 \le R(t) \le 1\\
&F(0) = 0 \quad \operatorname*{Lim}_{t\to\infty} F(t) = 1\\
&R(0) = 1 \quad \operatorname*{Lim}_{t\to\infty} R(t) = 0
\end{aligned} \tag{2.40}$$

From the above expressions at $t = 0$ the failure probability is given as zero and reliability is given as 1. That means all the components are survived at $t = 0$ and there are no failures less than this time. Similarly, as the time tends to infinity the failure probability approaches to 1 and reliability approaches to zero. That means no component will be survived or all the components will be failed by the infinity time. Here infinity means very long time.

Example 2.8 Consider the failure data of a component follows exponential distribution function whose PDF is given as follows:

$$f_T(t) = \lambda e^{-\lambda t} \quad \text{for } 0 \leq t \leq \infty$$

where λ is parameter of the distribution and is a constant which represents the failure rate of a component. Find the CDF and reliability function?

$$F(t) = P(T \leq t) = \int_0^t f(t)\mathrm{d}t \quad \text{for } 0 \leq t \leq \infty$$

$$= \int_0^t \lambda e^{-\lambda t}\,\mathrm{d}t = 1 - e^{-\lambda t}$$

$$R(t) = P(T > t) = \int_t^\infty f(t)\mathrm{d}t \quad \text{for } 0 \leq t \leq \infty$$

$$= \int_t^\infty \lambda e^{-\lambda t}\,\mathrm{d}t = e^{-\lambda t}$$

The PDF, CDF, and reliability functions of exponential distribution function are shown in Figs. 2.27, 2.28 and 2.29.

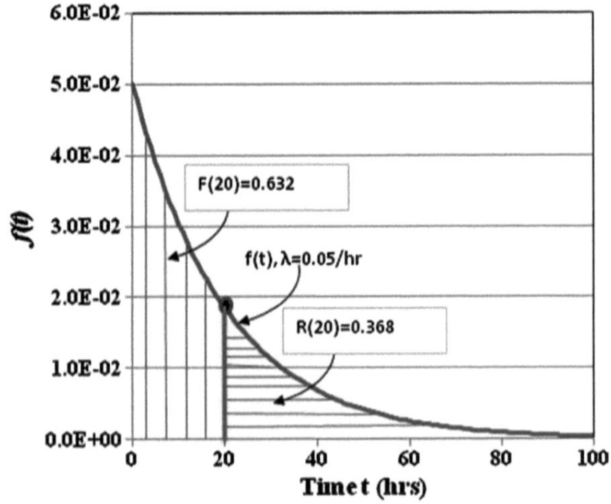

Fig. 2.27 PDF of exponential distribution

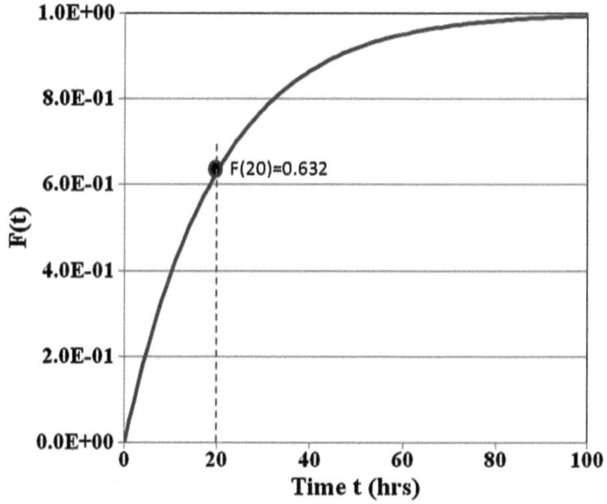

Fig. 2.28 CDF of exponential distribution

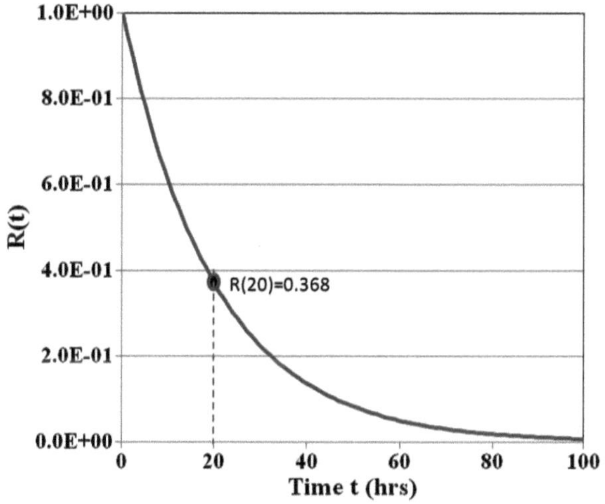

Fig. 2.29 Reliability function of exponential distribution

2.5 Measures of Reliability

Once the failure data of a component is available one can estimate the different parameters from the data. These parameters indirectly give the measure of reliability. These are namely

- Mean time to failure
- Median time to failure
- Mode
- Variance (standard deviation)

These measures are briefly explained below.

2.5.1 Mean Time to Failure

The mean value represents the average time to failure or expected time to failure of a component. This is same as the mean value of a continuous random variable as discussed earlier. This can be mathematically represented as follows:

$$\mathrm{MTTF} = E(t) = \int_0^\infty t f(t)\mathrm{d}t \quad \text{for } 0 \le t \le \infty \tag{2.41}$$

Mean Time to Failure (MTTF) can also be represented as a function of reliability by simplifying the Eq. 2.41 as follows:

$$\mathrm{MTTF} = \int_0^\infty t f(t)\mathrm{d}t \quad \text{for } 0 \le t \le \infty$$

$$f(t) = -\frac{\mathrm{d}R(t)}{\mathrm{d}t}$$

$$\mathrm{MTTF} = \int_0^\infty t\left[-\frac{\mathrm{d}R(t)}{\mathrm{d}t}\right]\mathrm{d}t$$

$$= -\int_0^\infty t\,\mathrm{d}R(t)$$

$$\mathrm{MTTF} = -[tR(t)]_0^\infty + \int_0^\infty R(t)\,\mathrm{d}t \quad \text{(integration by parts)}$$

$$\text{at} \quad t = 0 \quad R(0) = 1$$

$$t = \infty \quad R(\infty) = 0$$

$$\therefore \mathrm{MTTF} = \int_0^\infty R(t)\,\mathrm{d}t \tag{2.42}$$

The integration of above expression will be simpler compared to the previous expression which is a function of PDF.

2.5.2 Median Time to Failure

Median time to failure represents the median value of the failure data. In other words by this time 50 % of the components under operation would have failed or only 50 % of the component would survive. This can be derived from the cumulative distribution function or reliability function. Since the 50 % of the components would have failed by this time, hence, probability of failure or reliability at median time will be 0.5. This can be represented as

$$F(t_{50}) = P(T \leq t_{50}) = 0.5$$
$$R(t_{50}) = P(T > t_{50}) = 0.5 \quad (2.43)$$

It divides the distribution into two halves, with 50 % of the failures occurring before the median time to failure and 50 % occurring after the median value. Median is preferred to the mean when the distribution is highly skewed.

2.5.3 Mode

The mode of the failure data represents the most likely observed failure time. This can be calculated from the PDF of the data. For the continuous distributions wherever the $f(t)$ attains maximum in the entire range that value will be the mode for that data. This also can be represented as follows:

$$f(t_{\text{mod}e}) = \max_{0 \leq t \leq \infty} f(t)$$
$$\frac{df(t)}{dt} = 0 \quad \text{at } t = t_{\text{mod}e} \quad (2.44)$$

This means at the peak value slope of the PDF curve is zero. It is applicable only if the PDF has both +ve and −ve slope. For a monotonically decreasing function t_{mode} will be lower limit of t. Similarly, for the continuously increasing function t_{mode} will be upper limit of t. In the case of normal distribution MTTF, median time to fail and mode all are same. Whereas in the case of lognormal distribution (distribution which is skewed to the right) the reliability measures will appear in the following order:

$$t_{\text{mode}} < t_{\text{median}} < t_{\text{mean}}$$
$$t_{\text{mode}} < t_{50} < \text{MTTF}$$

The same is shown in Fig. 2.30.

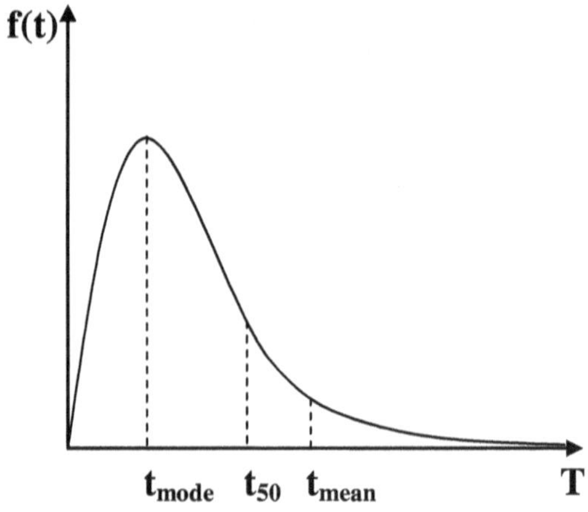

Fig. 2.30 Reliability measures for lognormal distribution

Even if two reliability functions have the same mean, their reliabilities may be quite different for the same operating time.

2.5.4 *Variance*

It is a measure of the spread or dispersion of the failure times about the mean. An average squared distance, a failure time will be from the MTTF. Square root of the variance will give standard deviation.

$$\begin{aligned}\sigma^2 &= \int_0^\infty (t - \mathrm{MTTF})^2 f(t)\,\mathrm{d}t \\ &= \int_0^\infty t^2 f(t)\,\mathrm{d}t - \mathrm{MTTF}^2 \end{aligned} \quad (2.45)$$

Example 2.9 Consider the time to failure data of a component follows exponential distribution function whose PDF is given as follows:

$$f_T(t) = \lambda e^{-\lambda t} \quad \text{for } 0 \le t \le \infty$$

Find the MTTF, median, mode, and variance?

$$F(t) = P(T \leq t) = \int_0^t f(t)\mathrm{d}t \quad \text{for } 0 \leq t \leq \infty$$

$$= \int_0^t \lambda e^{-\lambda t}\,\mathrm{d}t = 1 - e^{-\lambda t}$$

$$R(t) = e^{-\lambda t}$$

$$\text{MTTF} = \mu$$

$$\mu = \int_0^\infty tf(t)\mathrm{d}t = \int_0^\infty R(t)\mathrm{d}t$$

$$\mu = \int_0^\infty e^{-\lambda t}\mathrm{d}t = \frac{1}{\lambda}$$

at $t = t_{50}$

$$F(t_{50}) = 0.5 = 1 - e^{-\lambda t_{50}}$$

$$e^{-\lambda t_{50}} = 0.5$$

$$\ln(e^{-\lambda t_{50}}) = \ln(0.5)$$

$$-\lambda t_{50} = -0.693$$

$$t_{50} = \frac{0.693}{\lambda}$$

at $t = 0$

$$f(0) = \lambda = f_{\max}$$

$$\therefore t_{mode} = 0$$

$$\text{Variance} = \sigma^2$$

$$\sigma^2 = \int_0^\infty t^2 f(t)\mathrm{d}t - \mu^2$$

$$= \int_0^\infty t^2 \lambda e^{-\lambda t}\mathrm{d}t - \mu^2$$

$$\sigma^2 = \frac{1}{\lambda^2}$$

$$\sigma = \frac{1}{\lambda}$$

Example 2.10 The failure distribution of a component is defined by

$$f_T(t) = \frac{4t^3}{\mathrm{a}} \quad \text{for } 0 \le t \le 100\,\mathrm{h}$$

(a) Find the constant 'a'
(b) Compute the MTTF
(c) Find the design life for a reliability of 0.99.
(d) Find the mode for the given function.

Solution:

(a) According to the axioms of probability the area under the PDF curve should be equal to '1' in its defined range.

$$f_T(t) = \frac{4t^3}{\mathrm{a}} \quad \text{for } 0 \le t \le 100\,\mathrm{h}$$

$$\int_0^{100} f_T(t)\,\mathrm{d}t = 1$$

$$\int_0^{100} \frac{4t^3}{a}\,\mathrm{d}t = 1$$

$$\left[\frac{4t^4}{4a}\right]_0^{100} = 1$$

$$\frac{10^8}{a} = 1$$

$$\therefore a = 10^8$$

$$f_T(t) = \frac{4t^3}{10^8} \quad \text{for } 0 \le t \le 100\,\mathrm{h}$$

(b) MTTF can be calculated as follows:

$$\mathrm{MTTF} = \mu = \int_0^{\infty} tf(t)\mathrm{d}t$$

$$\mu = \int_0^{100} tf(t)\mathrm{d}t$$

$$= \int_0^{100} t\frac{4t^3}{10^8}\mathrm{d}t$$

$$\mu = \frac{4}{10^8}\int_0^{100} t^4\mathrm{d}t$$

$$\mu = \frac{4}{10^8}\left[\frac{t^5}{5}\right]_0^{100}$$

$$= \frac{4}{10^8}\frac{10^{10}}{5} = \frac{400}{5}$$

$$\mu = 80\,\mathrm{h}$$

(c) Find the design life for a reliability of 0.99. We know that

$$R(t) = \int_t^{\infty} f(t)\mathrm{d}t$$

$$= \int_t^{100} \frac{4t^3}{10^8}\mathrm{d}t$$

$$= \int_t^{100} \frac{4t^3}{10^8}\mathrm{d}t$$

$$= \frac{4}{10^8}\int_t^{100} t^3\mathrm{d}t$$

$$= \frac{4}{10^8}\left[\frac{t^4}{4}\right]_t^{100}$$

$$= \frac{4}{10^8}\left[\frac{10^8 - t^4}{4}\right]$$

$$R(t) = \frac{1}{10^8}\left[10^8 - t^4\right]$$

But $R(t) = 0.99$

$$\frac{1}{10^8}\left[10^8 - t^4\right] = 0.99$$

$$t^4 = 10^8 - 0.99 \times 10^8 = 0.01 \times 10^8 = 10^6$$

$$t = 10^{6/4} = 31.623\,\mathrm{h}$$

Hence, the design life of the component for the reliability of 0.99 is 31.623 h.

(d) Find the mode for the given function

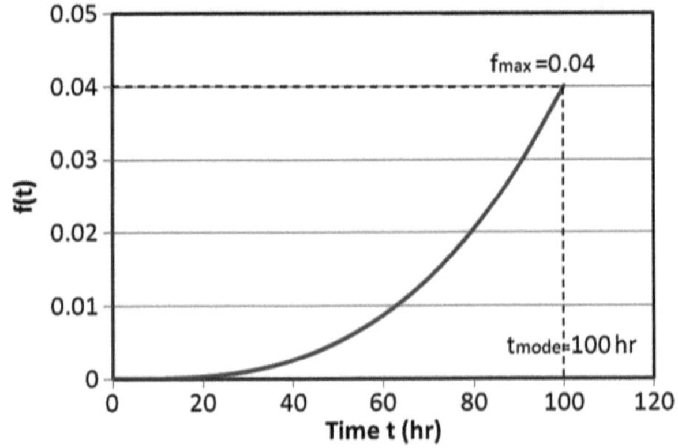

Fig. 2.31 Probability density function of T

The mode of the function will be time at which PDF will be maximum. From the density function it is clear that it is a continuously increasing function of time. Hence, the f_{max} will occur at its upper limit, i.e., at t = 100 h. Hence mode for the present function is 100 h. This can be seen in the PDF curve as shown in Fig. 2.31.

2.6 Hazard Rate Function

One more important function that is used often in reliability is hazard rate function. This can be sometimes used as failure rate depending on the context of use. It gives the instantaneous failure rate of a component. In simple terms it is nothing but number of failures per unit time. The hazard rate function can be derived based on the conditional probability concepts. Assume that a component has survived up to time 't.' Now what is the probability that the component will fail immediately after time t. This can be probabilistically represented as below. Consider the component will fail in a finite time interval Δt, then probability that the component will fail in the time interval t and $t + \Delta t$ can be given as

$$P[t \leq T \leq t + \Delta t] = F(t + \Delta t) - F(t) \tag{2.46}$$

But the component has survived till time t, this can be represented with conditional probability as follows:

$$P[t \le T \le t + \Delta t | T \ge t] = \frac{P[t \le T \le t + \Delta t \cap T \ge t]}{P[T \ge t]}$$

$$\because P(A|B) = \frac{P(A \cap B)}{P(B)}$$

$$P[t \le T \le t + \Delta t | T \ge t] = \frac{P[t \le T \le t + \Delta t]}{P[T \ge t]}$$

$$P[t \le T \le t + \Delta t | T \ge t] = \frac{F(t + \Delta t) - F(t)}{R(t)} \tag{2.47}$$

Now divide the above function with the interval size Δt and set the limits as $\Delta t \to 0$, then this can be written as

$$h(t) = \operatorname*{Lim}_{\Delta t \to 0} \frac{F(t + \Delta t) - F(t)}{R(t)} \times \frac{1}{\Delta t}$$

$$= \operatorname*{Lim}_{\Delta t \to 0} \frac{F(t + \Delta t) - F(t)}{\Delta t} \times \frac{1}{R(t)}$$

$$= \operatorname*{Lim}_{\Delta t \to 0} \frac{\Delta F(t)}{\Delta t} \times \frac{1}{R(t)} = \frac{\mathrm{d}F(t)}{\mathrm{d}t} \times \frac{1}{R(t)}$$

$$\text{but} \quad f(t) = \frac{\mathrm{d}F(t)}{\mathrm{d}t}$$

$$\therefore \quad h(t) = \frac{f(t)}{R(t)} \tag{2.48}$$

Further simplifying the above function

$$h(t) = \frac{f(t)}{R(t)}$$

$$\text{but} \quad f(t) = -\frac{\mathrm{d}R(t)}{\mathrm{d}t}$$

$$\therefore h(t) = -\frac{\mathrm{d}R(t)}{\mathrm{d}t} \times \frac{1}{R(t)}$$

$$h(t)\,\mathrm{d}t = -\frac{\mathrm{d}R(t)}{R(t)}$$

$$\int_0^t h(t)\,\mathrm{d}t = \int_0^t -\frac{\mathrm{d}R(t)}{R(t)} = -\int_0^t \frac{\mathrm{d}R(t)}{R(t)}$$

$$\int_0^t h(t)\,\mathrm{d}t = -[\ln(R(t))]_0^t \quad \text{at } t = 0 \;\; R(0) = 1$$

$$\ln(R(t)) = -\int_0^t h(t)\,\mathrm{d}t$$

$$\therefore R(t) = e^{-\int_0^t h(t)\,\mathrm{d}t} \tag{2.49}$$

From the above reliability function is a function of hazard rate function. The hazard function itself is a function of time. The hazard rate function can be decreasing function of time or can be constant or it can be increasing function of time. If the hazard rate function is constant then the reliability function will take the following form:

$$\begin{aligned} h(t) &= \lambda = \text{Constant} \\ R(t) &= e^{-\int_0^t h(t)\,\mathrm{d}t} = e^{-\int_0^t \lambda\,\mathrm{d}t} = e^{-\lambda t} \end{aligned} \tag{2.50}$$

This is same as the reliability function of an exponential distribution. This can be derived as follows:

$$\begin{aligned} h(t) &= \frac{f(t)}{R(t)} \\ f(t) &= \lambda e^{-\lambda t} \quad \text{and} \quad R(t) = e^{-\lambda t} \\ h(t) &= \frac{\lambda e^{-\lambda t}}{e^{-\lambda t}} = \lambda \end{aligned} \tag{2.51}$$

Hence, if the component failure data follows exponential distribution then its hazard rate is constant and is equivalent to its distribution parameter λ, i.e., failure rate. As discussed earlier hazard rate is a function of time and it can be decreasing or constant or increasing function of time. Based on this there are different types of hazard rate models existed. This has got beautiful application in life characteristic curve of components and is explained in the following section.

2.7 Life Characteristic Curve

Life characteristic curve (bathtub curve) is the graph between the hazard rate and the component lifetime. There are three stages of failures in the life of a product: early stage (infant mortality), operating stage, and wear out stage as shown in Fig. 2.32 which is called life characteristic curve or bathtub curve because of its shape.

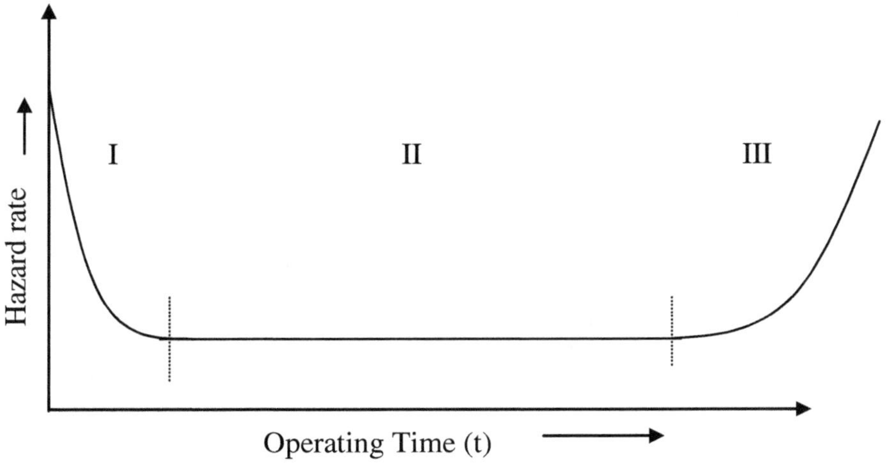

Fig. 2.32 Life characteristic curve

(1) Early failure region (infant mortality, burn-in)
(2) Useful life region (hazard rate constant)
(3) Wear out region.

When the equipment is put into use for the first time any inherently weak parts normally fail soon. Thus early hazard rate is very high. But once the weak parts are replaced the hazard rate falls and fairly constant and finally hazard rate rises again as parts start to wear out.

The region (1) suggests that no item be used unless it has survived this period. Some of the reputed manufacturers sell only those components which have survived this period. The region (2) is useful life period where hazard rate is governed by chance failure and is fairly constant. The region (3) indicates that the component should be replaced or scrapped. Table 2.4 gives the differences between the three phases.

Example 2.11 Times to failure of a component are obtained as follows:

Time interval	No. of failures
0–20	95
20–40	50
40–60	23
60–80	12
80–100	6
100–120	4
>120	10

Table 2.4 Different phases of bathtub curve

	Phase I	Phase II	Phase III
Characterized by	Decreasing failure rates	Constant failure rate	Increasing failure rate
	Infant mortality, burn-in	Useful life	Wear out
Caused by	Manufacturing defects, welding flaws, cracks, defective parts, poor quality control, contamination, poor workmanship	Environment, random loads, human error, chance events	Fatigue, corrosion, aging, friction, cyclical loading
Reduced by	Burn-in testing, screening, quality control acceptance testing	Redundancy, excessive strength	Derating, preventive maintenance, parts replacement

(a) Find the probability density function $f(t)$, cumulative distribution function $F(t)$, reliability function $R(t)$, and the hazard rate function $h(t)$?

(b) In which phase of the bathtub curve the above data fits?

Solution:

To solve the problem we first list down all the relevant formulas

$$f(t) = \frac{\mathrm{d}F(t)}{\mathrm{d}t} = \lim_{\Delta t \to 0} \frac{F(t+\Delta t) - F(t)}{\Delta t}$$

$$F(t) = \frac{N_f(t)}{N}$$

$$R(t) = \frac{N_s(t)}{N} = 1 - F(t)$$

$$h(t) = \frac{f(t)}{R(t)}$$

where

$N_f(t)$ is Total number of components failed till time 't'

$Ns(t)$ is Total number of components survived till time 't'

N is the Total number of components

- From the Table each time interval has lower and upper limit and are denoted as t_L and t_U. In the first interval $t_L = 0$ and $t_U = 20$ h.
- Interval size $\Delta t = t_U - t_L = 20$ h
- Total number of components (N) = 200
- For the first interval total number of components failed till $t = 0$ is 'Zero' and total number of components survived till $t = 0$ is 200.

Table 2.5 Calculation of CDF and reliability

t_L	$N_f(t_L)$	$N_s(t_L)$	$F(t_L) = \frac{N_f(t_L)}{N}$	$R(t_L) = \frac{N_s(t_L)}{N} = 1 - F(t_L)$
0	0	200	0/200 = 0.000	200/200 = 1.000
20	95	105	95/200 = 0.475	105/200 = 0.525
40	145	55	145/200 = 0.725	55/200 = 0.275
60	168	32	168/200 = 0.840	32/200 = 0.160
80	180	20	180/200 = 0.900	20/200 = 0.100
100	186	14	186/200 = 0.930	14/200 = 0.070
120	190	10	190/200 = 0.950	10/200 = 0.050
>120	200	0	200/200 = 1.000	0/200 = 0.0000

- For the second interval total number of components failed till $t = 20$ is 95 and total number of components survived till $t = 20$ is 105. This is shown in Table 2.5 below.
- $F(t)$ and $R(t)$ are calculated at each and every point of time and are also tabulated in Table 2.5 and are also shown in Figs. 2.33 and 2.34, respectively.
- $f(t)$ is calculated by using the following formula

$$f(t) = \frac{F(t + \Delta t) - F(t)}{\Delta t}$$

where t is the lower limit of the interval and Δt is the interval size.
Here $f(t)$ is calculated for the entire interval, i.e., average value for the interval. This is because we do not have the enough failure information at each and every point of the interval. Calculation is shown in Table 2.6. This can be seen in Fig. 2.35.
- Similarly $h(t)$ is also calculated (Table 2.7) for the entire interval, which will represent average value for the interval (Fig. 2.36).

Fig. 2.33 Cumulative distribution function

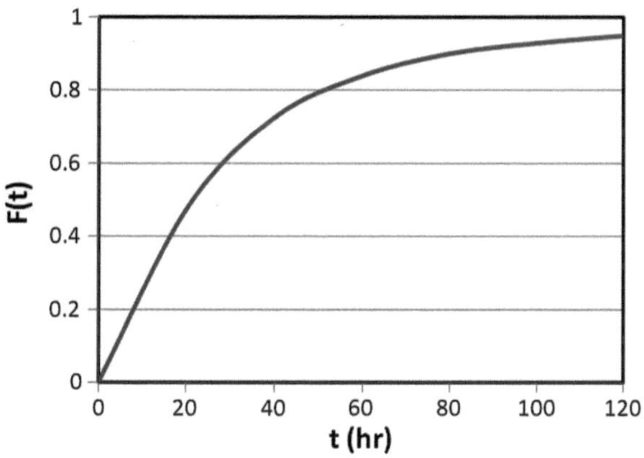

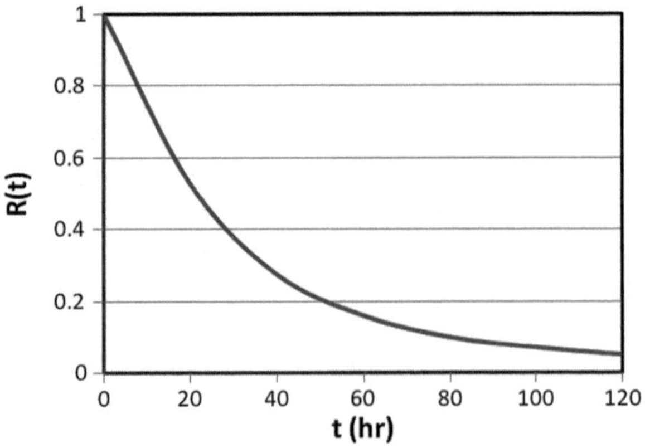

Fig. 2.34 Reliability function

Table 2.6 Calculation of PDF

Time interval	t_L	t_U / $t_L + \Delta t$	$F(t_L)$	$F(t_L + \Delta t)$	$f(t) = \frac{F(t_L+\Delta t)-F(t_L)}{\Delta t}$
0–20	0	20	0.000	0.475	(0.475 − 0)/20 = 0.02375
20–40	20	40	0.475	0.725	(0.725 − 0.475)/20 = 0.0125
40–60	40	60	0.725	0.840	(0.84 − 0.725)/20 = 0.00575
60–80	60	80	0.840	0.900	(0.9 − 0.84)/20 = 0.003
80–100	80	100	0.900	0.930	(0.93 − 0.9)/20 = 0.0015
100–120	100	120	0.930	0.950	(0.95 − 0.93)/20 = 0.001

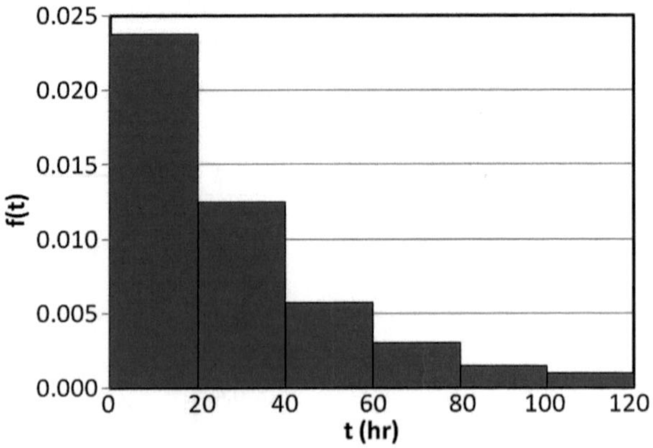

Fig. 2.35 Probability density function interval wise

- From Fig. 2.36 it is clear that the hazard rate is decreasing with respect to time. Hence, the failure data will represent the phase I in the bathtub curve, i.e., infant mortality region.

Table 2.7 Calculation of $h(t)$

Time interval	$f(t)$	$R(t)$	$h(t) = \frac{f(t)}{R(t)}$
0–20	0.02375	1.000	0.02375/1.0 = 0.02375
20–40	0.01250	0.525	0.0125/0.525 = 0.02381
40–60	0.00575	0.275	0.00575/0.275 = 0.0209
60–80	0.00300	0.160	0.003/0.160 = 0.0188
80–100	0.00150	0.100	0.0015/0.1 = 0.0150
100–120	0.00100	0.070	0.001/0.07 = 0.0143

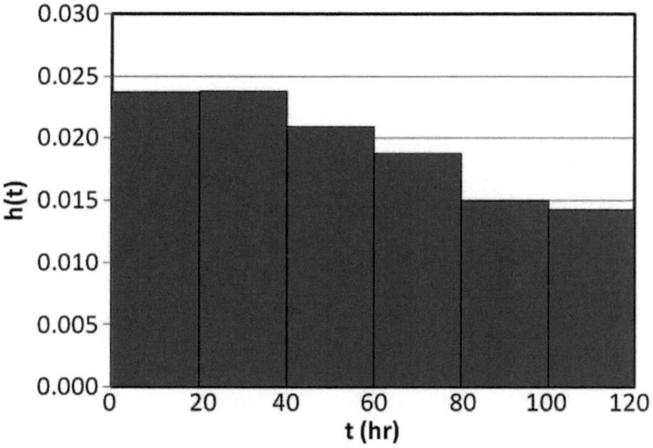

Fig. 2.36 Hazard rate function intervalwise

References

1. Ang AHS, Tang WH (1984) Probability concepts in engineering planning and design: decision, risk and reliability, vol 2. Wiley, New York
2. Montgomery DC, Runger GC (1999) Applied statistics and probability for engineers. Wiley, New York
3. Ross SM (1987) Introduction to probability and statistics for engineers and scientists. Wiley, New York
4. Weibull W (1951) A statistical distribution of wide applicability. J Appl Mech 18:293–297
5. Ebeling CE (1997) An introduction to reliability and maintainability engineering. Tata McGraw-Hill, New Delhi
6. Elsayed EA (1996) Reliability engineering. Prentice Hall, New Jersey
7. Misra KB (1992) Reliability analysis and prediction. Elsevier, Amsterdam

Chapter 3
Risk Analysis of Nuclear Power Plants

3.1 Introduction

Post Fukushima accident [1], a new problem has arisen for the nuclear community with respect to the establishment of new power plants in the country. There is public agitation wherever the government wants to establish the new plants. This is because of the wrong impression given about the risk from the NPPs. In the earlier days, people did not have much awareness about the NPPs and also kind of risk that is expected from it. Nowadays, the awareness is better but at the same time people have wrong opinion about the nuclear accidents. People started asking different questions about the risk, for example,

1. What is the risk expected from the NPP during its normal operation and during an accident?
2. Whether the accident affects the public near by the plant?
3. Is it necessary to live with this kind of risk?
4. Finally, the question boils down to cannot we sustain without the NPPs?

Now if one examines these questions carefully the answer to the first question is that during the normal and accidental conditions the risk from the NPP is defined in terms of dose absorbed by a person located at a distance from the plant. There is always risk involved in any plant operation (thermal, hydraulic, chemical etc.). But how much risk is acceptable from a plant? A simple answer to this question is that it is very important to see that the risk from the plant should be less than the risk associated with the natural hazards (Earthquakes, Tsunami, Storm, Flooding, striking of asteroids etc.). One more question arises that already we are living in a society where so much of risk is involved with respect to the natural hazards, then why to invite these additional human-made risks?

To answer these questions, one can look at the risk caused by the natural hazards as discussed in Chap. 1. From these discussions, it is evident that human life is always threatened by several hazards (natural as well as human made). Here the

A.K. Verma et al., *Risk Management of Non-Renewable Energy Systems*,
Springer Series in Reliability Engineering, DOI 10.1007/978-3-319-16062-7_3

question comes cannot we avoid these risks? The answer could be No, but at the same time one can take precautions to minimize these risks. As discussed earlier, risk is associated with two terms one is likelihood of occurrence of the event and the other term is consequences. If we want to minimize the risk, we have to minimize either the likelihood of occurrence of the events or consequences or both the terms. In case of natural hazards, one cannot have control over their likelihood of occurrence but the consequences can be minimized by taking appropriate precautions. Whereas in the case of human-made hazards, one can have control over both the terms.

Now we examine the risk from NPPs. Even though the risk from the NPPs is less but still the question arises cannot we sustain without the NPPs? The answer to this is simple we can sustain, but the establishment and operation of nuclear and radiation facilities, and the use of radioactive sources, contribute to the economic and social progress of the country. However, while undertaking such activities, safety of the workers, the general public, and the environment has to be ensured. To understand the different aspects of the risk of NPPs, first it is very important to know about the plant. A brief introduction to a typical NPP is given in the following section.

3.2 Nuclear Power Plants

In a typical pressurized heavy water reactor (PHWR) [2], a reactor core consists of a calandria, coolant channels, fuel bundles, fuel pins, and moderators as shown in the Fig. 3.1. During normal power operation or shutdown condition, the coolant should be circulated continuously inside the reactor core so as to remove heat or decay heat from the core. In any case, if heat is not removed from the core, the fuel temperature increases and if it attains the melting point of the fuel, core melt takes place. Hence, in the case of NPPs, core getting damaged is called an accident. Once the core gets damaged, the fission product noble gases release from the core to the containment.

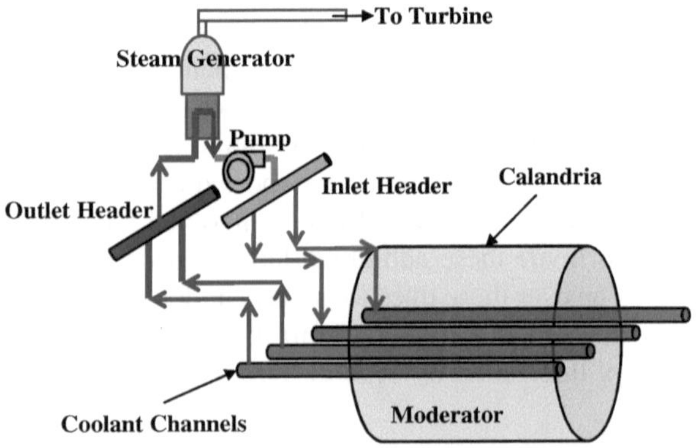

Fig. 3.1 A typical pressurized heavy water reactor

In the absence of the engineered safety features of the containment and through the leakage paths in the containment, the radioactivity may get released into the atmosphere. Once the radioactivity gets released into the atmosphere, it gets transported and diffused depending on the weather conditions and can affect the people depending on the concentration. From the above, one can observe that whenever there is an accident kind of situation in a NPP, the radioactivity will not be directly released into the atmosphere; initially, it will be contained in the reactor core, then released into the containment, and if there is any leakage, then it will finally get released into the atmosphere.

In the context of NPPs, an undesirable event or accident is the core getting damaged and its consequences will be in terms of radioactivity release into the atmosphere and its effects on the public [3, 4]. In this case, risk can be reduced by reducing the accident frequency as well as consequences. In NPPs, accident frequency can be reduced by reducing the frequency of occurrence of events that may lead to a loss of control over a nuclear reactor core, nuclear chain reaction, and by increasing the reliability of the safety systems. Now question comes how do we ensure safety of the plant? Safety of a plant can be ensured by the achievement of safety objectives.

Example 3.1 In a NPP, let the frequency of occurrence of core damage is 10^{-6}/year and the containment leakage probability is 10^{-2}. Let a person is standing at a distance of 1.5 km from the release point of radioactivity, then the consequence can be in terms of absorbed dose rate by the person. Let the absorbed dose rate for a particular amount of radioactivity release is 1 mSv. Then the risk of a person in terms of absorbed dose rate at 1.5 km distance can be calculated as 1×10^{-8} mSv/year.

3.3 Safety Objectives of NPPs

The general safety objective of NPPs is to protect people and the environment from harmful effects of radiation by establishing and maintaining effective defences against radiological hazards in nuclear installations [5]. The general safety objective applies to all the nuclear power plants in all the stages of its lifetime. The different stages of NPPs include planning, siting, designing, manufacturing, construction, commissioning and operation, and also decommissioning. This general safety objective forms a basis for two other safety objectives that deal with radiation protection and technical aspects of design.

3.3.1 Radiation Protection Objective

The radiation protection objective is provided to ensure that radiation exposure from the nuclear power plant in all its operational states is kept well below the prescribed limits and as low as reasonably achievable (ALARA) [5].

3.3.2 Technical Safety Objectives

The technical safety objectives are provided to ensure that proper measures are taken care to prevent accidents and also to mitigate their consequences if at all they occur in the nuclear power plants. This can be achieved by considering all the possible accidents in the design of NPPs, including those of very low probability. These technical safety objectives further provide basis for the safety criteria and safety goals as given below:

- Dose acceptance criteria for the design basis events (DBE) and
- Safety goals for beyond design basis accidents (BDBA)

In order to confirm that these safety criteria and goals are met, safety analyses are performed in NPPs. There are two types of safety analyses namely, deterministic safety analysis and probabilistic safety assessment. These are explained further in the following subsections.

3.3.2.1 Dose Acceptance Criteria

Deterministic safety analysis should demonstrate that the dose received by the individual at the plant boundary should be within the limits specified. According to the Atomic Energy Regulatory Board (AERB), India [6] these limits for design basis accident are specified as follows:

- Effective dose should not exceed 100 mSv or
- Child's thyroid dose should not exceed 500 mSv.

According to Canadian Nuclear Safety Commission (CNSC), the committed whole-body dose for a period of 30 days should not exceed [7]:

- 0.5 mSv for any anticipated operational occurrence (AOO); or
- 20 mSv for any design basis accident (DBA).

3.3.2.2 Qualitative Safety Goals

Two qualitative safety goals have been established in order to limit the risks posed by nuclear power plant operation on the public. These are given as follows:

- No significant additional risk to the life and health of individuals from the consequences of nuclear power plant operation; and
- Societal risks from nuclear power plant operation to life and health should not significantly add to other societal risks and should be less than the risks from other plants where electricity is generated with different technologies.

3.3.2.3 Quantitative Safety Goals

To achieve the above-mentioned qualitative safety goals, quantitative safety goals are established for practical applications, the two quantitative safety goals are:

- Core damage frequency [7]
  - The sum of frequencies of all dominating accident sequences (significant core degradation) should be less than 10^{-5} per reactor year.
- Large early release frequency
  - The sum of frequencies of all event sequences that can lead to a release of more than 10^{14} Becquerel of cesium-137 to the environment should be less than 10^{-6} per reactor year.

3.4 Safety Analyses

There are two types of safety analyses namely deterministic safety analysis and probabilistic safety assessment. This analysis will tell whether the safety objectives are achieved in the design of a nuclear power plant or not. With this analysis it is possible to identify all sources of exposure, to evaluate radiation doses received by the workers in the plant as well as by the public, and the potential effects on the environment. The safety analysis generally examines:

a. Normal operational modes of the plant;
b. Anticipated operational occurrences;
c. Design basis accidents;
d. Event sequences that may lead to a severe accident.

With this analysis, one can establish whether the engineering design withstands the postulated initiating events, accidents or not, it is also possible to demonstrate the effectiveness of the safety systems and safety-related systems, and requirements for emergency response can be established.

3.4.1 Deterministic Safety Analyses

Safety is the main concern in Nuclear Power Plants (NPP). The general safety objective of NPPs is that it should be designed and operated in a manner that will protect individuals, society, and the environment from harm. This objective relies on the establishment and maintenance of effective defences against radiological hazards in NPPs. One can achieve these safety objectives by performing safety analysis. Safety analysis involves deterministic and probabilistic analysis [5]. These analyses identify all sources of exposure, in order to evaluate potential radiation doses to workers at the plant and to the public, and to evaluate potential effects on the environment.

3.4.2 Probabilistic Safety Assessment

Probabilistic safety assessment (PSA) [3, 8] is an analytical technique for assessing the risks by integrating diverse aspects of design and operation of a nuclear power plant. Risk can be defined as the product of likelihood of occurrence of an accident and the consequences from that accident. In the context of a nuclear power plant, core damage of the reactor represents an accident and release of radioactivity in the public domain and its effects on them will be the consequence. In order to prevent the occurrence of an accident, various engineered safety features are designed.

An accident situation occurs when an initiating event is coupled with the unavailability of one or more engineered safety features. The accident frequency can be minimized by reducing the frequency of an initiating event and improving the availability of safety systems. This can be achieved by incorporating good design practices and selection of reliable components. The consequence of an accident that may occur can be minimized by providing barriers (e.g., reactor containment and exclusion zone) for arresting the release of radioactivity and reaching the public.

PSA differs from traditional deterministic safety analysis in that it provides a methodological approach to identifying accident sequences that can follow from a broad range of initiating events and it includes the systematic and realistic determination of accident frequencies and consequences. PSA has been shown to provide important safety insights in addition to those provided by deterministic analysis. In international practice, three levels of PSA (Fig. 3.2) are generally recognized [3, 9, 10, 11].

In Level 1 PSA [12], the design and operation of the plant are analyzed in order to identify the sequences of events that can lead to core damage and the core damage frequency is estimated. Level 1 PSA provides insights into the strengths and weaknesses of the safety-related systems and procedures in place or envisaged as preventing core damage.

In Level 2 PSA [12], the chronological progression of core damage sequences identified in Level 1 PSA is evaluated, including a quantitative assessment of phenomena arising from severe damage to reactor fuel. Level 2 PSA identifies ways in which associated releases of radioactive material from fuel can result in releases to the environment. It also estimates the frequency, magnitude, and other relevant characteristics of the release of radioactive material to the environment. This analysis provides additional insights into the relative importance of accident prevention and mitigation measures and the physical barriers to the release of radioactive material to the environment (e.g., a containment building).

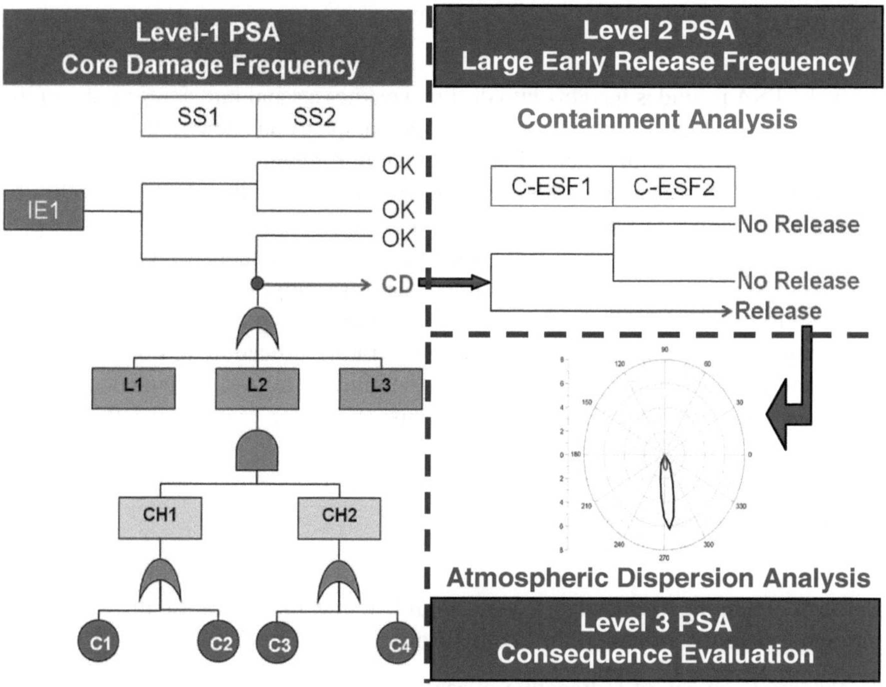

Fig. 3.2 Different levels of PSA

In Level 3 PSA [4], public health and other societal consequences such as the contamination of land or food from the accident sequences that lead to release of radioactivity into the environment are estimated. Level 1 PSA, Level 2 PSA, and Level 3 PSA are sequentially analyzed where the results of each assessment usually serve as a basis for the PSA at the next level.

To establish any plant, it is very much needed to estimate the risk of the plant. Unless it is proved that the risk from the plant is well within the regulatory limits, permission will not be given for commissioning of the plant. Probabilistic safety assessment (PSA) is an analytical technique to quantify the risk of a plant.

3.4.2.1 Objectives of PSA

The general objectives of a PSA are given as follows:

- To assess the level of safety of the plant and to identify the most effective areas for improvement.
- To assess the level of safety and compare to with explicit or implicit standards.
- To assess the level of safety to assist plant operation.

3.5 Level 1 PSA

A level 1 PSA provides insights into design weaknesses and into ways of preventing core damage, which in most cases is the precursor to accidents leading to major radioactive releases with potential health and environmental consequences [3].

In general, PSA can be used to assess the safety level of the plant, to identify the most effective areas for improvement, to compare it with standards, and to assist plant operation. It can be used to identify dominant accident sequences, to identify components, systems, and human actions important for safety, for design modifications, for evaluation of plant technical specifications, and limiting conditions of operations and for accident management (Emergency Operating Procedures). PSA can be performed at any stage of the plant life cycle either during the design stage or during the operating stage. In this section, we will discuss the concepts of level 1 PSA.

3.5.1 Scope

Before performing the analysis, one should define the scope of the study. The following parameters define the scope of any PSA study [3, 13].

1. Potential sources of radioactive releases
    a. Reactor core
    b. Spent fuel storage pool
    c. Spent fuel-handling facilities
    d. Radioactive waste storage tanks
2. Core damage states
3. Plant operational states
    a. Nominal full-power operation
    b. Reduced power operation
    c. Shutdown state
4. Initiating events
    a. Internal initiating events
        i. Loss of off-site power
        ii. Fires and floods generated inside the plant
    b. External initiating events
        i. Earthquakes
        ii. Floods
        iii. High winds
        iv. Aircraft crashes

3.5.2 *Procedure*

A general procedure for carrying out level 1 PSA [3] are given in the following steps.

- Information collection on design and operation of plant.
- Identification of sources of radioactive releases
- Selection of plant operating states
- Definition of core damage states or other consequences
  - Core damage
  - Partial Core damage
  - Core degradation state
- Identification of initiating events
- Event tree development
- System Modeling
  - Fault tree methodology
  - Data development and parameter estimation
  - Common cause failure analysis
  - Human reliability analysis
- Accident sequence modeling
- Event tree quantification
- Uncertainty and Sensitivity analysis
- Importance analysis

The above steps are explained in detail in the following subsections.

3.5.2.1 Information Collection on Design and Operation of Plant

The first and foremost step of the analysis is thorough understanding of the systems. This can be done through plant visits and discussions with designers, operating, and maintenance engineers. The information about the design aspects of the systems and operational practices in the plant can be collected from various sources such as

- Safety Analysis Reports
- Design Basis Reports
- Piping and Instrumentation diagrams
- Technical specifications
- Operating procedures
- Maintenance records

3.5.2.2 Defining the Scope of Analysis

In this step, one has to identify the different sources of radioactive releases in the plant, select the plant operating state, and define the core damage state as discussed in the beginning of the section. For example, the analysis can be done for a full-power operating plant with reactor core as the source of radioactive release and considering only core degradation. Like this, the analysis can be done by considering different options.

3.5.2.3 Identification of Initiating Events

No accident will take place in any plant without an initiating event (IE). Here, IE can be defined as "an event that creates a disturbance in the plant and has the potential to lead to core damage, in the absence of various mitigating systems (safety systems) in the plant." Hence, it is very much needed to identify all the possible initiating events that can happen in the plant. Unless the IEs are identified or postulated, it is not possible to design all kinds of safety systems that will be useful in mitigating the accidents. There are several approaches that can be followed for preparing the list of IEs. Here the aim is to make the list as complete as possible. The generally followed approaches are:

- Engineering evaluation
- Use of operational experience
- Reference to previous list
- Review of support systems
- Review of safety systems fault trees
- Precursor review

The accident initiators can be clubbed into two categories:

- Loss-of-coolant accident (LOCA) initiators
  - All events that directly cause loss of integrity of the primary coolant pressure boundary
- Transient initiators
  - Events that could create the need for a reactor power reduction or shutdown and subsequent removal of decay heat

3.5.2.4 Determination of Safety Functions

Once the IEs are identified it is important to identify the safety functions that need to be performed for each IE in order to prevent core damage. The following are the important safety functions for protecting against core damage:

- Control reactivity
- Remove core decay heat
- Maintain integrity of primary reactor coolant boundary
- Maintain primary reactor coolant inventory
- Protect containment integrity
- Scrub radioactive materials from containment atmosphere

3.5.2.5 Identification of Safety Systems

In this step, different safety systems that are required to perform safety functions are identified. The safety systems are classified into two categories

- Frontline systems (FLs)
  - The systems that directly perform a safety function, for example

    Reactor protection system
    High-/Low-pressure injection system
    Emergency feedwater system
    Pressurizer safety relief valves
    Reactor building spray injection system
    Reactor building cooling system

- Support systems (SSs)
  - The systems that are required for the proper functioning of the front line systems, for example

    Service water system
    Instrument air system
    Off-site AC power
    Diesel AC generators

The outcome of this step include

- Identification of safety functions for each IE and table of front line systems that perform each function
- A list of front line systems
- A list of support systems
- Dependency table among FLs and SSs
- Dependency table among SSs and SSs
- Identification of success criteria for each front line system of each IE

3.5.2.6 Grouping of the IEs

IEs can be grouped in such a way that all events in the same group impose essentially the same success criteria on the front line systems as well as the same special conditions. Also, IEs that evoke the same type of plant response but for which front line system success criteria are not identical. Examples of different IE groups are:

- LOCAs
  - Small LOCA
  - Medium LOCA
  - Large LOCA
- Main steam line break
- Feedwater line break
- Steam generator tube rupture etc.

3.5.2.7 Accident Sequence Modeling

An event sequence model provides sequences of events that, following an initiating event, lead either to a successful state or to a core damage state. Event sequences are expressed in terms of initiating events and success or failures of mitigating systems. There are different methods available for modeling event sequences, some of them are listed below:

- Event trees (ET)
- Cause consequence diagrams (CCD)
- Event sequence diagrams

Similarly, there are different methods available for modeling the systems as given below:

- Fault tree (FT)
- State space diagrams and Markov analysis
- Reliability Block Diagram (RBD)
- Go chart (GO)

In general, it is recommended to use combined event tree/fault tree method with the option of small event tree/large fault tree approach in which dependences between front line systems and support systems do not appear in the event trees. Whereas, in the case of large event tree/small fault tree approach the dependences between front line systems and support systems do appear in the event trees.

3.5.3 Event Tree Analysis

If successful operation of a system depends on approximately chronological, but discrete, operation of its units or subsystems (Units should work in a defined sequence for operational success), then an event tree method is used. This is used for complex systems like NPPs but not for simple systems. Event tree is a graphical representation of success or failure of safety systems in a chronological order when an initiating event occurs. Event tree will start with a specific Initiating event. Whenever an initiating event occurs in a NPP, the following activities should be performed in order to avoid accident kind of situation, i.e., shutting down the reactor, removing the decay heat from the core and maintaining the long-term subcriticality. To perform these activities, different safety systems are existed in the plant. Each initiating event will call for different safety systems depending on the type of event. To shutdown the reactor, reactor protection system (primary shut-down system or secondary shut down system) will be used. Depending upon the type of initiating event, different decay heat removal systems are designed. Similarly, there are different systems designed for maintaining the long-term sub-criticality in the core.

In the event trees, success of the safety systems is represented with upper arrow and failure of the same is represented with downward arrow. The outcome of the event trees will be representation of different sequences and state of the core. Each sequence will represent either success or failure of the safety systems along with occurrence of initiating event. Some sequences will result into safe state of the core whereas some sequences result into unsafe state of the core (core damage). With the help of event trees one can identify the dominating accident sequences which are leading to core damage kind of situation.

An event tree of a Class IV power supply failure of a typical PHWR is shown in the Fig. 3.3. The event tree is developed based on the actions to be performed during a given initiating event. Upon failure of Class IV power, reactor trips on 'No Primary Coolant Pump running'. High primary heat transport (PHT) Pressure trip will follow if the first trip parameter fails. This leads to the actuation of Reactor Protection System (RPS), initially with Primary Shutdown System (SDS-1) and with secondary shutdown system SDS-2, if SDS-1 fails. Emergency power supply (EPS), i.e., Class III is 6.6 kV system with 4 DG sets. If Class III is available the mode of decay heat removal and long-term reactivity control will be same as normally followed with decay heat removal systems (DHRS) such as Secondary Steam Relief System (SSR), Auxiliary Boiler Feed Water System (ABFWS), and Shut down Cooling System (SDCS).

If there is a failure in decay heat removal systems, core cooling will be achieved through valving in of fire water system (FWS). Class IV failure followed with complete loss of Class III failure leads to a Station Blackout scenario. During station black out scenario core cooling will be achieved through valving in of Fire water system.

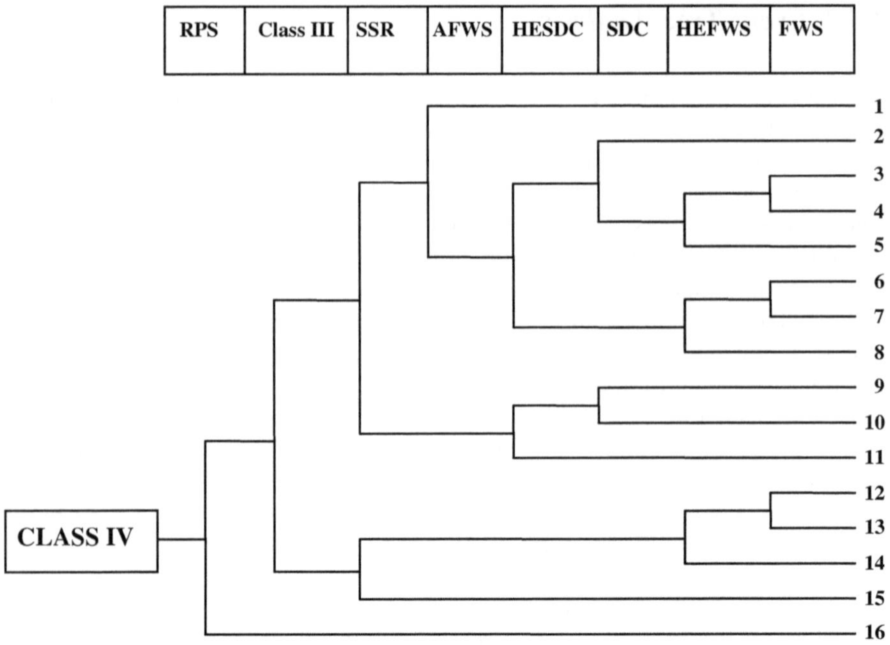

Fig. 3.3 Event tree for Class IV failure

In the above event tree, one can find different sequences that are enumerated. Each sequence represents success or failure of different safety systems along with the occurrence of the initiating event. The consequence of each sequence will differ from each other depending on the functioning of the safety systems. The consequences are usually defined based on the status of the core which in turn represented by fuel temperature. Now the question comes which are all the sequences that will lead to core damage kind of situation. To answer this question first one has to define the term "Core Damage." There are several definitions for this term depending on the type of reactor. Following are the some of the definitions provided [14].

IAEA: Core damage state is determined by criteria such as

- The limiting peak cladding temperature is below 1204 °C (LOCA)
- The amount of fuel element cladding that reacts chemically with water or steam does not exceed 1 % of the total amount of zircalloy in the reactor
- The core remains coolable to cooling during and after the accident
- Possibility for core relocation following an accident

CANDU 'Pickering A': CANDU defines various fuel damage categories (FDC). Categories FDC1 and FDC2 are included in severe core damage state.

- FDC1 (estimated frequency: 5.0 per 10 million years) involves rapid loss of the structural integrity of the core. Event sequences in this category involve a rapid reactivity excursion (at least 1 mk/s) followed by failure to shut down the reactor. The reactivity excursion could arise from a large loss-of-coolant accident (LOCA).
- FDC2 (estimated frequency: 1.3 per 10 thousand years) involves loss of the structural integrity of the core, but over a longer period than applies for FDC1. The onset of significant core damage might occur within 0.6–36 h after accident initiation. Event sequences in this category involve failure to remove decay heat from the core.

IPHWR: Reactor state brought about by the accident conditions with loss of fuel geometry or resulting in crossing of design basis limits or acceptance criteria limits for one or more parameters.

- The parameters to be considered include: fuel clad strain, fuel clad temperature, primary and secondary system pressures, fuel enthalpy, clad oxidation, % of fuel failure, H_2 generation from metal water reaction, radiation dose, time required for operator to take emergency mitigatory actions. This definition does not indicate the critical values of the parameters for defining core damage. However, accident sequences involves failure of shutdown systems, failure of long-term subcriticality, and loss of heat sink when the reactor is at either high pressure or low pressure have been considered to result in severe core damage.

To determine the status of the core, one has to perform thermal hydraulic studies of each and every sequence. Out of these sequences, very few sequences will fall under core damage category. These sequences are called dominating accident sequences. In the above event tree, the following sequences are identified as dominating accident sequences.

- Class IV-RPS
- Class IV-Class III-FWS

From the above sequences, core damage can occur through any of the sequences. Hence, core damage frequency can be mathematically defined as follows.

$$\text{CDF} = \text{Frequency of Accident Sequence 1} + \text{Frequency of Accident Sequence 2}$$

The above expression can be generalized to all the initiating events and can be written as follows:

$$\text{CDF} = \sum \text{Dominating Accident Sequences} \tag{3.1}$$

In order to quantify the value of core damage frequency, one should know the frequency of occurrence of all the initiating events and the safety systems failure probabilities. The frequency of IEs can generally be obtained from the plant operating experience data or by performing system reliability analysis (fault tree methods). The failure probabilities of systems can be obtained by performing system reliability analysis. There are different techniques available for performing system reliability analysis but in general fault tree method is used for very complex systems like NPPs. The fault tree method is explained in detail in the subsequent sections with example problems.

Example 3.2 Consider the event tree of Class IV failure as shown in the Fig. 3.3. Probability of failure of RPS is 1×10^{-6}, Class III is 1×10^{-3}, and that of FWS is 2×10^{-3}. The frequency of occurrence of Class IV failure is 0.8/yr. The following sequences are leading to core damage situation.

- Class IV-RPS
- Class IV-Class III-FWS

Find the frequency of occurrence of core damage due to Class IV failure.

Solution:

- Class IV failure frequency = 0.8/year
- RPS failure probability = 1×10^{-6}
- Class II failure probability = 1×10^{-3}
- FWS failure probability = 2×10^{-3}
- Accident Sequences (AS)
  - AS1 = Class IV-RPS
  - AS2 = Class IV-Class III-FWS
- Core damage frequency (CDF)

$$\begin{aligned}
\text{CDF} &= \lambda_{\text{AS1}} + \lambda_{\text{AS2}} \\
\lambda_{\text{AS1}} &= \lambda_{\text{Class IV}} P_{\text{F}}(\text{RPS}) \\
&= 0.8 \times 1 \times 10^{-6} \\
\lambda_{\text{AS1}} &= 8 \times 10^{-7}/\text{year} \\
\lambda_{\text{AS2}} &= \lambda_{\text{Class IV}} P_F(\text{RPS} \cap \text{FWS}) \\
&= \lambda_{\text{Class IV}} P_F(\text{RPS}) P_F(\text{FWS}) \\
&= 0.8 \times 1 \times 10^{-3} \times 2 \times 10^{-3} \\
\lambda_{\text{AS2}} &= 1.6 \times 10^{-5}/\text{year} \\
\text{CDF} &= \lambda_{\text{AS1}} + \lambda_{\text{AS2}} \\
&= 8 \times 10^{-7} + 1.6 \times 10^{-5} \\
\text{CDF} &= 1.608 \times 10^{-5}/\text{year}
\end{aligned}$$

- From the above it can be seen that CDF due to Class IV power supply failure is estimated as 1.608×10^{-5}/year.

3.5.4 *Fault Tree Analysis*

A Fault Tree [15, 16] is a graphical representation of component failures relating to a particular system failure. This technique was originally developed by Bell Telephone Laboratory in 1961 to evaluate the control system of a missile. Fault tree actually looks like an inverted tree with the system failure as the top event and branches down to the root causes which is generally represented as subcomponent failures. One starts with the top event as any undesirable event of system failure and analysis proceeds by asking "what could cause this?" In answering this question, one generates other events and their logical combinations which are pictorially displayed. Analysis terminates when events are encountered which cannot or need not be developed further. These events are called basic events. The combination of events are logical operation like "AND," "OR." This can be expressed by Boolean expression and top event probability is calculated in terms of probabilities of basic events.

3.5.4.1 Evaluation of Fault Trees

- Qualitative Evaluation: In this, we determine the minimal cut sets of the fault tree (minimum number of components failure will lead to the system failure).
- Quantitative Evaluation: In this, we determine the probability of the occurrence of the top event, unreliability or reliability associated with the top event can be determined.

3.5.4.2 Advantages of Fault Tree

(1) It provides a pictorial display of failure consequence chains which is very useful in the marginal assessment of reliability of a complex system.
(2) It gives an insight into the week parts or critical parts of the investigated system and gives a chance to improve the reliability by suitable design modifications.
(3) The logic diagram of a fault tree is useful as an aid for system trouble shooting if the considered failure mode (undesired event) occurs.

The generally used logic symbols in a fault tree are listed below [17]:

AND gate - a logic gate where an output event occurs only when all the input events have occurred.

OR gate - a logic gate where an output event occurs if at least one of the input events have occurred.

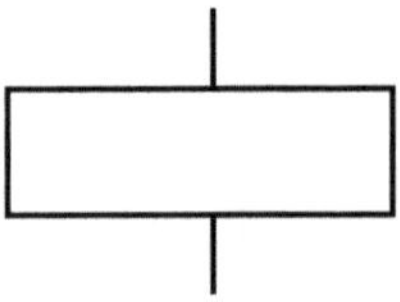

Resultant event - a fault event resulting from the logical combination of other fault events and usually an output to a logic gate.

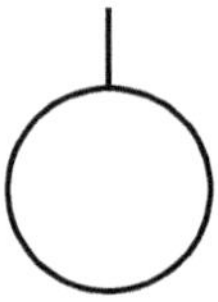

Basic event - an elementary event representing a basic fault or component failure.

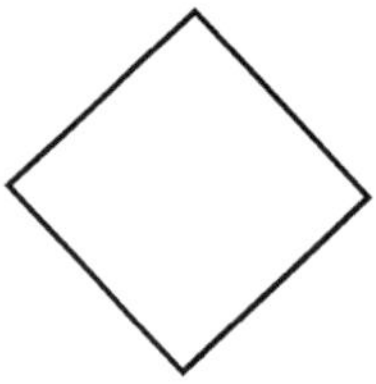

Incomplete event - an event that has not been fully developed because of lack of knowledge or its unimportance.

We will examine how to develop fault trees for simple systems to complex systems.

Case 1: Series System

Now consider a simple series system composed of two components as shown in the Fig. 3.4a. The fault tree for the system can be developed by defining the failure criteria for the given system. In this case, failure criterion is if any one of the components fails the system will fail. The fault tree is shown in the Fig. 3.4b.

The probability of failure can be calculated mathematically as follows:

$$P_F = P(A_F \cup B_F) = P(A_F) + P(B_F) - P(A_F \cap B_F)$$

Considering the components are independent (failure of one component does not affect failure of the other component), then

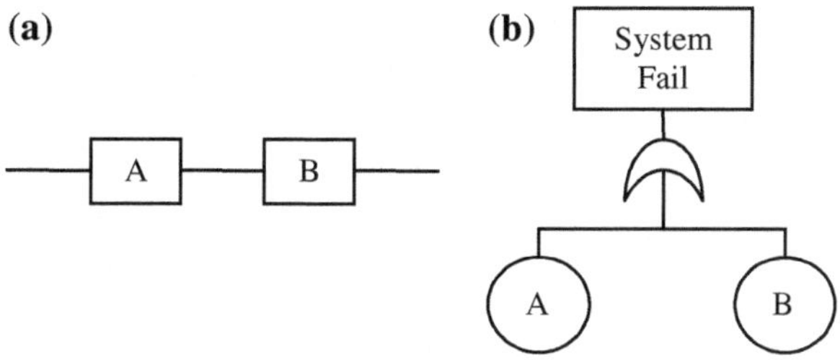

Fig. 3.4 **a** Series system. **b** Fault tree for series system

$$P_F = P(A_F \cup B_F) = P(A_F) + P(B_F) - P(A_F)P(B_F)$$
$$\text{Let } P(A_F) = F_1 \; P(B_F) = F_2$$
$$P_F = F_1 + F_2 - F_1F_2$$

Reliability of the system can be calculated as follows:

$$\text{We know that } F_1 = 1 - R_1 \quad F_2 = 1 - R_2 \quad P_{\mathrm{F}} = 1 - P_S = 1 - R_S$$
$$1 - R_S = 1 - R_1 + 1 - R_2 - (1 - R_1)(1 - R_2)$$
$$R_S = R_1R_2$$

If the component failure probabilities are very low then the simultaneous occurrence of component failure can be neglected. This approximation is called rare event approximation. Since the probability of simultaneous occurrence of multiple component failures is very rare. By adopting this assumption, the probability of failure of the series system can be rewritten as follows:

$$P_F = F_1 + F_2 - F_1F_2$$
$$\text{Assume} \quad F_1F_2 \cong \text{Very low} \cong 0$$
$$P_F = F_1 + F_2$$

As an example, assume the following component failure data

$$F_1 = 1 \times 10^{-3} \quad F_2 = 1 \times 10^{-2}$$
$$P_F = F_1 + F_2 - F_1F_2$$
$$P_F = 1 \times 10^{-3} + 1 \times 10^{-2} - 1 \times 10^{-3} \times 1 \times 10^{-2}$$
$$P_F = 1.1 \times 10^{-2} - 1 \times 10^{-5} = 1.101 \times 10^{-2} \cong 1.1 \times 10^{-2}$$

In the above calculation since the product of failure probabilities of both the components is in the order of 10^{-5} this is not contributing much to the system failure probability. Hence, this contribution can be neglected. Here main contribution is coming from individual component failure probabilities. Care should always be taken when this kind of approximation is imposed. Because, if the failure probabilities are very high, assume that they are in the order of 10^{-1} then if one impose this condition then the system failure probability may sometimes go beyond 1 and is violation of probability axioms. Normally, if we consider NPP systems whose component failure probabilities should be maintained very low for better availability of the plant. Hence, this approximation can be applied.

The failure probability of the system can also be obtained by finding out the cut sets and minimal cut sets (MCS) for the system. Here, cut sets can be defined as all the combination of component failures leading to system failure is called cut sets. Whereas, minimal combination of component failures leading to system failure is called minimal cut sets. This MCS can be obtained from cut sets of the system. For the simple series system as discussed above, the cut sets can be obtained as:

$$A_F$$
$$B_F$$
$$A_F, B_F$$

This means if component *A* fails system fails or component *B* fails system fails or if both the component fails simultaneously then also system fails. These are all possible combinations of component failures that will lead to system failure. If system is failing just because of single component failure then the combination of both component failures simultaneously is need not to consider and then MCS for the above system can be represented as below:

$$A_F$$
$$B_F$$

These are called single-order cut sets because single component failure leads to the system failure. System failure probability can be calculated directly by summing up all the cut set probabilities.

$$P_F = P(A_F) + P(B_F)$$
$$P_F = F_1 + F_2$$

Case 2: Parallel System

Now consider a parallel system composed of two components as shown in the Fig. 3.5a. The fault tree for the system can be developed by defining the failure criteria for the given system. In this case, the failure criterion is if both the components fail simultaneously then only the system will fail. The fault tree is shown in the Fig. 3.5b.

The probability of failure can be calculated mathematically as follows:

$$P_F = P(A_F \cap B_F)$$

Considering the components are independent (failure of one component does not affect failure of the other component), then

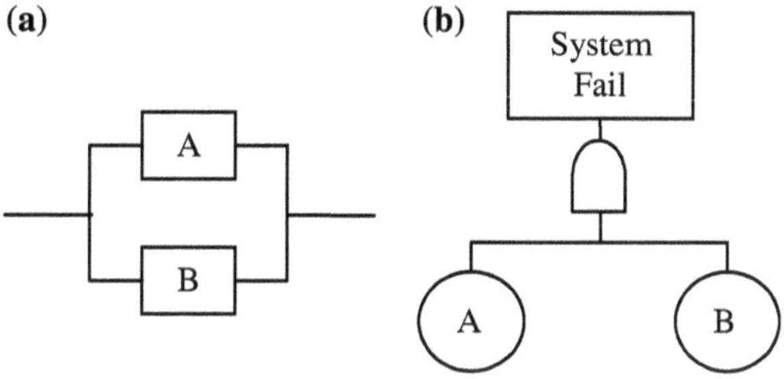

Fig. 3.5 **a** Parallel system. **b** Fault tree for parallel system

$$P_F = P(A_F)P(B_F)$$
$$\text{Let } P(A_F) = F_1 \quad P(B_F) = F_2$$
$$P_F = F_1F_2$$

Reliability of the system can be calculated as follows:

$$\text{We know that } F_1 = 1 - R_1 \; F_2 = 1 - R_2 \; P_{\text{F}} = 1 - P_{\text{S}} = 1 - R_{\text{S}}$$
$$1 - R_{\text{S}} = (1 - R_1)(1 - R_2)$$
$$R_{\text{S}} = R_1 + R_2 - R_1R_2$$

For the above system, both cut sets and MCS are same because system will fail when both the components fail simultaneously and is given as follows:

$$A_F, B_F$$

This is called the second-order cut set, because for the system to fail two components should fail simultaneously. Probability of failure can be calculated as follows

$$P_F = P(A_F)P(B_F)$$
$$P_F = F_1F_2$$

Case 3: Series and Parallel Combination
Now consider a system composed of both series and parallel combination as shown in the Fig. 3.6. In this case failure criterion is if both the components in

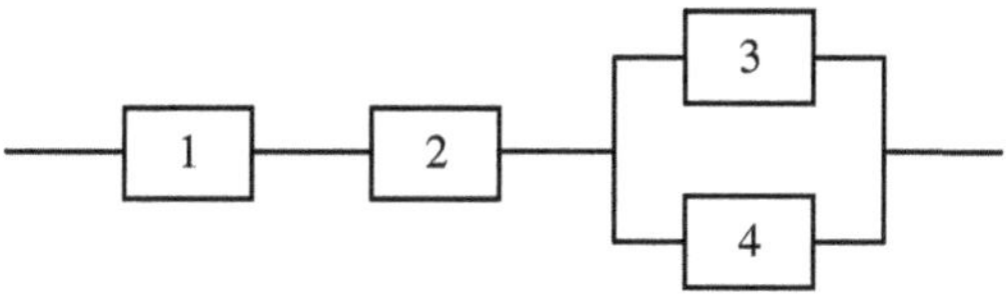

Fig. 3.6 Series and parallel combination system

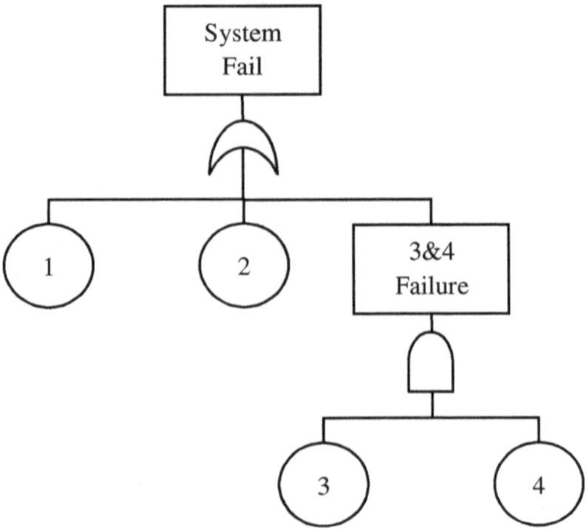

Fig. 3.7 Fault tree for series and parallel combination system

parallel fail simultaneously then only the system will fail. The fault tree is shown in the Fig. 3.7.

For the above system, cut sets are given in the Table 3.1 and the MCS list for the same system is given in the Table 3.2.

Probability of failure can be calculated as follows:

$$P_F = P(1_F) + P(2_F) + P(3_F)P(4_F)$$
$$P_F = F_1 + F_2 + F_3F_4$$

3.5.5 Common Cause Failures

Common cause failure (CCF) [18, 19] is the failure of more than one component or system simultaneously due to the same common cause. Hence, these failures defeat

Table 3.1 Cut sets for the system

1st order	2nd Order	3rd Order	4th Order
1	1, 2	1, 2, 3	1, 2, 3, 4
2	1, 3	1, 2, 4	
	1, 4	2, 3, 4	
	2, 3		
	2, 4		
	3, 4		

Table 3.2 MCS for the system

1st order	2nd Order
1	3, 4
2	

the redundancy or diversity that is employed to improve the reliability of the systems. In the absence of common cause failures, separate components of redundant system or diverse methods of providing the same function, are regarded as independent so that the unavailability of the system is essentially the product of the unavailabilities of the separate components or diverse systems. However, a common cause failure arises from some cause that fails more than one system, or more than one component of a system, simultaneously [13]. Thus, the effect of common cause failures results in higher unavailability of the function with respect to the situation of true independence.

3.5.5.1 Common Cause Groups

Now the question arises how the CCFs are treated in Level 1 PSA analysis. These are generally treated at fault tree/event tree level which will have effect on the system failure probabilities. In implementing CCFs in the analysis, first one should identify root causes of failure of components and determine the groups of components that are susceptible to the cause of failure. Depending on the root causes that affect the components, there are three groups of components that are listed below:

- *Root causes that affect similar equipment*: This group contains similar components. Because similar components will have same design, common manufacturing processes, same installation, same maintenance, and testing procedures. Due to the above features, the components under this group will be affected by some root cause.
- *Root causes that affect equipment operated according to the same procedures*: In the plant, there are some components for which emergency or normal operating procedures are same. Hence, due to some common operator error all these components could fail. This group may contain dissimilar components.
- *Root causes that affect equipment in the same location*: CCFs can also be attributable to harsh environmental conditions caused by energetic events such as fire, flood, earthquakes, explosions, missiles etc. and nonenergetic events such as contamination, vibration, moisture, corrosion, high temperature etc. These events affect similar components, since these events are caused by human errors in design, installation, and maintenance activities. CCFs can be avoided providing barriers such as walls, fire doors within the plant.

Some examples of common cause groups of a typical NPP are listed below

1. Active process water system of a typical NPP contains four pumps. All the pumps may fail to run simultaneously due to the same design, manufacturing processes, same installation, same testing, and maintenance procedures.
2. Similarly, diesel generators of the Class III power supply system of a NPP will fall under the same CCF group. The failure modes can fail to start or fail to run. The root causes can be attributed to same design and manufacturing processes as mentioned above.

3.5.6 Common Cause Failure Models

Once the CCF groups are identified, the next step is to implement the failure models into the system fault trees. There are different models available for treating the CCFs. These models can be categorized based on the number of parameters, their assumptions regarding the cause, coupling mechanism, and the impact of common cause failures. Based on the number of parameters, there are single and multiple parameter models available [19, 20]. Under the single parameter model, Beta factor (β) model is used whereas, Alpha factor (α) and Multiple Greek Letter (MGL) models fall under the category of multiple parameter models. These models are in general used in level 1 PSA analysis. There are some more models available in the literature but are not discussed here. Refer EPRI NP-5613 [19] for full list of models. The parameters of these models are estimated based on the operating experience data. Some of the models are explained as follows:

3.5.6.1 Beta Factor Model

The beta factor model is a single parameter model, i.e., in addition to the total component failure probability, it uses one parameter to calculate the CCF probabilities. In this model, beta (β) is the parameter that is used to represent the contribution from common cause failures. Here β represents the fraction of the component failure rate that can be associated with common cause events shared by other components in that group. This model is developed based on the assumption that whenever a common cause event occurs, all the components in that common cause component group are assumed to fail. Based on this assumption, the model can be mathematically represented as follows. The total failure probability of a component can be divided into two parts: one is contribution from individual component failure and the other is failure contribution from component failure sharing with other components. This is given as

$$\begin{aligned} Q_t &= Q_{\mathrm{I}} + Q_{\mathrm{CCF}-m} \\ Q_I &= (1-\beta)Q_t \\ Q_{\mathrm{CCF}-m} &= \beta Q_t \end{aligned} \tag{3.2}$$

This implies that

$$\beta = \frac{Q_{\mathrm{CCF}-m}}{Q_I + Q_{\mathrm{CCF}-m}}$$

where

Q_t Total failure probability of one component

Q_I Independent failure probability of the single component

$Q_{\mathrm{CCF}-m}$ Probability of basic event failure involving m specific components, and m is the number of components in that common cause group

The Eq. (3.2) can be generalized and can be written for m components involving failure of k components ($k \leq m$),

$$Q_k = \begin{cases} (1-\beta)Q_t & k = 1 \\ 0 & 2 \leq k < m \\ \beta Q_t & k = m \end{cases} \quad (3.3)$$

where
Q_k Probability of basic event involving k specific components.

Consider a three-component parallel configuration system as shown in the Fig. 3.8a and all the components are similar components (valves, pumps etc.).

The fault tree shown in Fig. 3.8b is developed without considering the common cause failures among the components. Now consider the above three components belonging to one common cause group and they will fail together due to some common cause. Then total failure probability of each component can be written as follows:

$$\begin{aligned} A_F &= A_I + CC_{AB} + CC_{AC} + CC_{ABC} \\ B_F &= B_I + CC_{BA} + CC_{BC} + CC_{ABC} \\ C_F &= C_I + CC_{CA} + CC_{CB} + CC_{ABC} \end{aligned} \quad (3.4)$$

where
A_I Independent failure contribution = Q_1
CC_{AB} Common cause contribution in which two components fail due to one common cause = Q_2
CC_{ABC} Common cause contribution in which three components fail due to one common cause = Q_3

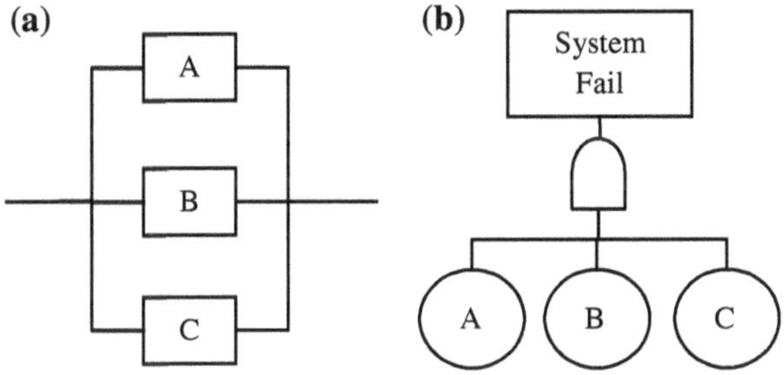

Fig. 3.8 **a** 3 Component parallel system. **b** Fault tree for 3 component parallel system

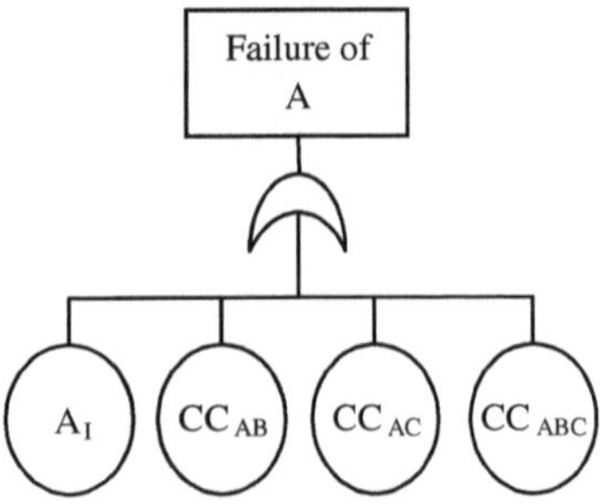

Fig. 3.9 Fault tree representing component A failure

The total failure of the individual component can be represented with a fault tree as shown in Fig. 3.9.

In case of beta factor model, whenever a common cause event occurs, all the components in that common cause component group are assumed to be failed. Hence, in the present case, failure of two components due to common cause is not applicable. Therefore, the total failure probability of each component can be rewritten as

$$\begin{aligned} A_F &= A_{\mathrm{I}} + CC_{ABC} \\ B_F &= B_{\mathrm{I}} + CC_{ABC} \\ C_F &= C_{\mathrm{I}} + CC_{ABC} \end{aligned} \tag{3.5}$$

This can be represented in the fault tree as shown in the Fig. 3.10. Now the total system failure can be represented in a fault tree as shown in the Fig. 3.11. For the present system after taking the contributions from both independent and common cause failures, MCS can be given as follows:

$$\{A_{\mathrm{I}}, B_{\mathrm{I}}, C_{\mathrm{I}}\}$$
$$\{CC_{ABC}\}$$

Hence the system failure probability can be calculated as

$$F_{\mathrm{sys}} = A_I \times B_I \times C_I + CC_{ABC} = Q_{\mathrm{I}}^3 + Q_3 \tag{3.6}$$

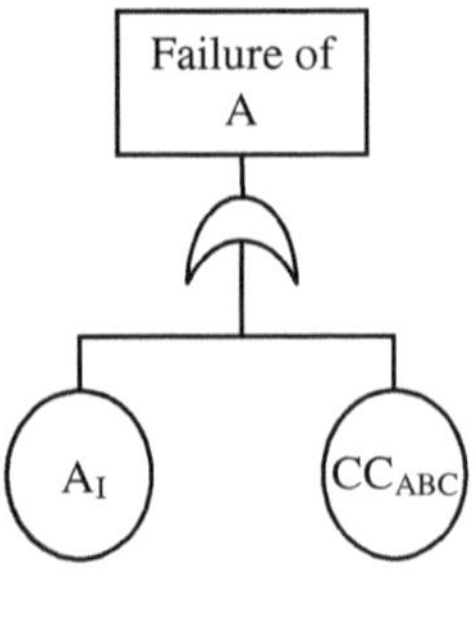

Fig. 3.10 Fault tree representing component A failure

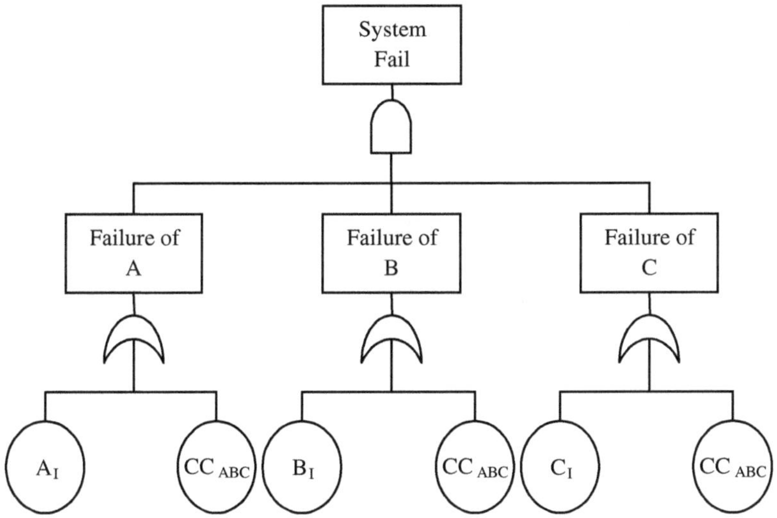

Fig. 3.11 Fault tree representing common cause and independent failures

From the beta factor model

$$Q_1 = (1 - \beta)Q_t$$
$$Q_3 = \beta Q_t$$

where

Q_t Total failure probability of each component and is considered to be same for all the components.

Example 3.3 Consider the example of a check valve in a system. There are three redundant loops in the system each loop has a check valve in its path. Consider these three components fall under one common cause group due to their same design, manufacturing processes, functioning etc.

- The failure mode of the component fails to function.
- The failure probability of each component is 2.3×10^{-3} and all the components are identical.
- Consider a beta factor of 0.1, then according to beta factor model, the independent failure probability of each component can be calculated as

$$\begin{aligned} Q_I &= (1 - \beta)Q_t \\ &= (1 - 0.1) \times 2.3 \times 10^{-3} \\ Q_I &= 2.07 \times 10^{-3} \end{aligned}$$

- And the common cause contribution can be calculated as

$$\begin{aligned} Q_{\mathrm{CCF}} &= \beta Q_t \\ &= 0.1 \times 2.3 \times 10^{-3} \\ Q_{\mathrm{CCF}} &= 2.3 \times 10^{-4} \end{aligned}$$

- From the above, total failure probability of each component can be verified as

$$\begin{aligned} Q_t &= Q_I + Q_{\mathrm{CCF}} \\ &= 2.07 \times 10^{-3} + 2.3 \times 10^{-4} \\ Q_t &= 2.3 \times 10^{-3} \end{aligned}$$

3.5.6.2 Alpha Factor Model

The alpha factor model is a multiple parameter model [19, 21]. This model, in addition to the total component failure probability, uses multiple parameters to calculate the CCF probabilities. In this model, different alpha factors ($\alpha_1, \alpha_2, \alpha_3 \ldots \alpha_k$) are used to represent the contribution from common cause failures. Here α_k represents the ratio of the probability of failure events involving any k components over the total failure probability of all failure events in a given common cause group. The alpha factor model defines common cause failure probabilities from a set of alpha factors and the total component failure probability, Q_t. The basic event probabilities can be expressed in terms of Q_t and the alpha factors as follows [13]:

$$Q_k^{(m)} = \frac{k\alpha_k^{(m)}}{\binom{m-1}{k-1}\alpha_t} Q_t \tag{3.7}$$

where

$$\alpha_t = \sum_{k=1}^{m} k\alpha_k^{(m)}$$

In terms of the basic event probabilities, the alpha factor parameters are defined as:

$$\alpha_k^{(m)} = \frac{\binom{m}{k} Q_k^{(m)}}{\sum_{k=1}^{m} \binom{m}{k} Q_k^{(m)}} \quad \text{and} \quad \sum_{k=1}^{m} \alpha_k^{(m)} = 1 \tag{3.8}$$

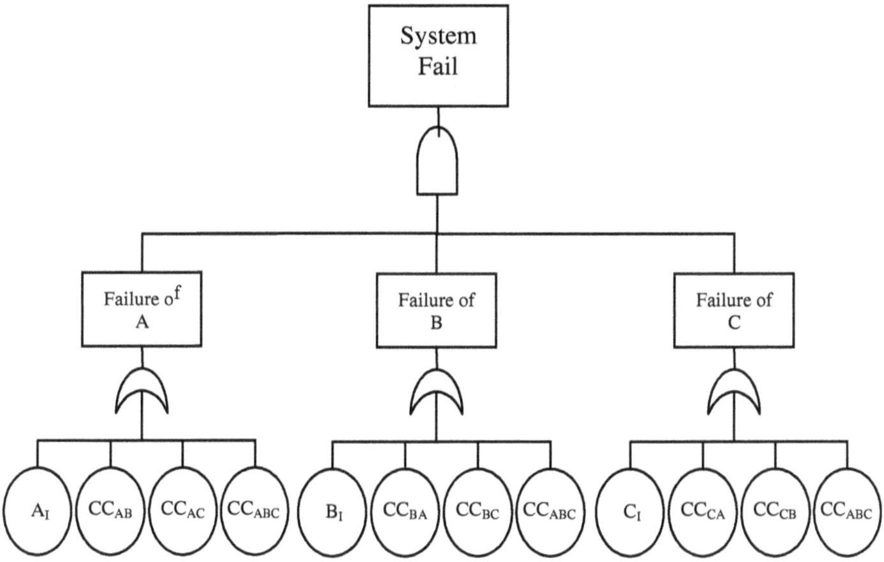

Fig. 3.12 Fault tree representing common cause and independent failures

where $\binom{m}{k}Q_k^{(m)}$ is the probability of events involving k component failures in a common cause group of m components, and the denominator is the sum of such probabilities. In other words, $\alpha_k^{(m)}$ is the ratio of the probability of failure events involving any k components over the total failure probability of all failure events in a group of m components.

Consider the three-component parallel configuration system as shown in the Fig. 3.8a. The fault tree of the system is also shown in the Fig. 3.8b. The total failure probability of each component can be expressed in terms of individual failure probability and common cause contribution along with the other components in the group as given in Eq. 3.4. Now the total system failure can be represented in a fault tree as shown in the Fig. 3.12. For the present system, after taking the contributions from both independent and common cause failures, MCS can be given as follows:

$$\{A_{\mathrm{I}}, B_{\mathrm{I}}, C_{\mathrm{I}}\};$$
$$\{A_{\mathrm{I}}, CC_{BC}\}; \{B_{\mathrm{I}}, CC_{AC}\}; \{C_{\mathrm{I}}, CC_{AB}\};$$
$$\{CC_{ABC}\}$$

Hence, the system failure probability can be calculated as

$$\begin{aligned} F_{\text{sys}} &= A_I \times B_I \times C_I \\ &+ A_I \times CC_{BC} + B_I \times CC_{AC} + C_I \times CC_{AB} + CC_{ABC} \\ &= Q_1^3 + 3Q_1Q_2 + Q_3 \end{aligned} \tag{3.9}$$

From the alpha factor model (m = 3)

$$\begin{aligned} Q_1 &= \frac{\alpha_1}{\alpha_t} Q_t \quad \alpha_t = \alpha_1 + 2\alpha_2 + 3\alpha_3 \\ Q_2 &= \frac{\alpha_2}{\alpha_t} Q_t \quad Q_3 = \frac{3\alpha_3}{\alpha_t} Q_t \end{aligned} \tag{3.10}$$

Example 3.4 Consider the same example of a check valve as discussed in Example 3.3. Consider the alpha factors as listed below:

- $\alpha_1 = 0.95$
  $\alpha_2 = 0.035$
  $\alpha_3 = 0.015$

- In this case, the number of components are three, i.e., m = 3
- Then according to the alpha factor model, the independent failure probability of each component can be calculated as

$$Q_k^{(m)} = \frac{k\alpha_k^{(m)}}{\binom{m-1}{k-1}\alpha_t} Q_t$$

for $m = 3,\ k = 1$

$$Q_1^{(3)} = \frac{1\alpha_1^{(3)}}{\binom{3-1}{1-1}\alpha_t} Q_t = \frac{\alpha_1}{\alpha_t} Q_t$$

where $\alpha_t = \sum_{k=1}^{m} k\alpha_k^{(m)} = 1\alpha_1 + 2\alpha_2 + 3\alpha_3$

$$\alpha_t = 0.95 + 2 \times 0.035 + 3 \times 0.015 = 1.065$$

$Q_t = 2.3 \times 10^{-3}$

$$\therefore \quad Q_1 = \frac{0.95}{1.065} \times 2.3 \times 10^{-3} = 2.0516 \times 10^{-3}$$

for $m = 3,\ k = 2$

$$Q_2^{(3)} = \frac{\alpha_2}{\alpha_t} Q_t$$

$$Q_2 = \frac{0.035}{1.065} \times 2.3 \times 10^{-3} = 0.0756 \times 10^{-3}$$

for $m = 3,\ k = 3$

$$Q_3^{(3)} = \frac{3\alpha_3}{\alpha_t} Q_t$$

$$Q_3 = \frac{3 \times 0.015}{1.065} \times 2.3 \times 10^{-3} = 0.0972 \times 10^{-3}$$

- From the above, the total failure probability of each component can be verified as

$$\begin{aligned} Q_t &= Q_1 + 2 \times Q_2 + Q_3 \\ &= 2.0516 \times 10^{-3} + 2 \times 0.0756 \times 10^{-3} + 0.0972 \times 10^{-3} \\ &= 2.3 \times 10^{-3} \end{aligned}$$

3.5.6.3 Multiple Greek Letter (MGL) Model

The MGL model is further extension of the beta factor model [20]. To account for higher order redundancies and to implement the different component failure combinations other than the failure of all the components apart from beta factor other parameters are introduced. As is the case of other models this also consists of total component failure probability Q_t which includes the effects of not only independent but also common cause contributions for all component failure and a set of failure fractions. The parameters are defined based on the size of the group. For example for a four-component group, MGL model has four parameters and they are expressed as follows:

$$\begin{aligned} Q_t &= \sum_{k=1}^{m} \binom{m-1}{k-1} Q_k^m \\ &= Q_1^{(4)} + 3Q_2^{(4)} + 3Q_3^{(4)} + Q_4^{(4)} \\ \beta &= \frac{3Q_2^{(4)} + 3Q_3^{(4)} + Q_4^{(4)}}{Q_1^{(4)} + 3Q_2^{(4)} + 3Q_3^{(4)} + Q_4^{(4)}} \\ \gamma &= \frac{3Q_3^{(4)} + Q_4^{(4)}}{3Q_2^{(4)} + 3Q_3^{(4)} + Q_4^{(4)}} \\ \delta &= \frac{Q_4^{(4)}}{3Q_3^{(4)} + Q_4^{(4)}} \end{aligned} \tag{3.11}$$

where

β Conditional probability that the cause of a component failure will be shared by one or more additional components, given that a specific component has failed.

γ Conditional probability that the cause of a component failure that is shared by one or more components will be shared by two or more additional components, given that two specific components have failed.

δ Conditional probability that the cause of a component failure that is shared by two or more components will be shared by three or more additional components, given that three specific components have failed.

For a general case,

$$Q_k^{(m)} = \frac{1}{\binom{m-1}{k-1}} \prod_{l=1..k} \rho_l (1 - \rho_{k+1}) Q_t \tag{3.12}$$

where

$$\rho_1 = 1,\ \rho_2 = \beta,\ \rho_3 = \gamma, \ldots,\ \rho_{m+1} = 0$$

The Eq. 3.13 expresses the probability of multiple component failures due to common cause, Q_k, in terms of the MGL parameters, for a four-component group:

$$\begin{aligned} Q_1^{(4)} &= (1-\beta)Q_t \\ Q_2^{(4)} &= \frac{\beta(1-\gamma)}{3} Q_t \\ Q_3^{(4)} &= \frac{\beta\gamma(1-\delta)}{3} Q_t \\ Q_4^4 &= \beta\gamma\delta Q_t \end{aligned} \tag{3.13}$$

3.5.7 Component Failure Probability Models

In the fault trees, system failures are represented with logical combination of component failures and it terminates when events are encountered which cannot be developed further. These events are called basic events. These basic events represent either failure of the component or different failure modes of the components. In quantifying the failure probability of the system, one should have the information about the failure data as well as failure models of the components.

There are different failure models [3] existed depending on the type of component that is being used in the plant for evaluating the failure probability of the components. These components also differ by the different systems in the plant. The systems can be either continuously operating systems or can be standby systems.

3.5.7.1 Operating Systems

These systems are continuously operating systems. The failure probability of these systems is defined as the probability that the system will fail to function for a given period of time (mission time). However, the system failure probability depends on the component failure probability. These operating systems generally contain two types of components viz., repairable components and nonrepairable components.

3.5.7.2 Repairable Components

As the name implies, these components can be repaired during operation of the system when they fail. This means components can be repaired without taking the total system out of service. If there is no redundant component and the failed component is not accessible for repair, then the component should be considered as nonrepairable component. While modeling these components, the above conditions should be considered. The failure probability or unavailability of online repairable components can be calculated as follows:

$$P_F = \frac{\lambda_0 T_R}{1 + \lambda_0 T_R} \tag{3.14}$$

where

λ_0 is the operating failure rate
T_R the mean time to repair (MTTR)

In the above expression, the parameters can be estimated based on the plant operating experience data. The operating failure rate (λ_0) can be calculated as follows:

$$\lambda_0 = \frac{\text{Number of observed failures}}{\text{Total observed time}} \tag{3.15}$$

Similarly, MTTR can be estimated based on the observed individual times for repair.

Example 3.5 Find the failure probability of a pump for the following information. The pump is required to run during its operation. The failure mode of this pump is considered as failed to run. The mean time to repair of the pump is given as 8 h and the failure rate of the pump is given as 1.3×10^{-5}/h

Solution:

Mean time to repair = 8 h
Failure rate = 1.3×10^{-5}/h

$$P_F = \frac{\lambda_0 T_R}{1 + \lambda_0 T_R}$$
$$= \frac{1.3 \times 10^{-5} \times 24}{1 + 1.3 \times 10^{-5} \times 24}$$
$$P_F = 3.12 \times 10^{-4}$$

3.5.7.3 Nonrepairable Components

As the name implies, these components cannot be repaired during the operation of the system. Unless the system is taken out of service, these components cannot be repaired. The failure probability of these components can be calculated as follows:

$$P_F = 1 - e^{-\lambda_0 T_M} \quad (3.16)$$

where
λ_0 is the operating failure rate
T_M is the mission time

Example 3.6 Find the failure probability of a valve for the following information. The valve is required to be open during its operation. The failure mode of this valve is the failure to remain open. The mission time is given as 48 h and the failure rate of the valve is given as 2.3×10^{-6}/h

Solution:

Mission time = 48 h
Failure rate = 2.3×10^{-6}/h

$$P_F = 1 - \mathrm{e}^{-\lambda_0 T_M}$$
$$= 1 - \mathrm{e}^{-2.3 \times 10^{-6} \times 48}$$
$$P_F = 1.10 \times 10^{-4}$$

3.5.7.4 Standby Systems

In general, these systems will be in standby mode and will come into operation on demand. Hence, the demand failure probability is the measure of reliability for these systems. Once these systems come into operation, they have to operate for a given mission time. Two types of models can be used for these types of components: one is constant failure probability per demand or unavailability as a function of standby time, period of testing, and testing duration. Depending on how a component is tested, there are three types of components existed for standby systems.

3.5.7.5 Periodically Tested Standby Components

Normally, these components are tested periodically and will be in standby mode. During the testing if they are found to have failed, they will be repaired. Also these components may be subjected to periodic maintenance. For these components, the unavailability will have contributions from hardware failure, testing, repair, and scheduled maintenance. These are mathematically expressed as below [3, 22].

$$P_F = F_{\mathrm{HF}} + F_{\mathrm{TO}} + F_{\mathrm{RO}} + F_{\mathrm{SM}} \tag{3.17}$$

where

F_{HF} Failure probability due to hardware failure
F_{TO} Failure probability due to test outage
F_{RO} Failure probability due to repair outage
F_{SM} Failure probability due to scheduled maintenance

The above terms are further mathematically expressed as follows:

$$\begin{aligned} F_{\mathrm{HF}} &= 1 - \frac{1 - e^{-\lambda_S T}}{\lambda_S T} \\ F_{\mathrm{TO}} &= \frac{\tau}{T} q_0 \\ F_{\mathrm{RO}} &= \lambda_S T_R \\ F_{\mathrm{SM}} &= f_m T_m \end{aligned} \tag{3.18}$$

where

λ_S Standby failure rate
T Component test period
τ Average test duration
q_0 Override unavailability
T_R Mean time to repair
f_m Frequency of scheduled maintenance
T_m Mean time of scheduled maintenance

Example 3.7 Find the failure probability of a pump for the following information. Test interval of the pump is 1 month. Assume override unavailability is zero. Test duration is 30 min. Failure rate of the pump in standby mode is 4 × 10^{-6}/h. Frequency of scheduled maintenance is 4 times/year. Mean time of scheduled maintenance 72 h. Mean time to repair is 8 h.

Solution: Given data

$$\begin{aligned}
\lambda_S &= 4 \times 10^{-6}/\text{h} \\
T &= 1 \text{ month} = 720 \text{ h} \\
\tau &= 30 \text{ min} = 0.5 \text{ h} \\
q_0 &= 0 \\
T_R &= 8 \text{ h} \\
f_m &= 4/\text{year} = 4.57 \times 10^{-4}/\text{h} \\
T_m &= 72 \text{ h} \\
P_F &= F_{\text{HF}} + F_{\text{TO}} + F_{\text{RO}} + F_{\text{SM}} \\
F_{\text{HF}} &= 1 - \frac{1 - e^{-\lambda_S T}}{\lambda_S T} \\
&= 1 - \frac{1 - e^{-4\times10^{-6}\times720}}{4 \times 10^{-6} \times 720} \\
F_{\text{HF}} &= 1.44 \times 10^{-3} \\
F_{\text{TO}} &= \frac{\tau}{T} q_0 = 0 \\
F_{\text{RO}} &= 4 \times 10^{-6} \times 8 = 3.2 \times 10^{-5} \\
F_{\text{SM}} &= f_m T_m \\
&= 4.57 \times 10^{-4} \times 72 \\
F_{\text{SM}} &= f_m T_m \\
&= 4.57 \times 10^{-4} \times 72 \\
F_{\text{SM}} &= 3.2904 \times 10^{-2} \\
P_F &= F_{\text{HF}} + F_{\text{TO}} + F_{\text{RO}} + F_{\text{SM}} \\
P_F &= 1.44 \times 10^{-3} + 0 + 3.2 \times 10^{-5} + 3.2904 \times 10^{-2} \\
P_F &= 3.44 \times 10^{-2}
\end{aligned}$$

3.5.7.6 Untested Standby Components

These components are not tested but they will be in standby mode. The failure probability of these components can be expressed as a function of standby failure rate and fault exposure time (T_p). This can be mathematically expressed as follows [12]:

$$F = 1 - \frac{1 - e^{-\lambda_S T_P}}{\lambda_S T_P} \tag{3.19}$$

where
λ_S Standby failure rate
T_p Fault exposure time

In this expression, the fault exposure time (T_p) is defined as the time during which failure can occur but state of the component is unknown. This time is usually set to the plant life time (40 years). However, if a standby system is called for functioning, then at that time the status of these components can be detected whether they are working or failed. In this case, the fault exposure time is equivalent to the mean time to challenge the system to which it belongs.

3.5.7.7 Continuously Monitored Standby Components

Some components are continuously monitored even though they belong to the standby systems. Since they are continuously monitored, the failure of the component can be detected as soon as it occurs. The failure probability of these components is similar to those of online repairable component. However, in the present case, one uses standby failure rate as one of the parameters of the failure model whereas in the later case operating failure rate is used in the failure model. This is mathematically expressed as follows.

$$F = \frac{\lambda_S T_R}{1 + \lambda_S T_R} \tag{3.20}$$

where
λ_S Standby failure rate
T_R Mean time to repair

3.5.8 Estimation of Parameters of Failure Models

To estimate the failure probability of the components based on their respective failure models, one should have the information about the parameters that are entering into the models. This information one can get from the operating experience data or generic data sources. The plant-specific data sources include plant design, operating, maintenance records, test records, and calibration records. From these plant records, one can extract component failure data, component repair data, component test data, and component maintenance data. These data include the number of failures, failure modes, total exposure time, duration of component repair (includes detection time and waiting time), frequency of tests, test duration times, and frequency of maintenance and maintenance duration times. Whenever plant-specific data is not available, one can use the generic data sources to estimate the parameters of the failure models.

3.5.8.1 Standby Failure Rate/Operating Failure Rate

In estimating the standby or operating failure rate from the plant-specific data sources, following steps can be used.

- First identify the total component population in the plant for which the failure rate is to be estimated
- Find out the time period during which the component failures have to be considered
- Count the total number of failures (*N*) in the total component population
- Find out the total operating time or stand by time (*T*) of all the component population for the time period considered. Then the plant-specific failure rate (operating/standby) can be found from the following expression [12, 23].

$$\lambda = \frac{N}{T} \tag{3.21}$$

3.5.8.2 Mean Time to Repair

The mean repair time can be estimated from the observed repair times of a particular type of component. This can be estimated as sum of the repair times divided by the total number of repair actions. The repair time should consider both detection time as well as waiting times. This can be expressed as

$$T_R = \frac{\sum_{i=1}^{n} T_{Ri}}{n} \tag{3.22}$$

where

T_{Ri} Repair time for *i*th time

n Total number of repair actions

3.5.8.3 Test Frequency and Test Duration

In general, technical specifications will provide the information about the time between two surveillance tests. If the tests are more frequent than those specified in the technical specification, then the extra tests should also be considered in the test frequency estimation. Similarly, the test duration (τ) can be found as sum of the total test duration in a certain period of time divided by the number of test operations. This can be expressed as

$$\tau = \frac{\sum_{i=1}^{n} \tau_i}{n} \tag{3.23}$$

where

τ_i Test duration for ith time

n Total number of test operations

3.5.8.4 Maintenance Parameters

The maintenance parameters such as maintenance frequency and maintenance time can be estimated similar to those of test frequency and test duration.

3.5.8.5 Initiating Event Frequency

The frequency of an initiating event can be calculated as follows:

$$\begin{aligned} f_{\mathrm{IE}} &= \frac{N}{T} \\ N &= N_1 + N_2 + \cdots \\ T &= T_1 + T_2 + \cdots \end{aligned} \tag{3.24}$$

where

N Total number of occurrences

T Total operating years

N_1, N_2 Number of occurrences in plant 1, plant 2…

T_1, T_2 Total operating years of plant 1, plant 2…

Care should be taken while calculating the total operating years. This value differs from the calendar years. This is illustrated as follows: consider plant 1 has operated for three full calendar years. The year-wise operating hours of the plant is given as follows: in the first year, the plant has operated for 5000 h, in the second year it has operated for 6000 h, and in the third year it has operated for 8000 h out of 8760 h.

Hence, the total operating hours will be 19,000 h out of 26280 h. From this, one can find out ratio between actual operating hours (19,000 h) and total number of hours (26,280 h) which can be termed as availability factor. Now the total operating years of the plant 1 is calculated as

$$T_1 = \frac{19000}{8760} = 2.169\,\text{years}$$

This also can be calculated from the availability factor as follows:

$$A_1 = \frac{19000}{26280} = 72.298\ \%$$
$$T_1 = T_{\text{cal-years}} \times A_1 = 3 \times 0.72298 = 2.169\,\text{years}$$

Example 3.8 Find the frequency of occurrence of Class IV power supply failure from the following information. Consider two power plants (plant 1 and plant 2), number of occurrences in plant 1 is 5 and number of occurrences in plant 2 is 7. Operating years of plant 1 is 8.3 years and operating years of plant 2 is 9.5 years.

Solution: From the above information, the frequency of Class IV power supply failure can be calculated as 0.674/year as shown in the Table 3.3.

3.5.9 Parameter Estimation Using Bayesian Analysis

Proper estimation of parameters of the failure models are very important from reliability point of view of the systems as well as risk point of view of the plant. These parameters can be estimated directly from the operating experience data as described in the previous sections. However, getting plant-specific data for the new plants is not possible. In such cases, one should always depend on the similar existing plants data. These data sources are called generic data sources. By using this data, one can always estimate the system reliabilities or accident sequence frequencies or Core damage frequency. This is the situation when plant-specific data is not available. However, in some cases, very few data points will be available but one cannot make any kind of statistical inference from this information. At the same time, one should not omit this information, because this is some kind of a plant-specific data and it cannot be overlooked.

Table 3.3 Frequency calculation of Class IV power supply failure

$N_1 = 5$	$T_1 = 9.5$ years
$N_2 = 7$	$T_2 = 9.5$ years
$N = 12$	$T = 17.8$ years
$f = \frac{N}{T} = \frac{12}{17.8} = 0.674/\text{yr}$	

For example, from the generic sources it is found that failure rate of a check valve is higher. Suppose that the system failure probability is calculated by using this information and it also turned out to be higher (considering check valve is one of the dominant minimal cut sets). Consider that from the actual plant operating experience it is found that number of times check valve failed over a period of time is lesser. But this information by its own not sufficient for evaluating the failure rate of the component due to the lesser plants operating time. However, this data can be coupled with the generic data and can be used in calculations. In this way, the evidence collected from the plant is not omitted at the same time full credit is also not given to the generic data sources. For a small dataset (evidence), the updated value resembles the generic data to some extent. As the dataset becomes larger, the updated value departs more and more from the generic value, because the data contribute the dominant information.

There are two approaches for estimating the parameters viz., the Bayesian method and the frequentist, or classical, method. Both approaches use probability distributions to describe the behavior of random outcomes, such as a random number of initiating events or a random number of failures to start. The two approaches differ in the way they treat uncertainty of unknown parameters.

In the Bayesian approach [8], probability is a measure of uncertainty. In this methodology, each unknown parameter is assigned an initial prior probability distribution. Based on available data (operating experience), the analyst's prior belief about the parameter is updated, using Bayes' Theorem. The final outcome of this approach is the posterior distribution of the parameter. It is conditional on the observed data. The frequentist approach is quite different. The probability of a random event is defined as the long-term fraction of times that the event would occur, in a large number of trials. Probabilities are used only for random quantities, the possible data values. Probability distributions are never used to describe parameters, because the parameters are not random.

When estimating parameters for PSA, the Bayesian approach clearly works better, for two reasons. First, data from reliable equipment are typically sparse, with few or even zero observed failures. In such cases, it is reasonable to draw on other sources of information. The Bayesian approach provides a mechanism for incorporating such information as prior belief. Second, the Bayesian framework allows straightforward propagation of basic event uncertainties through a logical model, to produce an uncertainty on the frequency of the undesirable end state. To do this, it assigns a probability distribution to each of the unknown parameters, draws a random sample from each, and constructs the corresponding sample for the frequency of the undesirable end state. The frequentist approach cannot handle such complicated propagation of uncertainties except by rough approximations.

3.5.9.1 Bayesian Estimation

Bayesian approach can be used to estimate the parameters like failure rate, demand failure probability etc. The estimation of failure rate (λ) involves several steps. In the first step, the prior belief about the parameter (λ) is quantified by a probability distribution called as the prior distribution. This distribution will be restricted to the positive real values, because λ must be positive. The data (operating experience) are then collected, and the likelihood function is constructed. This is given by Eq. 3.25 for initiating events. It is the probability of the observed data, written as a function of λ.

$$P(X = x) = \frac{e^{-\lambda t}(\lambda t)^x}{x!} \tag{3.25}$$

Finally, the posterior distribution is constructed, by combining the prior distribution and the likelihood function through Bayes' theorem. This theorem says that

$$f_{\text{post}}(\lambda) \;\alpha\; \text{Likelihood}\,(\lambda) \times f_{\text{prior}}(\lambda) \tag{3.26}$$

Here, the symbol 'α' denotes "is proportional to." The posterior distribution shows the updated belief about the values of λ. It is a modification of the prior belief that accounts for the observed data.

3.5.9.2 Choosing a Prior

The first step in the Bayesian analysis is to choose a proper prior distribution for the parameter under estimation. The following sections consider estimation of λ using various possible prior distributions. The prior distribution can be discrete or it can be continuous in nature. In the case of discrete prior, the posterior distribution can be obtained easily by multiplying the prior probabilities with the evidence probabilities. In the case of continuous prior, the prior distribution can be either conjugate or nonconjugate. In the case of conjugate prior, the posterior distribution will be same as the prior distribution for some likelihood functions. Hence, the analysis with the conjugate prior is also simple. Whereas, in the case of nonconjugate prior for obtaining the posterior distribution, one should use either numerical integration or simulation methods. The prior distribution should accurately reflect prior knowledge or belief about the unknown parameter.

3.5.9.3 Estimation with a Conjugate Prior

In this case, the posterior distribution will be same as prior distribution for some particular likelihood functions. For example, if prior distribution is gamma function, then the posterior will also be gamma function if the likelihood function is of Poison distribution. Also the beta prior distributions are conjugate to the binomial data. Consider the following two-parameter gamma prior distribution,

$$f(\lambda) = \frac{\beta^{\alpha}}{\Gamma(\alpha)} \lambda^{\alpha-1} e^{-\lambda\beta} \tag{3.27}$$

Here, λ has units of 1/time and β has units of time, so the product '$\lambda\beta$' is unit less. The parameter β is called scale parameter. The other parameter, α, is unit less, and is called the shape parameter. The gamma function, $\Gamma(\alpha)$, is a standard mathematical function. If α is a positive integer, $\Gamma(\alpha)$, equals $(\alpha - 1)!$.

Let λ is a continuous random variable whose probability density function follows gamma distribution. In the case of gamma distribution, the mean or expected value $E(\lambda)$ and variance (σ^2) of the random variable can be written as follows

$$E(\lambda) = \frac{\alpha}{\beta} \qquad \sigma_{\lambda}^2 = \frac{\alpha}{\beta^2} \tag{3.28}$$

The posterior distribution is related to the prior distribution by

$$f_{\text{post}}(\lambda) \ \alpha \ P(X = x|\lambda) \times f_{\text{prior}}(\lambda) \tag{3.29}$$

The probability of the data is also called the likelihood, in which case it is considered as a function of the parameter λ for a given x. For Poisson data, it is given by Eq. 3.25. By removing the normalizing constants, the gamma p.d.f. can be given as follows:

$$f(\lambda) \ = \ \lambda^{\alpha-1} e^{-\lambda\beta} \tag{3.30}$$

The posterior distribution combines the gamma distribution and the Poisson likelihood, the result will be as follows:

$$\begin{aligned} f_{\text{post}}(\lambda) \ \alpha \ & \mathrm{e}^{-\lambda_i t} \frac{(\lambda_i t)^x}{x!} \lambda^{\alpha-1} \mathrm{e}^{-\lambda\beta} \\ \alpha \ & \lambda^{(x+\alpha)-1} \mathrm{e}^{-\lambda(t+\beta)} \end{aligned} \tag{3.31}$$

In the final expression, everything that does not involve λ has been absorbed into the proportionality constant. Hence, the posterior distribution of λ is again a gamma distribution. This is the meaning of conjugate: if the prior distribution is a member of the family (in this case, the gamma family), the posterior distribution is also a member of the same family. The update formulas are:

$$\begin{aligned} \alpha_{\text{post}} &= x + \alpha_{\text{prior}} \\ \beta_{\text{post}} &= t + \beta_{\text{prior}} \end{aligned} \tag{3.32}$$

Similarly for the case of beta distribution, it is conjugate for the binomial likelihood that is the posterior also a beta distribution. In this case, the number of failures, M, in N tests considered as evidence and it follows binomial distribution. The failure probability follows beta distribution and the probability density function of the beta distribution is given as follows:

$$p(f) = \frac{\Gamma(\alpha+\beta)}{\Gamma(\alpha)\Gamma(\beta)} f^{\alpha-1}(1-f)^{\beta-1} \quad (3.33)$$

$$\text{where} \quad f = \text{probability of failure}$$

Binomial distribution is as follows:

$$P(R = r) = \binom{n}{r} f^r (1-f)^{n-r} \quad (3.34)$$

Combining the prior with the binomial update in Bayes' Eq. 3.26 results in

$$\begin{aligned} p_{\text{post}}(f) &= \frac{\Gamma(N+\alpha+\beta)}{\Gamma(M+\alpha)\Gamma(N-M+\beta)} \\ &\quad \times f^{M+\alpha-1}(1-f)^{N-M+\beta-1} \\ \alpha_{\text{post}} &= M + \alpha_{\text{prior}} \\ \beta_{\text{post}} &= N - M + \beta_{\text{prior}} \end{aligned} \quad (3.35)$$

The mean and standard deviation of the posterior distribution are given as follows:

$$\begin{aligned} E(f) &= \frac{M+\alpha}{N+\alpha+\beta} \\ \sigma &= \frac{1}{(N+\alpha+\beta)} \frac{\sqrt{(M+\alpha)(N-M+\beta)}}{(N+\alpha+\beta+1)} \end{aligned} \quad (3.36)$$

3.5.9.4 Estimation with a Continuous Nonconjugate Prior

The posterior distribution in the case of discrete and conjugate priors can be easily obtained with the simple formulas. But it is not the case with continuous non-conjugate priors. Prior distribution can be any continuous distribution (exponential, normal, lognormal, Weibull etc.). The resulting posterior distribution is a continuous distribution, with no simple form because the posterior distribution does not have a simple analytical form.

3.5.9.5 Lognormal Prior

The following steps are followed in finding out the posterior distribution of the parameter. First the parameter to be estimated (Failure rate, demand failure probability) is identified. In the second step, one should fix the prior distribution; in this case, lognormal distribution is taken as the prior distribution. The parameters taken from the generic data are the median and error factor of the distribution. The error factor is used to find the lower limit and upper limit of the parameter in distribution as follows. One should make sure that these bounds should cover the entire range of the distribution.

$$\begin{aligned}\lambda_{\text{Lower}} &= \frac{\text{Median}}{\text{EF}} \\ \lambda_{\text{Upper}} &= \text{Median} \times \text{EF}\end{aligned} \tag{3.37}$$

Now divide the whole range into N intervals.

$$\delta = \frac{(\lambda_{\text{Upper}} - \lambda_{\text{Lower}})}{N} \tag{3.38}$$

Now the N values are found out as follows and i vary from 0 to N.

$$\lambda_i = \lambda_{\text{Lower}} + (i + 0.5) \times \delta \tag{3.39}$$

The probability density function for lognormal distribution is expressed as:

$$\begin{aligned}f(\lambda_i) &= \frac{1}{\sigma \lambda_i \sqrt{2\pi}} e^{\left[-\frac{1}{2}\left(\frac{\ln \lambda_i - \mu}{\sigma}\right)^2\right]} \\ \mu &= \ln(\text{Median}) \\ \sigma &= \frac{\ln(\text{EF})}{1.646}\end{aligned} \tag{3.40}$$

Let the likelihood function follow the Poisson distribution. Data required are number of failures, total operating time, and is same as the Eq. 3.25. Finally, one can find the posterior distribution using Bayes' theorem as given in the Eq. 3.26. In this case, instead of summation one should use integration in the denominator. This is illustrated in Fig. 3.13. As an example consider the failure rate λ follows lognormal distribution with median 1×10^{-3}/h and error factor as 3. Now consider the observed data as 5 failures in 10,000 h. For the given data, the prior and posterior distributions are shown in the Fig. 3.14. In the case, nonconjugate priors since the posterior distribution is not similar to the prior distribution, one has to find the posterior distribution after updating with the evidence. For this purpose, one can utilize the concepts of goodness of fit test like Chi-Square or Kolmogorov–Smirnov goodness of fit tests.

Example 3.9 Consider a case study on a Class IV power supply system of a typical NPP. The parameter to be estimated, is failure rate. The data collected from different units are shown in the Fig. 3.15. In general, the prior distribution can be taken from generic sources and the posterior can be obtained by combining the evidence

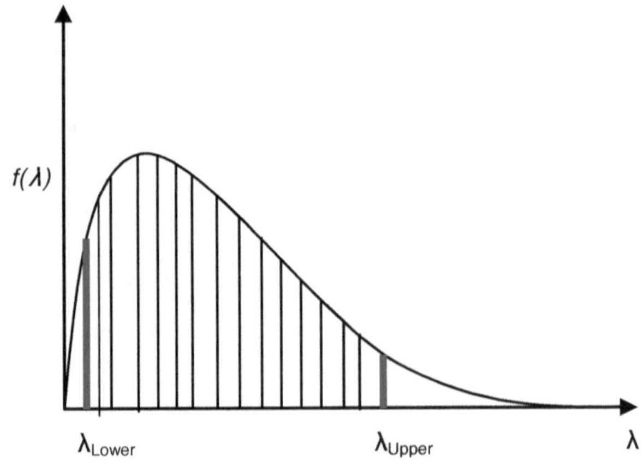

Fig. 3.13 Illustration of lognormal distribution

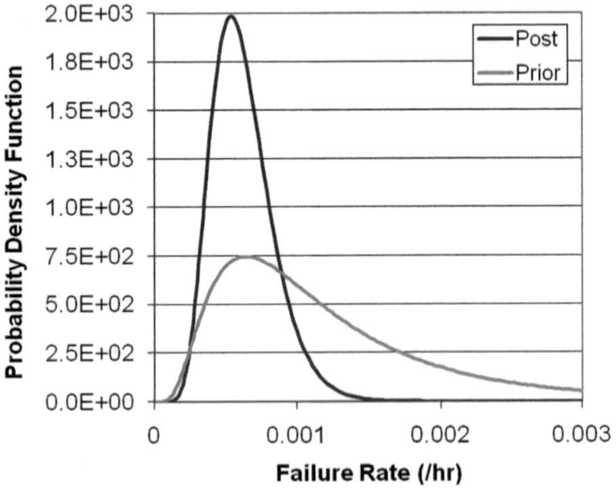

Fig. 3.14 Lognormal prior and posterior distributions

with the prior. For the purpose of illustration, in this case the data from year 1993 to 2005 has been utilized for obtaining the prior distribution. This data has been checked for different distributions (exponential, gamma, truncated normal, lognormal, and Weibull). With the help of KS test this can be verified. From the KS test it is found that the data best fits the Gamma distribution. Considering this as the prior distribution and evidence, the data from 2006 to 2008, the posterior distributions for different units can be evaluated.

Now considering gamma distribution as the prior distribution, the distribution parameters have been estimated from the data. The data from 2006 to 2008 has been taken as evidence and the likelihood function is considered as Poisson distribution. Since its conjugate prior the posterior distribution also follows the Gamma

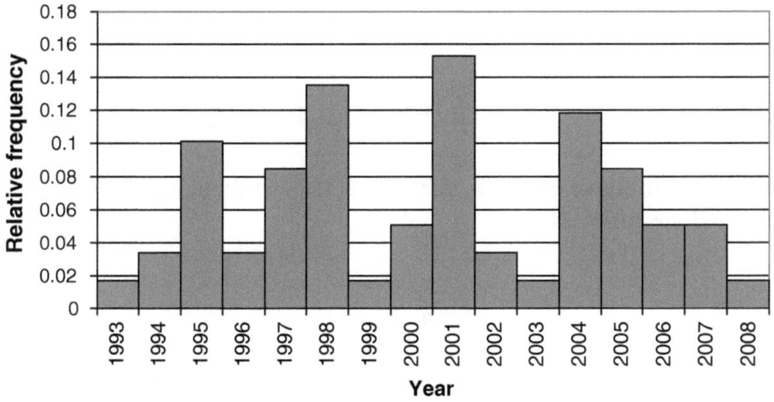

Fig. 3.15 Year-wise failure data

distribution. Hence, for the posterior distribution the posterior parameters have been calculated from the evidence and the posterior distribution has been developed. This has been done for individual units and also for the combined data. The corresponding prior and posterior distributions are shown in the Fig. 3.16 and are explained as follows. The mean and standard deviation have been calculated from the data and are given as 1.43 and 1.171/year, respectively. For the gamma distribution, these can be given as follows:

Fig. 3.16 Posterior distribution from different units

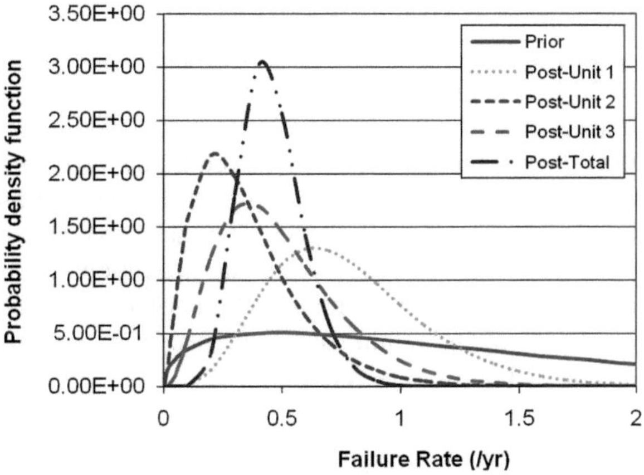

$$E(\lambda) = \frac{\alpha}{\beta} = 1.43$$

$$\sigma_\lambda^2 = \frac{\alpha}{\beta^2} \Rightarrow \sigma_\lambda = \sqrt{\frac{\alpha}{\beta^2}} = 1.171$$

By solving the above equation, one can get the distribution parameters of gamma function and are given as

$$\alpha = 1.49$$
$$\beta = 1.043$$

Now consider the following evidence. In unit 1 there were 2 events in 3 years, in unit 2 there were 1 event in 3 years, and in unit 3 there were 2 events in 3 years. The posterior gamma parameters have been found from the following formulas. In which x indicates number of events occurred and t indicates time period.

$$\alpha_{\text{post}} = x + \alpha_{\text{prior}} \qquad \alpha_{\text{post-unit 1}} = 1 + 1.49 = 2.49$$
$$\beta_{\text{post}} = t + \beta_{\text{prior}} \qquad \beta_{\text{post-unit 1}} = 3 + 1.043 = 4.043$$

3.5.10 Human Reliability Analysis

A component or system can fail due to several reasons. They mainly include design errors, manufacturing errors, operating errors, human errors etc. The history of accidents shows that the major failures occur due to the human errors. For example, in aerospace industry 90 % of the accidents are due to human errors and in the case of NPPs this is around 40 % [24]. Hence, it is very much important to consider human interactions while performing PSA. With the help of PSA one can identify the different human actions involved in the analysis and emphasize can be given to reduce these errors by following good operational practices as well as giving good training to the plant personnel. The main objective of Human Reliability Analysis (HRA) is to see that all the human interactions are properly accounted in the PSA and identify major human interactions which are contributing more to the risk of the plant and try to minimize the human error probability in operator actions. The details of HRA are covered in [24].

3.5.11 Uncertainty Analysis

The authenticity of PSA results is mainly dependent on the data that is being used in the analysis. If there is any variation in the data PSA results also vary. One can deal this variation of data and other aspects through the uncertainty analysis. Hence, uncertainty analysis is an important task of PSA. The objective of uncertainty analysis is [25]

- Identification of sources of uncertainty
- Propagation of uncertainty and
- Quantification of uncertainty

The uncertainty analysis will finally affect the results of PSA [26] such as accident sequence frequency, dominating accident sequences, and core damage frequency. While developing PSA model of a plant it is inevitable that there might be some underlying assumptions and simplifications used for representation of complex processes. These simplifications will generate uncertainties. There are three major categories of sources of uncertainties existed, namely [27]

- Completeness uncertainty
  - Whether complete list of initiating events have been considered
  - All the accident sequences covered or not
- Modeling uncertainty
  - Assumptions in the mathematical models
  - Numerical approximations
  - Coding errors and computational limits
- Input parameter uncertainties
  - Parameters used in the various models are not exactly known due to lack of data

Among all the above sources of uncertainty input parameter uncertainty can be quantified easily.

3.5.11.1 Propagation of Uncertainty

The input parameter uncertainty can be well treated with the help of PSA results such as core damage frequency (CDF) [17, 28]. In PSA, the CDF is represented with minimal cut sets at the plant level. This MCS is a function of initiating events and component failures. Component failures can be represented with corresponding failure probability models which is further function of different input parameters (failure rates, repair time, mission time, inspection intervals etc.). Hence, with the help of this PSA model, one can propagate uncertainty in the input parameters to the CDF level. This is explained with an example consider one of the accident sequences of the CDF as follows:

$$\mathrm{AS}_1 = \mathrm{IE}_1 \times C_1 \times C_2 \tag{3.41}$$

From the above it is evident that core damage can occur due to occurrence of initiating event IE_1 and simultaneous failures of component 1 (C_1) and component 2 (C_2). For the given values of IE frequency and component failure probabilities, one can obtain a constant value of accident sequence frequency. Assume the following data:

- IE frequency = 1×10^{-2}/year
- Failure probability of $C_1 = 2.3 \times 10^{-3}$
- Failure probability of $C_2 = 6 \times 10^{-4}$

For the above data

$$
\begin{aligned}
\mathrm{AS}_1 &= \mathrm{IE}_1 \times C_1 \times C_2 \\
&= 1 \times 10^{-2} \times 2.3 \times 10^{-3} \times 6 \times 10^{-4} \\
\mathrm{AS}_1 &= 1.38 \times 10^{-8}/\mathrm{year}
\end{aligned}
$$

However, if one assumes a variation in the failure probability of components, the frequency of accident sequence also vary. Assume the variation in the failure probability data as shown in Table 3.4

From the above it can be seen that failure probability of the components are not constant and they follow lognormal distribution. By using simulation techniques (Appendix), these uncertainties can be propagated from component level to accident sequence level.

3.5.12 Importance Analysis

The objective of importance analysis is to identify important accident sequences, systems failures, component failures, human interactions, and human errors which will have more effect on core damage frequency. There are two ways with which importances can be identified namely [3, 29],

- Qualitative importance analysis
- Quantitative importance analysis

In qualitative analysis, the importance of a component or system can be determined with the help of qualitative measures. As we know that minimal cut sets are used to quantify system failure probabilities or CDF. With the help of MCS, one can

Table 3.4 Variation in component failure probability

S. no	Component	Distribution	Median P_F	Error factor
1	C_1	Lognormal	2.3×10^{-3}	3
2	C_2	Lognormal	6×10^{-4}	3

identify which are all the components very important from CDF point of view. MCS may contain different orders of list like first order, second order etc. At the system level, there can be possible of first-order cut sets. However, in the case of CDF, it will start with second-order cut sets, because every accident sequence contains one initiating event and at least one component failure.

In the case of first-order cut set, the failure of one component leads to system failure; similarly, in the case of second-order cut sets, failure of both components simultaneously lead to failure of the system. Hence, if any first-order cut set appears in the MCS list then that component is very important from system failure probability point of view. Similarly, the next priority goes to second-order cut sets. This is just a qualitative way of identifying the importance of a component. Based on the second-order cut set if one wants to identify which component is more important, then one has to go for quantitative importance analysis.

In the case of quantitative analysis, there are different approaches available for quantification of importance of a components or systems. Some of them are listed below:

- Birnbaum importance measure
- Fussell Vesely importance measure

Birnbaum importance: This importance measure is defined as follows [3]:

$$I_B(C) = \mathrm{CDF}(1) - \mathrm{CDF}(0) \tag{3.42}$$

where CDF(1) is the value of CDF when the event is assumed to have occurred with a probability of one. CDF(0) is the value of CDF when the event is assumed not to have occurred, i.e., the occurrence probability of the event is considered as zero. Here the importance of a component is evaluated as follows: first evaluate the CDF by considering the component has failed with a probability of 1. Then reevaluate the CDF by considering the component is fully available, i.e., failure probability of the component is zero. Now taking the difference between these two cases will give importance of the component.

Fussell Vesely importance: This importance measure is defined as follows [3]:

$$I_{\mathrm{FV}}(C) = \frac{\sum \mathrm{MCS}_C}{\mathrm{CDF}} \tag{3.43}$$

where MCS_C is the minimal cut sets in which component C is existed. Here Σ represents summing up of all the cut sets where the component C is existed. Now one can calculate importance measure for all the components. Whichever the component has highest value that is very important from CDF point of view. For further information on the importance measures one may refer [29, 30].

3.6 Level 2 PSA

The level 2 PSA provides insights into the containment sequence modeling along with the dispersion of radio nuclides in the containment [12]. Level 2 PSA is dependent on the Level 1 PSA analysis of a given plant. As explained previously, the outcome of Level 1 PSA analysis is quantification of large number of accident sequence frequencies. One has to analyze each sequence to assess the accident progression, to know the containment response, and to identify the fission product releases. However, it is not practical or necessary to treat each one of these sequences. Hence, these large numbers of accident sequences will be grouped into a smaller number of plant damage states (PDS) in accordance with accident characteristics and containment response characteristics in Level 2 PSA analysis. This is the starting point of Level 2 PSA analysis.

After determining the plant damage states, the progression of accidents and impacts on containment behavior are examined probabilistically with containment event trees. The various end states of the containment event trees are grouped to a more manageable set of release categories for which distinct source terms are estimated. These distinct release categories define the conditions for estimation of consequences in terms of public risk.

The Level 2 PSA deals with frequency and magnitude of release to environment and consists of probabilistic and deterministic analysis elements. The probabilistic element consists of the development and quantification of containment logic models for each plant damage state. The deterministic element consists of calculating the release magnitude from the core, physical process of accident progression including containment response, and source term analysis of radionuclide releases to the environment for the representative events from each plant damage state using appropriate codes. Figure 3.17 shows the graphical representation of Level 2 PSA analysis mainly emphasizing on containment analysis and leakage from the containment.

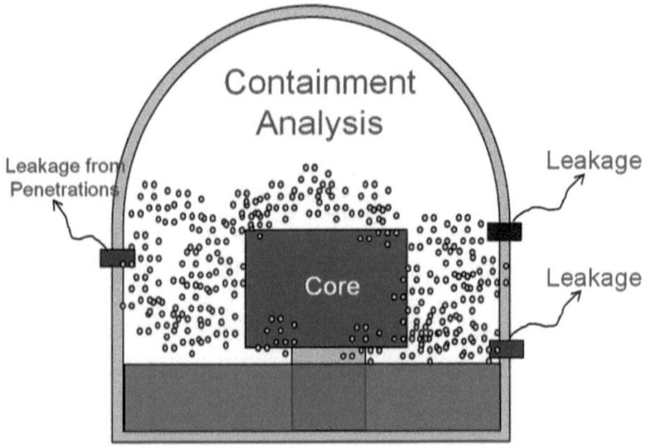

Fig. 3.17 Graphical representation of Level 2 PSA analysis

3.6.1 Objectives of Level 2 PSA

Some of the objectives of Level 2 PSA are [12]

- To gain insights into the progression of severe accidents and containment performance.
- To identify plant-specific vulnerabilities of the containment to severe accidents.
- To identify major containment failure modes and to estimate the corresponding releases of radionuclides.
- To provide a basis for plant-specific backfit analysis and evaluation of risk reduction options.
- To provide a basis for the evaluation of off-site emergency planning strategies.
- To provide a basis for the development of plant-specific Severe Accident Management (SAMG) strategies.

3.6.2 Steps in Level 2 PSA

The different steps involved in carrying out Level 2 PSA are shown in Fig. 3.18 and are further explained below.

1. Plant familiarization: The main purpose of plant familiarization is to identify the component, system, and operational data that may be useful in assessing the progression of sever accidents and the containment response. Identify the various plant features which are important to arresting the progression of sever accidents systems such as containment sprays, passive containment coolers, and suppression pool etc.

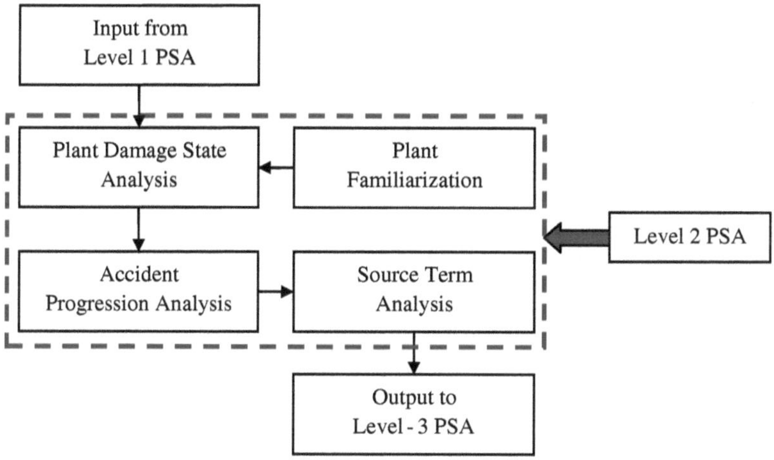

Fig. 3.18 Steps in performing Level 2 PSA

2. Interface with Level 1 PSA and Plant damage state analysis: As discussed earlier, the starting point for Level 2 PSA is Level 1 PSA analysis. In this step, various accident sequences from various initiating events will be grouped into smaller number of plant damage states in accordance with accident characteristics and containment response characteristics. One has to define the attributes for plant damage states. Once the PDS are identified, then the fission product release magnitude from each damage state will be determined.
3. In the second step, accident progression for severe accidents will be assessed deterministically.
4. In this step, containment analysis will be carried out to understand the containment performance features, to identify possible containment failure modes. Containment event trees will be developed to identify the spectrum of containment damage or release states. Success criteria of various containment Engineering Safety Features (ESFs) in various modes will be developed. Reliability analysis of containment ESFs and related systems will be carried out to quantify the failure probabilities of containment ESFs and other phenomena by detailed fault tree or other appropriate analytical methods.
5. Release categorization and source term analysis: In this step, different end states from the containment event trees will be grouped to form preliminary source term/release categories. And the source term analysis will be performed for the selected event sequences from this list of release categories. Final release categories will be determined based on the similarity in release magnitude and characteristics.
6. In the final step, various release category frequencies will be quantified and will estimate the large early release frequency (LERF) [12]. This will go as a input to the Level 3 analysis.

Some of the steps are further explained in detail in the following subsections.

3.6.3 Plant Damage States

The plant damage states represent the groups of accident sequences which generate similar loads on the containment, which have similar accident timelines and thereby resulting in a similar event progression and similar radiological source terms. The accident sequences can be grouped into various plant damage states on the basis that all the accident sequences in a PDS should have similar characteristics with respect to

- The magnitude of fuel damage,
- All other characteristics which can influence fission product release to containment such as mode of emergency core cooling system failure which can influence timing of the release of fission products from PHT during LOCA and associated thermal hydraulic phenomena and implications of the initiating event on the PHT pressure,

- The containment effectiveness with respect to the accident sequence, i.e., whether the event results in release in containment or outside the containment (containment bypass events).

Broadly plant damage states can be grouped into two main classes viz. Class-1 and Class-2. Class-1 includes all PDSs in which radioactive materials are initially released to the containment and Class-2 includes all PDSs in which the containment is bypassed and radiological releases go directly to the atmosphere. For PDSs, Class-1 containment event tree analysis will have to be performed while for Class-2 where the containment is bypassed, only source term analysis needs to be performed.

3.6.3.1 PDS Class 1

All the accident sequences which will result in release of radioactivity within the containment will fall in this class. From the Level 1 analysis, one can obtain the different core damage categories from different accident sequences. These accident sequences based on their level of core damage can be categorized into category 1, 2, 3 etc. Out of these categories, some of them will come under PDS Class 1 category. These are explained with example as follows:

Accident Sequence Category 1: The events in this category are those for which loss of core structural integrity is expected. The accident sequences in which initiating events are followed by failure to shutdown the reactor will fall in this category. That means both the primary and secondary shutdown system fail following the IE. Example of this category include

- LOCA initiators followed by failure of both shut down systems
- Loss of regulation system followed by failure of both shut down systems
- Small leak followed by failure of both shut down systems

Accident Sequence Category 2: The events in this category are those for which loss of core structural integrity is expected due to long-term subcriticality function fails. The accident sequence in which reactor is shutdown after the initiating event occurrence but unable to maintain the long-term subcriticality. Example of this category include

- Station black out scenario followed by failure of Gravity addition boron system

Accident Sequence Category 3: In this category, damage can arise from total loss of heat sink. These accident sequences will lead to overstressing of the reactor core or its component systems causing a loss of core structural integrity. Example of this category include

- Station black out followed by decay heat removal systems failure, i.e., failure of Class IV power supply system followed by failure of Class III power supply system and failure of Fire Water System which is the ultimate heat sink in case of station black out condition.

Accident Sequence Category 4: In this category, core damage can occur due to the events like loss-of-coolant accident followed by failure of ECCS as well as failure of moderator heat sink. Example of this category include

- Simultaneous rupture of pressure tube and calandria tube followed by ECCS failure.

Accident Sequence Category 5: In this category, the accident sequences can include large-break LOCA or small-break LOCA followed by failure of both ECCS injection recirculation systems. Early heatup, large fraction of fuel failures, and significant metal water reaction are expected in this category. Moderator heat sink is available as ultimate heat sink.

3.6.3.2 PDS Class 2

All those accident sequences that result in the release of radioactivity directly into the atmosphere bypassing the containment will fall under this classification. Example of this category includes

- Initiating events like steam generator tube rupture in case of PHWR result in some activity release into the atmosphere.
- Main steam line break outside the containment in case of advanced reactor will release radioactivity directly into the atmosphere.

3.6.4 Accident Progression

Thermal hydraulic analysis has to be carried out for the assessment of accident progression for the representative event sequences comprising the enveloping severe accidents by using suitable thermal hydraulic codes. Some of the severe accidents that can be analyzed are listed below:

- LBLOCA with failure of both shutdown systems
- SBO with failure of fire water system
- LBLOCA with failure of ECCS and moderator cooling

3.6.5 Containment Analysis

In general, containment is designed to withstand the temperature and pressure loading expected during severe accidents within design basis. Following are the steps used to perform containment analysis [12].

- Identification of containment performance features
- Containment Event Tree (CET) development by evaluation of containment performance, determination of probable failure modes of containment and containment ESFs.
- Development of containment release model
- Quantification of containment event trees and source term analysis

3.6.5.1 Containment Performance

In this analysis, it is necessary to know the details about the structural design of the containment such as type of containment, design pressure, and temperature and containment penetrations. These are shown in Table 3.5. This will help in determining the structural performance of the containment during severe accident conditions. During these conditions, containment should act as a leak tight barrier and this will characterize the performance of the containment.

In order to characterize the containment integrity, two basic models are being used in PSA studies such as threshold model and leak before break model. In the threshold model, if the containment pressure exceeds a certain threshold pressure then it is said to have failed which will result in a large rupture with rapid blow down of the containment atmosphere to the environment. Whereas, in the leak before break model, containment leakage is expected to occur before a major rupture. In general, leakage begins at a lower pressure and increases up to ultimate capability pressure at which point a larger failure may occur.

3.6.6 Containment Event Tree Development

Containment event trees are helpful in identifying the important failure modes of containment and containment ESFs that can lead to releases of fission product outside the containment. Fault tree analysis will be helpful in estimating the failure probabilities of the containment ESFs. As in the case of Level 1 PSA, event trees where the end states are grouped into various plant damage states, the end states of

Table 3.5 Design considerations of containment

Containment type	• Steel
	• Concrete
	Prestressed
	Reinforced
Containment penetrations	• Piping penetrations
	• Electrical penetrations
	• Vent lines etc.

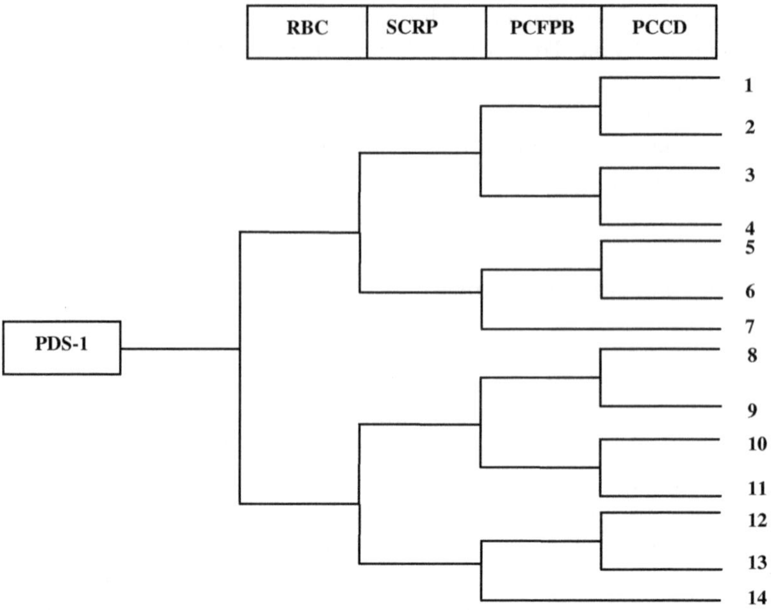

Fig. 3.19 Containment event tree for a typical NPP

the containment event tree are categorized in terms of their consequence/release potential. The containment event tree is developed based on the understanding of the accident progression in the containment and various functions required to be performed to prevent releases to environment. Containment event trees have to be developed in detail for all the plant damage states containing severe accidents within design basis and other plant damage states involving different LOCA scenarios. Figure 3.19 shows the typical CET for a typical NPP. This further explained in a case study at the end of the present chapter.

For beyond design basis accidents involving overpower transient and early core disassembly, due to the uncertainties involved in the likely energetic reactions containment failure can be assumed for conservative purpose. For other beyond design basis accidents involving core disassembly, containment failure occurs late into the event. During these accidents as the moderator starts boiling and fuel channels get uncovered, large amount of hydrogen gets generated. For BDBAs involving late containment failure, CETs can be developed considering the containment isolation function. If containment isolation gets actuated successfully in the beginning of the accident, means of attenuation for iodine like plate out and natural decay of some of the fission products take place during the available time frame before the containment fails. Thus at the time of containment failure, the radioactivity level in primary containment will be less in lower magnitude of radionuclide release to atmosphere on late containment failure.

For all other severe accidents within design basis, the CET can be developed using the following steps

- Identification of containment failure modes
- Identification of progression of events with regard to containment and containment ESF systems
- Quantification of the failure probabilities of the containment ESFs by fault tree analysis
- Grouping of CET end states into release categories according to the magnitude of the radionuclide release from containment.

After the occurrence of a plant damage state, the containment isolation function acts first. Consequences of primary containment isolation failure are more severe than secondary containment isolation failure in terms of radionuclide releases to environment. If primary containment isolation fails then the radionuclide release will be very high as most of the radionuclide in the PC will be released to environment. If the secondary containment isolation fails, whatever radionuclide leaks from PC to SC, gets diluted before it is released to environment through stack. Also containment ESFs like PCFPB and SCRP help to reduce the cumulative radio nuclide release due to SC isolation failure. The possibility of local hydrogen detonation exists for severe accidents within design basis about 1 h after the accident. During the accident conditions, there is a probability of excess leakage from the containment due to some previous degradation in the EPs.

After successful isolation, containment depressurization and cooling function is carried out by RB coolers. PCFPB and SCRP systems are used for radionuclide management in primary and secondary containments respectively.

3.6.7 *Reliability Analysis of Containment ESFs*

Some of the containment ESFs that are used in a typical NPP are listed below:

- Containment isolation system
- Secondary containment recirculation and purge system
- Primary containment filtration and pump back system
- Primary containment controlled discharge system
- Post accident hydrogen management system etc.

Fault tree analysis has to be carried out to estimate the failure probabilities of each system.

3.6.8 *Containment Failure Modes*

The function of a containment in a severe accident is to contain fission products. A containment building houses a nuclear reactor, coolant pumps and other equipment.

Table 3.6 Different failure modes of containment

S. No.	Failure mode	Failure mechanism
1	Direct bypass	• Interfacing systems LOCA • Steam generator tube rupture • Externally initiated
2	Isolation failure	System failure
3	Vapor explosion	• Rapid pressurization • Blast loads
4	Over pressurization	• Noncondensable gases • Direct energy transfer
5	Over temperature	• Direct contact by core debris • Core concrete interactions
6	Combustion	• Detonation • Deflagration
7	Concrete penetration	• Basement penetration • Support failure

Temperature rise threatens the containment because of the associated increase in the pressure of the containment atmosphere. The heat sources in the severe accident phenomena are decay heat, heat generated in metal water reaction, hydrogen combustion. In addition to the containment failure due to over temperature and pressure, possibility of direct bypass and failure to isolate are also considered as part of failure modes. In case of containment bypass, the fission products directly leak to atmosphere without entering containment. The generally considered containment failure modes are given in Table 3.6 [12]. Due to the over pressurization of containment, leak may occur through cracks, through penetrations, and leakage due to permeability of concrete. These are briefly explained as follows:

Leakage through cracks: The cracks may get formed during the accidental overpressure in the containment. These cracks may result in leakage of gas from the containment under accident condition. To estimate the leakage the crack size, location and the number of such cracks have to be estimated as a function of pressure.

Leakage through penetrations: Leakage through the holes made for the penetrating pipes, airlock and cables is most important. Normally, the holes made for these openings are fitted with metallic sleeves. The pipes pass through this sleeve. There is a bellow which is welded on the pipe on one end and the metallic sleeve on the other end to prevent any leakage between the gaps. Though efforts are made to compact the concrete around the metallic sleeve, still a gap between the concrete wall and the sleeve cannot be ruled out. This gap may get accentuated during the over pressurization condition.

Leakage due to Permeability of Concrete: Reinforced concrete is a porous material. In some of the containments, the inside wall is coated with epoxy paint to minimize the leakage due to permeability. There are relations existed to estimate the instantaneous discharge rate through a porous medium to the local pressure gradient and the permeability at that point.

3.6.9 Release Categorization and Source Term Analysis

The CETs produce a large number of end states, some are identical in terms of key release attributes. These end states are grouped together. These groupings are referred to as release categories. The release categories group CET end states that would be expected to have similar radiological characteristics and potential off-site consequences. The source terms/release groups are generally defined on the basis of attributes that affect fission product releases and accident consequences specific to the plant like amount and timing of release, location of release, energy of release, duration of release etc.

Source term is generally known as the amount of radio nuclides that can be released from a NPP in an accident. A quantitative estimation of the source term is of importance for assessing the effectiveness of safety design features for accident management in the immediate post accident phase and for the planning of post accident emergency measures in the public domain. In the most simplified selection of radio nuclides for severe accident analysis only two groups of radio activities are included i.e., Noble gases and Iodines.

3.6.10 Frequencies of Release Categories

By combining the frequencies of plant damage states and their associated uncertainties resulting from Level 1 PSA with the conditional probabilities of various containment-related system failures, the release frequency of each release category can be calculated [12].

3.7 Level 3 PSA

A Level 3 PSA [4] provides insights into the relative importance of accident prevention and mitigative measures expressed in terms of the adverse consequences for the health of the public, and the contamination of land, air, water and foodstuffs. Finally, a Level 3 PSA provides insights into the relative effectiveness of emergency response planning aspects of off-site accident management, and into the economic impacts. The general flowchart of the Level 3 analysis is shown in the Fig. 3.20. These are explained in the following steps.

3.7.1 The Input and Output

In this study, the input to the analysis include radioactivity release to the atmosphere and the output include

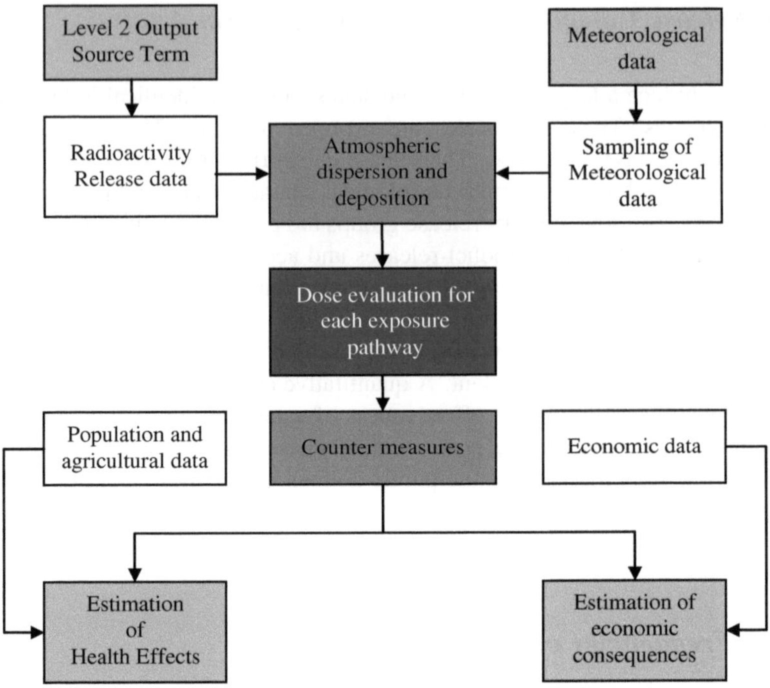

Fig. 3.20 Different steps of Level 3 PSA

- The individual risks of early and late fatal health effects,
- The societal risk of early health effects,
- Collective risks,
- Risks of unacceptable land contamination.

From the Level 2 PSA for a given accident scenario, the source term that is released to the atmosphere can be estimated. The radionuclide that is released to the atmosphere depending on the meteorological conditions existed at that point of time will get dispersed and deposited in the environment. Based on the type of exposure pathways, the dose absorbed can be calculated by using dose conversion factors. Similarly individual and societal risk can be calculated by using appropriate risk conversion factors. These are further discussed in the following subsections.

3.7.2 Source Term

The starting point for Level 3 PSA is output obtained from Level 2 PSA in terms of source term which is nothing but the radionuclide released to the environment [4]. In general radionuclide release to the environment can be either to the aquatic environment or to the atmosphere. However, during the accidents the releases to the aquatic environment will have very small contribution to the total risk from the NPPs compared to the releases to the atmosphere. It means the total risk contribution from the NPPs will be from the releases to the atmosphere. Hence, in Level 3 PSA much concern is given to the releases to the atmosphere compared to that of aquatic environment. In general, the source term includes

- The quantity of release
- Isotopic composition of released radionuclides, together with their
- Physical and chemical characteristics,
- The heat content of the plume,
- The time profile of the release,
- The release height, and
- Frequency of the release

If source terms from Level 2 PSA are not available, reference source terms available in the literature can be used. In determining the particulate material that is deposited in the respiratory tract if inhaled and in the environment, the following parameters are important namely, particle size and chemical form of the released aerosol. If the data on the chemical and physical form of the released radionuclide are not available, in general, the radionuclide can be assumed to be released in 1 μm activity median aerodynamic diameter (AMAD) aerosol. This is, in addition, to the release of noble gases and the fraction of iodine released in an organic or elemental form.

3.7.3 Meteorological Data and Its Sampling

Meteorological data is required for carrying out Level 3 analysis [4]. In general, the meteorological data collected on an hourly basis for at least one year will be considered in the analysis. This data must be obtained from the meteorological station nearest to the release point. This is called as source meteorology. Most of the NPPs have their own meteorological stations in the plant itself. The data is represented as sequence. Each sequence consists of six parameters namely

- Wind velocity
- Wind direction
- Stability category
- Rain fall rate etc.

The data can be collected on minute bases as well. However, the data collected on hourly bases is used in the analysis considering that each hour's conditions can be regarded as unique and it can be taken as starting point for a sequence of hourly weather data. In a year there are 8760 h and the meteorological data so collected will have 8760 possible sequences. Theoretically, one has to consider all the sequences and estimate risk. But it is not practicable and not necessary to consider each and every sequence. Instead, one can collect one or 2 year's data and can sample from that data. The sampling should be done in such a way that one should be able to select a truly representative weather sequence. The sequence which is selected by sampling will have certain probability of occurrence.

There are different sampling techniques available namely, random sampling, cyclic sampling, and stratified sampling etc. In the random sampling, sequence is selected randomly on the basis of uniform random number generator. In the cyclic sampling, the sequence is selected based on the fixed time interval between the samples. With these methods, there is a possibility that one can omit significant and infrequent weather sequences. To avoid this, nowadays stratified sampling is the preferable method. In this method, the data is divided into strata and the sampling is done from each strata.

3.7.4 Atmospheric Dispersion and Deposition

Normally, radionuclides will be released to the atmosphere in the form of fine aerosol or gas. This will create a plume and will be released into the atmosphere with an initial temperature and velocity which may be different from that of ambient conditions. The plume will have vertical motion due to the initial velocity and buoyancy due to the initial temperature. Buoyancy can be either positive or negative. The plume will rise until its momentum lost and this vertical rise of plume is called plume rise and this will change the effective height of release. In nuclear power plants, during normal operating conditions, the released radionuclide will not have much effect of buoyancy due to the lesser difference of initial temperature with that of ambient. Whereas, during accidental releases the radionuclides will have thermal energy due to the higher temperatures and buoyancy effect will be more.

Once the plume enters the atmosphere it will be carried away (transport) with the wind along the wind direction that is presented at that time of release. During this process, the plume will be expanded both in horizontal and vertical directions due to diffusion as well as turbulent eddies present in the atmosphere. The process in which both transport and diffusion are involved is termed as dispersion. In this process, transport of the plume is happening due to the wind velocity and dispersion of the plume is taking place due to the turbulent eddies in the horizontal (downwind and crosswind) and vertical directions.

The purpose of atmospheric dispersion analysis is to estimate the concentration of the dispersing plume at different points from the point of release (concentration profile). There are different models existed for predicting the concentrations at along the downwind direction, some of them are listed as below:

- Statistical model
- Gradient Theory Model (*K*-Model)
- Similarity model

Among these models statistical and *K*-models are being used widely. Sutton's model is a statistical model that is based on the statistical properties of turbulence. Gradient theory models are based on partial differential equations involving diffusion coefficients, velocities etc. *K* theory models are well suited for long distances varying from few hundreds to thousands of kilometers. For short distances, empirical models are being used widely which use double or triple Gaussian distribution of concentration. The Gaussian distribution is based on the statistical models and also it is an analytical solution for certain simplified *K* theory models. The dispersion model based on statistical and *K* theory model are illustrated in the following subsections.

3.7.4.1 Gaussian Plume Model

The Gaussian plume model is originally developed by Sutton, Pasquell and Gifford [31]. The following assumptions are made in the Gaussian plume model:

- The emission is assumed to be continuous and the rate of emission is not variable with time.
- The mass emitted from the source is assumed to remain in the atmosphere and is not undergoing any kind of depleting mechanisms.
- Meteorological conditions are assumed to be constant (not change with time) till the plume reaches the receptor.
- Crosswind and vertical concentration distributions follows Gaussian distribution.

Gaussian plume model is based on the Gaussian distribution that is used in the statistics for representing the probability density function of a random variable. Here the concentration of the plume is represented with this model. The Gaussian probability density function of a random variable can be represented as follows:

$$f_X(x) = \frac{1}{\sigma_X\sqrt{2\pi}} e^{-\frac{1}{2}\left(\frac{x-\mu_X}{\sigma_X}\right)^2} \tag{3.44}$$

where

X Random variable

μ_X Mean value of the random variable

σ_X Standard deviation of the random variable

Now if Eq. 3.44 is multiplied with the amount of radionuclide released, one can get the distribution of concentration of the plume at a particular distance. The modified equation representing the concentration can be given as Eq. 3.45.

$$\chi(y) = Q\frac{1}{\sigma_y\sqrt{2\pi}} e^{-\frac{1}{2}\left(\frac{y-\mu_Y}{\sigma_Y}\right)^2} \tag{3.45}$$

Fig. 3.21 Representation of plume dispersion

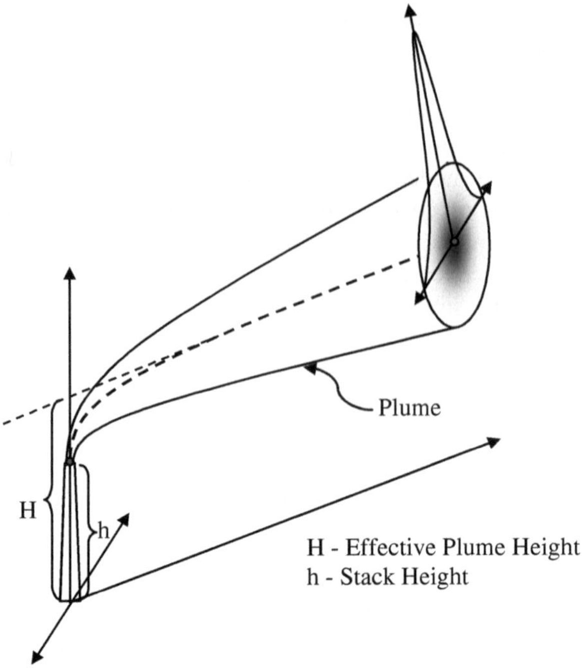

where Q is the amount of radionuclide released. Considering the Eq. 3.45 represents concentration profile at a particular distance from the source of release in the crosswind direction, then the maximum concentration will be at the center of the distribution in this case it is at $\mu_Y = 0$. Here σ_Y represents the spread of the plume around the center point. More the σ_Y value more the spread. The plume model and its parameters are shown in the Fig. 3.21 and the concentration profile at a particular distance in the crosswind direction is shown in the Fig. 3.22.

The concentration at any point for an instantaneous point source from the point of release can be given by triple Gaussian expression as follows:

$$\chi(x,y,z,t) = Q\frac{1}{\sigma_x\sqrt{2\pi}}e^{-\frac{1}{2}\left(\frac{x}{\sigma_x}\right)^2}\frac{1}{\sigma_y\sqrt{2\pi}}e^{-\frac{1}{2}\left(\frac{y}{\sigma_y}\right)^2}\frac{1}{\sigma_z\sqrt{2\pi}}e^{-\frac{1}{2}\left(\frac{z}{\sigma_z}\right)^2}$$
$$\chi(x,y,z,t) = Q\frac{1}{(2\pi)^{3/2}\sigma_x\sigma_y\sigma_z}e^{-\frac{1}{2}\left(\frac{x}{\sigma_x}\right)^2}e^{-\frac{1}{2}\left(\frac{y}{\sigma_y}\right)^2}e^{-\frac{1}{2}\left(\frac{z}{\sigma_z}\right)^2} \quad (3.46)$$

where x, y, and z are the coordinate points at which the concentration is required and σ_X, σ_Y, and σ_Z are the standard deviation of concentration along the respective axes and Q is the quantity of radioactive material released (Bq). In the present case, the origin is at the plume level at the release point. Similarly, for the case of continuous point source, the concentration profile can be given as

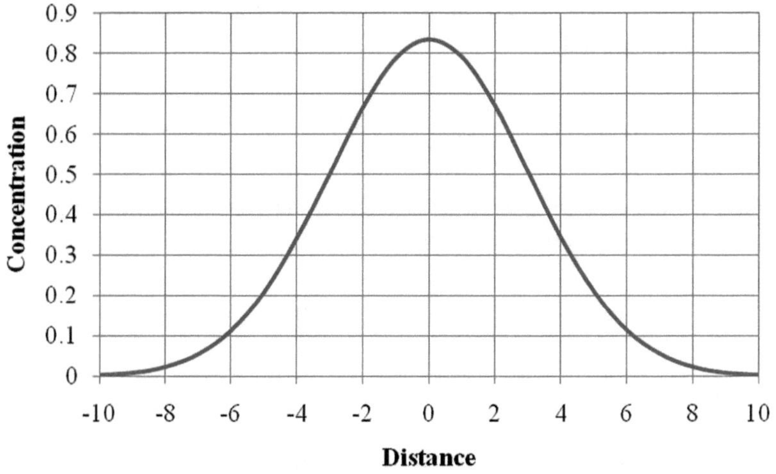

Fig. 3.22 Variation of concentration along crosswind direction

$$\chi(x,y,z) = Q' \frac{1}{U} \frac{1}{\sigma_y\sqrt{2\pi}} \mathrm{e}^{-\frac{1}{2}\left(\frac{y}{\sigma_y}\right)^2} \frac{1}{\sigma_z\sqrt{2\pi}} \left\{ \mathrm{e}^{-\frac{1}{2}\left(\frac{z-H}{\sigma_z}\right)^2} + \mathrm{e}^{-\frac{1}{2}\left(\frac{z+H}{\sigma_z}\right)^2} \right\}$$
$$\chi(x,y,z) = \frac{Q'}{2\pi U \sigma_y \sigma_z} \mathrm{e}^{-\frac{1}{2}\left(\frac{y}{\sigma_y}\right)^2} \left\{ \mathrm{e}^{-\frac{1}{2}\left(\frac{H-z}{\sigma_z}\right)^2} + \mathrm{e}^{-\frac{1}{2}\left(\frac{H+z}{\sigma_z}\right)^2} \right\} \tag{3.47}$$

where
Q^1 is the source in Bq/s
χ is in Bq/m^3
U is wind velocity in X direction in m/s
H is the effective height (stack height + plume rise)

In this case, the origin is assumed to be at the base of the stack. The first exponential term describes the distribution of the concentration along the Y direction, i.e., horizontal crosswind direction. With respect to the Z direction, there are two exponential terms: the first one represents the direct distance from the plume centerline to the point of interest and the second exponential term takes care of the reflected plume from the ground. Now Eq. 3.47 can be modified to get the concentration at different points of interest. For example, the case of concentration on the ground level can be obtained as

$$\chi(x,y,z) = \frac{Q'}{2\pi U \sigma_y \sigma_z} \mathrm{e}^{-\frac{1}{2}\left(\frac{y}{\sigma_y}\right)^2} \mathrm{e}^{-\frac{1}{2}\left(\frac{H}{\sigma_z}\right)^2} \qquad z = 0 \tag{3.48}$$

For the case of concentration on the ground level and beneath, the plume centerline can be obtained as

$$\chi(x, y, z) = \frac{Q'}{2\pi U \sigma_y \sigma_z} e^{-\frac{1}{2}\left(\frac{H}{\sigma_z}\right)^2} \qquad y = 0,\ z = 0 \tag{3.49}$$

Finally, for the case of ground level release, the concentration along the plume centerline on ground level can be obtained as

$$\chi(x, y, z) = \frac{Q'}{2\pi U \sigma_y \sigma_z} \qquad y = 0,\ z = 0,\ H = 0 \tag{3.50}$$

One can also calculate the concentration per unit release of radio nuclide. This can be represented as χ/Q. This is called dispersion factor per unit release. The above expressions do not take into account the modifications of the concentration due to physicochemical processes radioactive decay, build up, depletion, or washout etc. These effects can be introduced as correction factors by which the concentrations are to be multiplied.

3.7.4.2 Gradient Theory or *K*-Model

In this theory, the dispersion of a plume is represented with velocity and diffusion coefficients as follows:

$$\frac{\partial C}{\partial t} = -\left(u\frac{\partial C}{\partial x} + v\frac{\partial C}{\partial y} + w\frac{\partial C}{\partial z}\right) + \frac{\partial}{\partial x}\left(K_x\frac{\partial C}{\partial x}\right) + \frac{\partial}{\partial y}\left(K_y\frac{\partial C}{\partial y}\right) + \frac{\partial}{\partial z}\left(K_z\frac{\partial C}{\partial z}\right) \tag{3.51}$$

where

C	Concentration of the plume
t	time
u, v, w	wind velocities along X, Y, and Z directions
K_x, K_y, and K_z	diffusion coefficients along X, Y, and Z directions

Equation 3.51 can be simplified with the following assumptions

- Cross and vertical velocities are zero, i.e., $v = w = 0$
- Turbulent diffusion across the x axis can be neglected compared to the advection by the mean wind u, $K_x = 0$

With the above-mentioned assumptions, the Eq. 3.51 becomes

$$\frac{\partial C}{\partial t} = -\left(u\frac{\partial C}{\partial x}\right) + \frac{\partial}{\partial y}\left(K_y\frac{\partial C}{\partial y}\right) + \frac{\partial}{\partial z}\left(K_z\frac{\partial C}{\partial z}\right) \tag{3.52}$$

For stationary conditions and for constant diffusion coefficients along Y and Z directions, Eq. 3.52 becomes

$$\begin{aligned} \frac{\partial C}{\partial t} &= 0 \\ u\frac{\partial C}{\partial x} &= K_y\frac{\partial^2 C}{\partial y^2} + K_z\frac{\partial^2 C}{2} \end{aligned} \tag{3.53}$$

The analytical solution of Eq. 3.53 is similar to Gaussian plume model. Many models assume concentration profiles will have Gaussian shape and downwind transport goes along a straight line.

3.7.4.3 Atmospheric Stability and Sigma Functions

In the Gaussian plume model, the concentration profile depends on the wind velocity, standard deviations in Y and Z direction at a given distance in the downwind direction. Wind velocity can be estimated from the meteorological data. Now the question comes how to get the information about standard deviations. These standard deviations in general increase with downwind distance and also they depend on the atmospheric stability.

The atmospheric stability depends on the temperature profile of the atmosphere and it gives a measure for the vertical acceleration of the atmosphere. There are three main stability categories namely, unstable, neutral, and stable categories. These are briefly explained as follows [31]:

- *Neutral*
  - Vertical acceleration is zero
  - Temperature decrease with height at a rate equal to 0.98 °C for every 100 m
- *Unstable*
  - Positive vertical acceleration
  - Temperature decrease with height at a rate more than −0.98 °C/100 m
- *Stable*
  - Vertical motions are damped causing turbulence to decrease
  - Temperature decrease with height at a rate more than −0.98 °C/100 m

In the Pasquill model, the unstable and stable categories are further divided and it will lead to a total of six stability categories. These are divided into six classes from A to G.

Table 3.7 Pasquill stability categories [31]

Wind speed (m/s at 10 m)	Daytime conditions			Nighttime conditions	
	Strong sun	Moderate sun	Cloudy	>4/8 clouds	Clear sky
<2	A	A–B	B	E or F	F or G
2–3	A–B	B	C	E	F
3–5	B	B–C	C	D	E
5–6	C	C–D	D	D	D
>6	C	D	D	D	D

- Class A: Extremely unstable
- Class B: Moderately unstable
- Class C: Slightly unstable
- Class D: Neutral
- Class E: Slightly stable
- Class F: Moderately stable
- Class G: Very stable

This categorization is done based on wind speed along with the incoming solar radiation during day time and cloudiness during night time. The Pasquill stability categories are shown in Table 3.7.

The sigma functions can be well represented by power law. i.e.,

$$\begin{aligned} \sigma_y &= A_y x^{B_y} \\ \sigma_z &= A_z x^{B_Z} + C_Z \end{aligned} \tag{3.54}$$

where x is the downwind distance in '*m*'

A, *B*, and *C* are coefficients tabulated for each stability classes

σ_z cannot increase without any upper limit. The dispersion of the plume in vertical plane is restricted below by the ground and above by the inversion layer which acts as an atmospheric lid. Optimum value of σ_z can be used as 5000 m.

3.7.4.4 Wind Speed Correction with Height

Standard height for national networks recommended by World meteorological Organization is 10 m above ground. Anemometers are used to measure the wind speed. In the surface layer, wind speed increases with height and this can be represented with power law (empirical relations) as follows:

$$\frac{U_z}{U_1} = \left(\frac{z}{z_1}\right)^P$$
$$P = \frac{n}{2-n}, n = \text{sut tan parameter} \tag{3.55}$$
$$n = \begin{cases} 0.2 & \text{Unstable} \\ 0.25 & \text{Neutral} \\ 0.5 & \text{Stable} \end{cases}$$

where U_1 is reference level wind speed at z_1 (10 m)

Example 3.10 Find the wind speed at a height of 100 m from the ground considering the unstable conditions of the atmosphere. The reference wind speed at 10 m height is 5 m/s.

Solution: From the power law for wind speed correction.

$$\frac{U_z}{U_1} = \left(\frac{z}{z_1}\right)^P$$
$$P = \frac{n}{2-n}$$
$$n = 0.2 \text{ Unstable}$$
$$P = \frac{0.2}{2-0.2} = \frac{0.2}{1.8}$$
$$z_1 = 10 \text{ m}$$
$$U_1 = 5 \text{ m/s}$$
$$U_z = U_1\left(\frac{z}{z_1}\right)^P$$
$$z = 100 \text{ m}$$
$$U_{100} = 5\left(\frac{100}{10}\right)^{0.11} = 6.44 \text{ m/s}$$

3.7.4.5 Plume Rise

There are two types of plumes one is buoyant plumes and the other one is non-buoyant plumes. In case of buoyant plumes, the effluent discharge temperature is much higher than the ambient temperature. In the later case, the effluent temperature is equal to near ambient temperature. In nuclear industry, one generally come across nonbuoyant plumes. Bluff bodies can cause a downwash of the effluents. Stack should be of a height at least 2.5 times the height of the tallest adjacent structure within 20 stack heights radius (these are called tall stacks) in order to prevent downwash. The body of stack itself may cause down wash. To avoid this effect, the exit velocity of the effluent should be at least 1.5 times of wind speed. The plume rise from tall stacks can be calculated based on the following relations. For neutral and unstable conditions, this can be given as

$$\Delta h = 3\left(\frac{W_0}{U}\right)D_i \tag{3.56}$$

where

W_0 effluent exit velocity (m/s),
U wind speed at stack height (m/s), and
D_i internal stack diameter (m).

For stable conditions the plume rise can be given as

$$\Delta h = 1.5\,S^{-1/6}\left(\frac{F_m}{U}\right)^{1/3}$$

$$F_m = W_0^2\left(\frac{D}{2}\right)^2$$

$$S = 8.7 \times 10^{-4} \quad \text{for} \quad E$$

$$S = 1.75 \times 10^{-3} \quad \text{for} \quad F$$

In the case of plume rise from short stacks, if the exit velocity is greater than 5 times the horizontal wind speed at the height of release, the effluent should be treated as elevated release and the above expressions can be used for plume rise.

Example 3.11 Find the plume rise from the stack of height 100 m. The stack diameter is given as 1 m and the effluent exit velocity is 3 m/s. The wind speed at 10 m height is 4 m/s. Consider the stability category as B.

Solution:

- The plume rise for the unstable condition can be given as

$$\Delta h = 3\left(\frac{W_0}{U}\right)D_i$$

where

W_0 Effluent exit velocity = 3 m/s
D_i Stack inner diameter = 1 m
U wind speed at stack height of 100 m

- Wind speed can be obtained as

$$\frac{U_z}{U_1} = \left(\frac{z}{z_1}\right)^P$$
$$P = \frac{n}{2-n}$$
$$n = 0.2 \qquad \text{Unstable}$$
$$P = \frac{0.2}{2-0.2} = \frac{0.2}{1.8} =$$
$$z_1 = 10\,\text{m}$$
$$U_1 = 4\,\text{m/s}$$
$$U_z = U_1\left(\frac{z}{z_1}\right)^P$$
$$z = 100\,\text{m}$$
$$U_{100} = 4\left(\frac{100}{10}\right)^{0.11} = 5.153\,\text{m/s}$$

- Now the plume rise can be calculated as

$$\Delta h = 3 \times \left(\frac{3}{5.153}\right) \times 1 = 1.75\,\text{m}$$

- Hence, the plume can rise to a height of 1.75 m above the stack.
- The effective height of the plume is (stack height + plume rise), i.e., 101.75 m.

3.7.4.6 Sector-Averaged χ/Q Values

In general, the radiological assessment for nuclear facilities will be done for longer duration. For the extended periods, the atmospheric conditions (wind velocity, wind direction and stability categories) will definitely change. To account for these changes in the calculations, one need to determine the sector-averaged χ/Q values [31]. This is done on monthly, quarterly, or annual basis. In this, the released radioactive material is distributed into different sectors around the plant depending on the wind direction and it will be further grouped into various groups based on the wind speed and stability category for an average release quantity Q. For this purpose, the meteorological data has to be collected for longer periods. The long-term sector-averaged χ/Q values can be calculated as follows:

$$\frac{\bar{\chi}}{Q} = 2\frac{1}{\sqrt{2\pi}}\frac{f_i}{\beta x}\sum_{jk}\frac{F_{jk}}{\sigma_{zj}\bar{u}_k}e^{\left[-\frac{1}{2}\left(\frac{H}{\sigma_{zj}}\right)^2\right]} \tag{3.57}$$

where

- x is the downwind distance from source to the midpoint of sector (m)
- β is the sector width ($2\pi/n$) (radians)
- f_i fraction of time wind blows towards sector-i
- F_{jk} fraction of f_i during which stability class j exists for wind class k in sector i
- σ_{zj} vertical diffusion coefficient for stability class j
- $\bar{u}_k$ Median wind speed for wind class k

For n = 16 sectors, Eq. 3.57 will be

$$n = 16$$

$$\beta = \frac{2\pi}{n} = \frac{2\pi}{16} = 22.5^\circ$$

$$\frac{\bar{\chi}}{Q} = \frac{2.032 f_i}{x}\sum_{jk}\frac{F_{jk}}{\sigma_{zj}\bar{u}_k}e^{\left[-\frac{1}{2}\left(\frac{H}{\sigma_{zj}}\right)^2\right]} \tag{3.58}$$

For 16 sectors, the sector angle will be 22.5° and is shown in the Fig. 3.23.

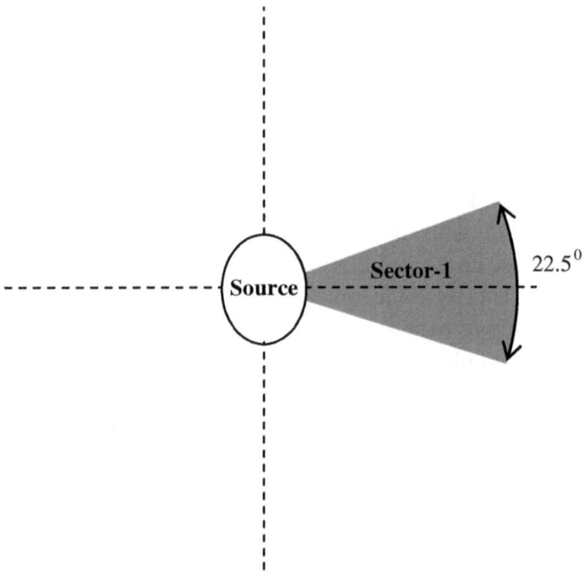

Fig. 3.23 Representation of sectors around plant

Table 3.8 Stability classes existed for each wind speed group

Wind speed (m/s)	Stability class	Duration (h)
0–2	F	50
2–4	B	60
	C	40
4–6	D	50

Example 3.12 The meteorological data around a plant for one month (720 h) shows that the wind blows in the direction of sector-1 for 200 h. The plant releases radionuclides at an average rate Q (Ci/s) from a 100 m stack. The wind speed variation is given as 0–2 m/s for 50 h, 2–4 m/s for 100 h and 4–6 m/s for 50 h. Stability classes existed for each wind speed group are given in Table 3.8:

Determine the monthly sector-averaged χ/Q for a downwind distance between 1 and 3 km in the sector 1.

Solution:

- f_1 fraction of time wind blows towards sector-1 = 200/720 = 0.278
- x = Downwind distance from source to the midpoint of sector-1 = 2000 m
- Vertical diffusion coefficient for different stability categories at distance 2 km is given in Table 3.9
- The exponential term is also calculated in the above Table 3.9 for a release height of 100 m.
- F_{jk} for each stability class and wind speed are calculated in Table 3.10
- Now sector-averaged χ/Q can be calculated as

Table 3.9 Vertical diffusion coefficient

Stability class	σ_z (m)	$\exp\left[-\frac{1}{2}\left(\frac{H}{\sigma_{zj}}\right)^2\right]$ (H = 100 m)
B	230	0.9098
C	120	0.7066
D	50	0.1353
F	22	3.26×10^{-5}

Table 3.10 Calculation of F_{jk}

Wind speed (m/s)	Stability class	Duration (h)	F_{jk}
0–2	F	50	50/200 = 0.25
2–4	B	60	60/200 = 0.30
	C	40	40/200 = 0.20
4–6	D	50	50/200 = 0.25
	Total duration = 200 h		$\Sigma F_{jk} = 1$

$$\frac{\bar{\chi}}{Q} = \frac{2.032 f_i}{x} \sum_{jk} \frac{F_{jk}}{\sigma_{zj} \bar{u}_k} e^{\left[-\frac{1}{2}\left(\frac{H}{\sigma_{zj}}\right)^2\right]}$$

$$= 2.032 \times \frac{200}{720} \times \frac{1}{2000} \begin{bmatrix} \frac{0.25}{1\ \text{m/s} \times 22\ \text{m}} \times 3.26 \times 10^{-5} + \\ \frac{0.3}{3\ \text{m/s} \times 230\ \text{m}} \times 0.9098 + \\ \frac{0.2}{3\ \text{m/s} \times 120\ \text{m}} \times 0.7066 + \\ \frac{0.25}{5\ \text{m/s} \times 50\ \text{m}} \times 0.1353 + \end{bmatrix}$$

$$\frac{\bar{\chi}}{Q} = 2.6073 \times 10^{-7} \text{s/m}^3$$

3.7.4.7 Depletion Mechanisms

In reality, the radionuclide content of the plume diminishes both by radioactive decay and through deposition mechanisms (dry and wet). Dry deposition is basically a surface effect, where by material in contact with the ground is removed through a number of processes (Particles may impact on surface projections and stick, gases may diffuse to leaf surfaces and be absorbed). Wet deposition is the removal of dispersing material as a result of either the interaction between the dispersing material and rain (snow, etc.) falling through it, or by its incorporation into rain clouds.

The effect of above-mentioned processes on plume concentration is very small compared to the input radionuclide release and it can be neglected in the calculation of downwind concentration. However, their effect can be seen as accumulation of radioactivity on vegetation and ground surface from where they can enter into the human body through food pathways or becomes a direct source of exposure to the radioactivity from the ground surface. Correction due to radioactive decay can be given as

1. For single radionuclide

$$F_r = e^{-\lambda t} \quad t = \frac{x}{U}$$

2. For mixed fission products

$$F_r = \left(\frac{t}{t_r}\right)^{-1.2}$$

where λ is the decay rate
x is the downwind distance,
U is the wind velocity,
t is the time elapsed, and
t_r is the reference time at which the initial activity is stated

The other correction factor is depletion due to dry deposition. In this case, depletion of particles take place by impaction and depletion of gases occur by absorption processes on surfaces (walls, foliage and ground). Gravitational settling of large particle is possible. Generally, noble gases do not deposit and aerosols, elemental iodine, organically bound iodine deposit. This can be represented with areal contamination $C_{A,\text{dry}}(\text{Bq/m}^2)$ over a given area as follows:

$$C_{A,\text{dry}} = v_d \frac{\bar{\chi}_i}{Q_i} Q_{\text{TOT}} \tag{3.59}$$

where
v_d Deposition velocity (m/s)
Q_{TOT} Total amount of material released over a period for which sector-averaged χ/Q is calculated.

The other deposition factor is due to wet deposition. Here depletion is due to precipitation process. Precipitation (rain, snow) can deplete the plume due to collection of effluent by the raindrops or snow particles as they fall through the plume. Falling rain drops collect the particles in their path. Soluble material gets scavenged by rain drops. The correction factor due to wet deposition can be calculated as

$$F_P = e^{-\Lambda \frac{x}{U}} \tag{3.60}$$

where
Λ washout coefficient (/s)
$\Lambda = aR^b$ a and b are constants
R rainfall rate

Example 3.13 Estimate the areal contamination due to dry deposition of Cs-137 (a) on the ground in the downwind direction of a 100 m stack that releases 10^9 Bq of Cs-137 if the average χ/Q value over the period of release is 2.6×10^{-7} s/m^3.

Solution:

- $Q_{\text{TOT}} = 10^9$ Bq
- $\frac{\bar{\chi}}{Q} = 2.6 \times 10^{-7}$ s/m^3
- Consider deposition velocity for Cs-137 on ground as 4×10^{-4} m/s
- The areal contamination over a given area can be calculated as follows:

$$
\begin{aligned}
C_{A,\text{dry}} &= v_d \frac{\bar{\chi}_i}{Q_i} Q_{\text{TOT}} \\
&= 4 \times 10^{-4} \times 2.6 \times 10^{-7} \times 2 \times 10^{9} \\
C_{A,\text{dry}} &= 0.208\,\text{Bq/m}^2
\end{aligned}
$$

3.7.5 Exposure Pathways and Dose Assessment

There are different pathways by which people can get exposed to the radiation caused by the radioactive material which is released to the atmosphere after an accident. They fall under two categories, i.e., external irradiation and internal irradiation and are listed as follows [4].

External irradiation:

- Cloud Shine: Irradiation caused by the radioactive material carried by the passing plume or cloud
- Ground Shine: Irradiation caused by the radioactive material deposited on the ground,
- Irradiation caused by the radioactive material deposited on skin and clothing,

Internal irradiation:

- Irradiation caused by the radioactive material inhaled directly from the passing plume,
- Irradiation caused by the radioactive material inhaled following resuspension of the ground deposit,
- Irradiation caused by the radioactive material ingestion: This is due to the ingestion of the food stuffs which are contaminated by the radioactive material that is deposited from the passing plume.

The above-mentioned pathways are shown in the Fig. 3.24. In calculating the dose absorbed by the humans all the exposure pathways have to be considered. The dose from a given a path can be calculated by converting the concentration of radioactive material in that particular path into the dose by using proper dosimetric models. The dose from a each exposure pathway is explained briefly in the following subsections.

3.7.5.1 Cloud Shine

This is an external irradiation caused by the radioactive material carried by the passing plume. In this pathway, the person on the ground will get exposed to the

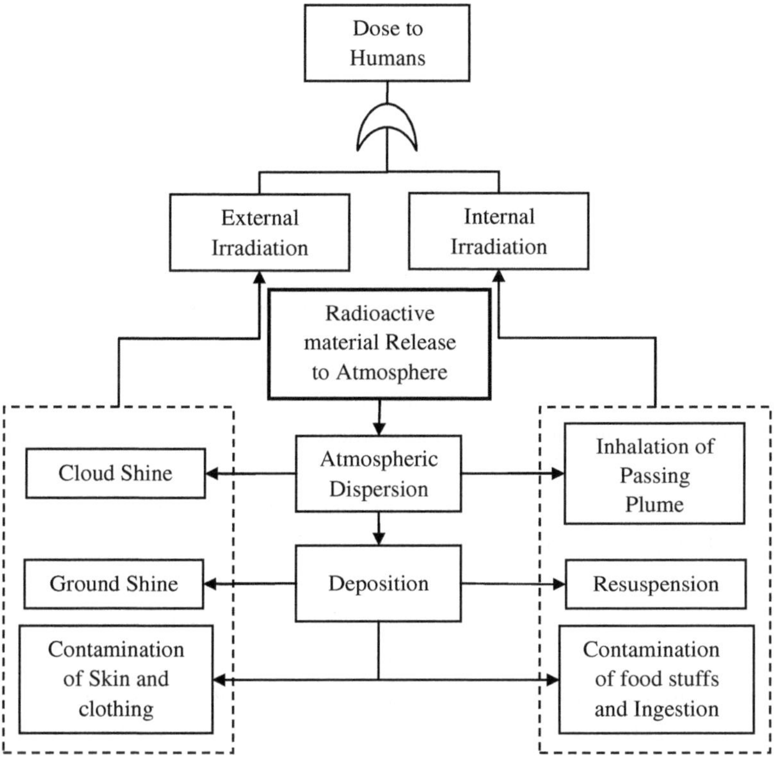

Fig. 3.24 Different radiation exposure pathways to humans

dose by both β and γ particles that are being emitted by the radionuclides which are carried by the passing cloud. Since β particles have short range in air (a few meters), the dose contribution from these particles can be neglected. The external gamma dose rate at a point from the passing plume can be calculated by using the following formula [31].

$$\begin{aligned} D_\gamma &= 5 \times 10^{-4} \phi \, E \, \mu_a \\ \phi &= B \, \frac{e^{-\mu R}}{4\pi R^2} C \mathrm{d}x \, \mathrm{d}y \, \mathrm{d}z \\ B &= 1 + k\mu R \\ k &= \frac{\mu - \mu_a}{\mu_a} \end{aligned} \tag{3.61}$$

where

D_γ gamma dose rate (μSv/h)

C concentration in the elemental volume dv (Bq/m^3)

B buildup factor for that photon energy

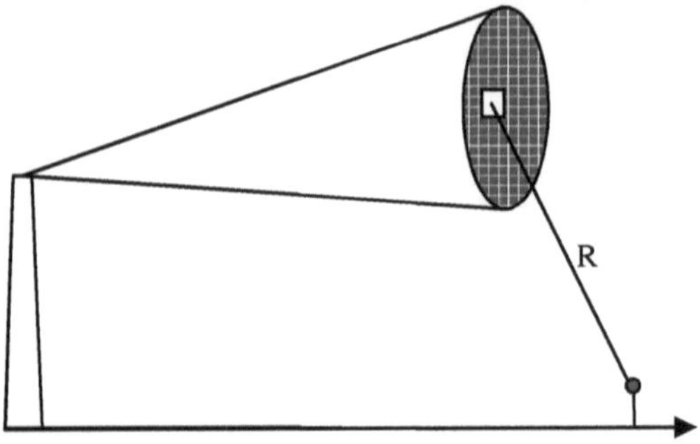

Fig. 3.25 Gamma dose from a passing plume

μ total attenuation coefficient in air (/m)
μ_a energy absorption coefficient in air (/m)
R distance between the receptor and elemental volume

The net gamma dose rate is obtained by integrating over entire plume. This is illustrated in Fig. 3.25 and this can be expressed as follows:

$$D_\gamma = 5 \times 10^{-4} \iiint B\,E\,\mu_a \frac{e^{-\mu R}}{4\pi R^2} C \mathrm{d}x\,\mathrm{d}y\,\mathrm{d}z \tag{3.62}$$

The above dose calculation does not take into account of shielding. Here shielding means, if the person is inside a building or staying indoor, the building will attenuate the γ rays and eventually reduces the dose absorbed. Hence, in dose calculation it is necessary to apply the shielding factor depending on the type of building. Here, shielding factor is defined as the ratio of interior dose to the exterior dose. Hence, this factor will always be less than 1. In general, it is assumed that some percentage of people is staying indoors in evaluating the dose.

3.7.5.2 Ground Shine

This is also an external irradiation caused by the radioactive material deposited on the ground. Here also both β and γ particles emitted by the radioactive material will contribute to the dose. However, the dose contribution due to β particles can be neglected as in the previous case. The dose received by the person depends on two factors as listed below:

- Duration of time the person is exposed and
- The time at which the radionuclides deposited on the ground.

As the time of exposure increases the dose received also increases. Also, radionuclides undergo radioactivity decay and with time it will result in a reduction of dose rate. The dose rate will also get reduced due to some natural weathering processes. The deposit from the exposed surfaces may be removed by these natural weathering processes. For example, depending on the properties of the soil and the weather, migration of radionuclides into soil may take place and also the deposits from the urban surfaces may be removed by the rainfall. These processes will eventually lead to reduced dose rates.

In this pathway, the dose is evaluated by multiplying dose per unit deposit conversion factor with the total deposit and is integrated over exposure time period. The pathway is often evaluated by multiplying the deposit by a dose per unit deposit conversion factor, integrated to appropriate time periods. These conversion factors also consider long-term removal mechanisms. As in the previous case, shielding factors are considered for people indoors. The shielding factors may range from 0.01 to 0.5 depending on type of shielding.

3.7.5.3 External Dose from Radioactive Material Deposited on Skin and Clothing

In this path, the individual is exposed to the radioactive material which is deposited on skin and clothing. This is an external irradiation. Unlike the other pathways in this case, both β and γ emitters contribute to the dose. While calculating the amount of radionuclide material deposited on skin and clothing, it is taken as a fraction of amount that is deposited on the ground surface. Shielding factors can be used for accounting the different types of clothing while calculating the β dose to the skin.

3.7.5.4 Inhalation Dose

Inhalation dose can be caused by either the person inhaling the passing plume directly or inhalation of the resuspended radioactivity material that is deposited on the ground surface. If a person inhales the passing plume containing the radioactive material it will give direct inhalation dose. This inhalation dose is a function of the time-integrated air concentration (TIC), breathing rate, and dose conversion factor (precalculated dose per unit activity inhaled). The inhalation dose can be obtained as a product of these factors. TIC can be calculated from the available plume models. The amount of radioactivity inhaled over some period of time depends on the breathing rate and it is a function of age of the person and level of physical activity. For adults, the typical breathing rate is 2.66×10^{-4} m^3/s. These factors are obtained from metabolic models.

The air concentration will be different depending on whether the person is inside or outside the building for short-term releases. In such cases, filtering factors can be applied to modify the inhaled dose. The filtering factor less than one is normally applied to the proportion of the population that is indoors. The inhalation dose can be mathematically expressed as follows [4].

$$\begin{aligned} I_{dose} &= \text{TIC} \times \text{BR} \times \text{Dose/unit activity inhaled} \\ \text{where TIC} &= \text{Time integrated concentration} \\ \text{BR} &= \text{Breathing Rate} \end{aligned} \tag{3.63}$$

Due to different deposition mechanisms, the radioactivity material will be deposited on the ground surface. This radioactivity material will become airborne over long periods of time after the initial deposition. Once it becomes airborne it will be available for inhalation. This will happen due to the action of the wind or human activities like driving vehicles or ploughing. A time-dependent resuspension factor is used to define the relationship between the air concentration and the amount of radioactivity material deposited on the ground surface. This factor can be defined as follows:

$$\text{Resuspension factor} = \frac{\text{Air concentration due to resuspension}}{\text{Initial surface concentration}}$$

The resuspension factors can be obtained from the actual measurements made in the environment. Dose calculation is similar to that explained in the previous case.

3.7.5.5 Ingestion Dose

The ingestion dose is caused due to the direct intaking of the food stuffs which are contaminated with the radioactive material. There are a wide range of food stuffs that are consumed by the human beings. There is a separate food chain for intake of radionuclides for each food stuff consumed. To calculate the ingestion dose, one should know the amount of radionuclides in food stuffs per unit deposition, the consumption rate of food, and the dose conversion factor (dose per unit activity ingested). The first factor that is concentration of radionuclides in food stuffs is time dependent whereas the other two factors that is consumption rate and dose conversion factor are age dependent. These are evaluated by using the dynamic food chain transport models. These are complex calculations and most of the consequence assessment codes do not implement ingestion dose calculations.

3.7.6 Countermeasures

Countermeasures represent the measures which are to be implemented in order to avoid or minimize human beings to get exposed to the radiation during accident kind of situation and thereby preventing deterministic effects as well as minimize the risks of stochastic effects. There are two types of countermeasures available depending on the implementation time and the mitigation effects [4].

3.7.6.1 Short-Term Counter Measures

- These are implemented either before or shortly after the radioactive material is released to the environment.
- This include
  - Sheltering

    Staying inside the buildings with windows and doors closed and ventilation systems turned off.
    It is intended to reduce

      cloud shine,
      ground shine,
      inhalation exposures, and
      contamination of clothing and skin while the plume is passing.

  - Evacuation

    evacuation of people within the area

  - Issuing of iodine tablets

    distribution of *tablets containing stable iodine*
    will saturate the thyroid gland with the nonactive isotope
    prevents the radioactive isotopes from being absorbed.

  - Decontamination of people

    changing the clothes
    taking shower to remove contamination from the skin

3.7.6.2 Long-Term Countermeasures

- These are designed to reduce the late health effects.
- It includes
  - relocation,
  - land decontamination, and
  - food bans.
- Other measures like

  changes to agricultural practices,
  deep ploughing,
  alternative feed, caesium binders,
  alternative crops, and
  alternative production (cheese instead of milk).

3.7.7 Health Effects

When the individual is exposed to ionizing radiation it will have an effect on the individual's health. There are two types of health effects: one is deterministic health effects and other one is stochastic health effects. These effects can be observed directly in the exposed individual (termed as somatic effects) or it can be observed in the descendants of the exposed person (known as hereditary effects) [4].

3.7.7.1 Deterministic Effects

Deterministic effects are sometimes referred as early effects. This is due to the reason that depending on the amount of radiation, the individual get exposed it will have early effects on individual. Either whole or part of the body can be exposed. If the doses are beyond certain level, then only these effects can be seen or else the effects may not be induced. This level of radiation dose is called threshold dose. The deterministic effects are evaluated based on a hazard function [Scott and Hahn]. The probability of an individual getting affected can be given by:

$$R = 1 - e^{-H} \tag{3.64}$$

where H is the cumulative hazard function that can be represented by a two-parameter Weibull distribution function as follows [4]:

$$H = \ln 2 \left(\frac{D}{D_{50}} \right)^{S} \quad D > T \tag{3.65}$$

where

D Avg. absorbed dose to the relevant organ

D_{50} Dose which causes the effect in 50 % of the exposed population

S Shape parameter, which characterizes the slope of the dose-risk function

T Threshold dose

3.7.7.2 Stochastic Effects

These effects can be observed either in the exposed population or in the descendents of the exposed population. The stochastic effects mainly include increased incidences of cancer. The probability of occurrence depends on the radiation dose. This effect will appear after longer period of time after the exposure. Hence it is often referred as late effects. In this, the risk of cancer incidence is given by a nonlinear dose response function which is a combination of both linear and quadratic terms of the radiation dose and is given as follows:

$$R = aD + bD^2 \tag{3.66}$$

where

D Absorbed dose to the organ of interest

a and b Risk coefficients, model parameters that quantify the risk per unit dose.

3.7.8 Population and Economic Data

Finally, the spatial distribution of the population, agricultural production, and economic data (land and housing values) are required

- To evaluate the health effects within the population caused due to the
  - external exposure (cloud shine, ground shine etc.)
  - internal exposure (inhalation and ingestion)
- To calculate the economic impact of implementing countermeasures, such as
  - relocation and
  - food bans.

3.7.9 Complementary Cumulative Frequency Distributions

Once all the results for all the weather conditions are compiled, it is common practice to produce a graph between consequence versus frequency of exceedance of a given consequence. This curve will be complementary to the cumulative distribution function. A sample CCFD curve is shown in the Fig. 3.26 [4].

3.8 Applications of PSA

PSA has been accepted all over the world as an important tool to assess the safety of a plant. PSA can be done during design stage and also during operational stage. It can be used to identify design weaknesses of various systems in the plant. PSA provides the quantitative estimate of risk which is useful for comparison alternatives in different design and engineering areas. It can provide dominant risk contributors and options to reduce the risk. Some important applications of PSA toward Safety Issues are explained below [3]:

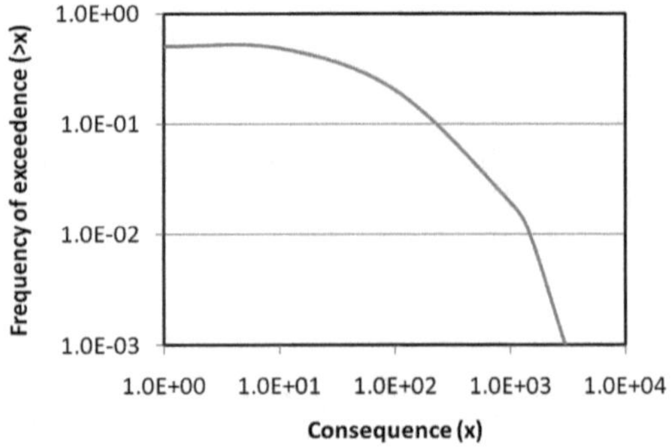

Fig. 3.26 A sample CCFD curve

- *Decision Making in operations*:
  Core Damage Frequency (CDF) value is an important parameter, which can provide risk insights. If CDF value exceeds the prescribed probabilistic safety criteria, that is termed as an unsafe condition. Also, efforts are always made to lower the CDF through different test and maintenance policies.
- *Risk Monitor*:
  One of the main applications of PSA is risk monitor. This is further explained in detail in Chap. 7.
- *Maintenance Strategies*:
  Risk achievement worth (RAW) is the best input for deciding maintenance policies. RAW and risk reduction worth (RRW) can be evaluated system wise and component wise. Components having higher RAW have to be maintained immediately, in order to minimize the CDF value. Similarly, component having higher RRW should be given attention from the design point of view, since it can enhance the reliability of the system.
- *Risk-Based In-Service Inspection*:
  The Risk Informed In-Service Inspection (RI-ISI) [29] programs aim at integrating traditional engineering evaluations with insights gained from PSA. The prime use of PSA is to obtain an estimate of risk and relegate it to various systems and down to components to obtain an idea of their importance in terms of contribution to the Risk. PSA tools can be effectively employed for analyzing the change in CDF whenever there is a change in inspection plans and thereby analyze for an optimum scheduling plan. Risk importance measures such as RAW, RRW, Fussell-Wessley etc. for various components and systems can be evaluated for risk-based inspection planning.
- *Review of Technical Specification*:
  The Technical Specifications are usually based on deterministic assessment and engineering judgement. Based on the PSA studies, technical specifications based on probabilistic considerations can be evolved to optimize the allowable outage time (AOT) and surveillance test interval (STI) for various Systems.
- *Emergency Operating Procedures and Risk Management*:
  The emergency operating procedures (EOPs) have been usually based on the considerations of failures in process systems only. EOPs based on dominating

accident sequences as identified in PSA can be effectively used in risk management.
- *Probabilistic Precursor analysis*:
  Probabilistic precursor analysis is very useful in identification of the precursors to a major accident. This is further explained in detail in Chap. 7.

In spite of the benefits, it is well recognized that PSA has its own limitations. The accuracy of the PSA depends on the uncertainties in aspects like data and models used in the analysis. In this chapter, different levels of PSA in the context of NPPs have been explained. The same procedure can be applied to the chemical plants also depending on the number of levels.

3.9 Case Study: Level 1, 2, and 3 PSA Analysis of a Typical NPP

The purpose of this case study is to understand the three levels of PSA applied to a typical NPP. The reactor core consists of calandria, coolant channels, fuel bundles, fuel pins, and moderator. During normal power operation or shutdown condition, coolant should be circulated continuously inside the reactor core so as to remove heat or decay heat from the core. There are different frontline systems and support systems to perform different safety functions during any transient. The outcome of the analysis is risk of the plant in terms of absorbed dose (msv/year) at a given distance from the plant due to some postulated initiating events.

3.9.1 Level 1 PSA

3.9.1.1 Identification of Initiating Events

Some of the initiating events considered along with their frequency of occurrence are shown in the Table 3.11.

Table 3.11 Different initiating events

S. No.	Initiating Event	Frequency (/year)
1	Small LOCA	3.25×10^{-2}
2	Large LOCA	6.12×10^{-4}
3	Mainsteam line break outside the containment	1.01×10^{-4}
4	Class IV power supply failure	1.41
5	Loss of feedwater system	1.45×10^{-1}
6	Active process water system failure	7.80×10^{-3}

For the purpose of illustration, large LOCA has been considered in the present case study.

3.9.1.2 Determination of Safety Functions and Safety Systems

Once the IEs are identified, it is important to identify the safety functions that need to be performed for each IE in order to prevent core damage. Some important safety functions for protecting against core damage are listed below:

- Shut down the reactor to control the reactivity
- Removing the decay heat from the core
- Maintaining long-term subcriticality in the core

Different safety systems (front line systems) are designed to perform different safety functions. For proper functioning of the front line systems, support systems are designed. Some of the safety systems and support systems are listed below.

- Frontline systems (FLs)
  - Reactor protection system
  - High/Low-pressure injection system
  - Emergency feedwater system
  - Pressurizer safety relief valves
- Support systems (SSs)
  - Service water system
  - Instrument air system
  - Diesel Ac generators

3.9.1.3 Grouping of the IEs

Depending on the safety systems to be evoked and same special conditions, IEs can be grouped. Example of different IE groups are

- LOCAs
  - Small LOCA
  - Medium LOCA
  - Large LOCA
- Mainsteam line break

3.9.1.4 Accident Sequence Modeling

Once the IEs are grouped, next step is the identification of different accident sequences and their quantification. For this purpose ,one uses event tree and fault tree methods.

3.9.1.5 Development of Event Trees

The event trees have been developed for all the identified initiating events. For the case of large loss-of-coolant accident (LOCA), the event tree is shown in the following Fig. 3.27. Any pipe rupture resulting in leak/breaks from main coolant system, greater than 10 % of the area of the largest diameter pipe is considered to result in Large LOCA. Upon identification of Large LOCA and reactor shutdown, emergency core cooling system (ECCS) will be initiated. The ECCS will be functioning first in high-pressure injection mode and later once the pressure comes down, it will function in low-pressure injection mode and these two are independent systems for core cooling. In case of failure of HPIS, LPIS will be initiated after the pressure falls below the set point.

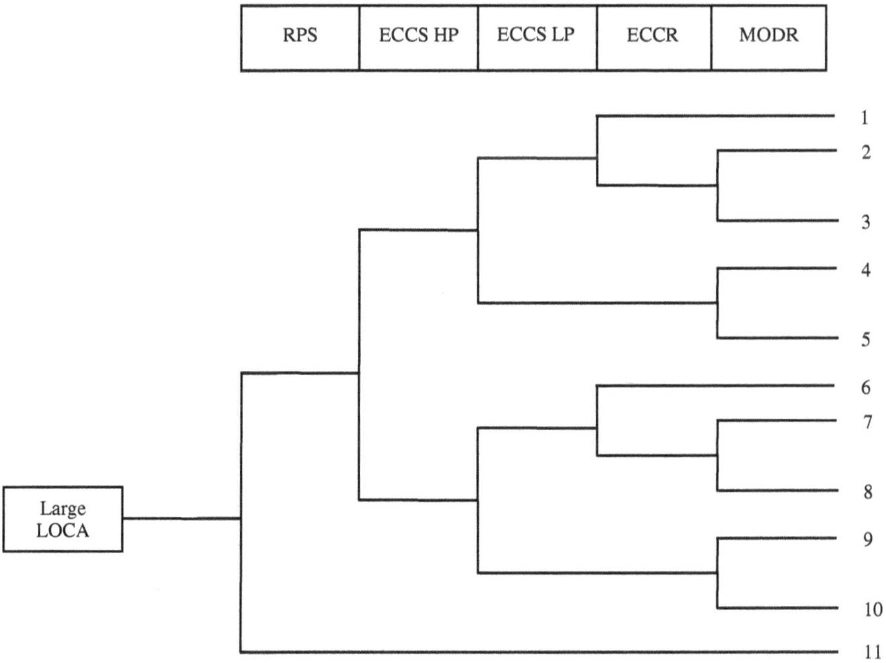

Fig. 3.27 Event tree of large LOCA

The spilled water gets collected in the cavity surrounding the core. This water is recirculated after cooling it in heat exchangers to continue long-term cooling of the core. The recirculation phase for ECCS can be started later after 3 days. If LPIS system fails to operate, there will not be any accumulation of water in the reactor cavity and recirculation in the normal mode will not be available. Credit for moderator as the ultimate heat sink has been considered. If moderator is also unavailable, it is expected to lead to core damage.

3.9.1.6 Identification of Dominant Accident Sequences

In the event tree, there are 11 accident sequences and out of these very few sequences lead to core damage state; these are called dominating accident sequences. Among these 5, 10, and 11 are core damage states. These are listed below

- Large LOCA-ECCSLP-MODR
- Large LOCA-ECCS HP-ECCSLP-MODR
- Large LOCA-RPS

3.9.1.7 Quantification of Accident Sequences

In order to quantify the accident sequences, one should have the information about IE frequency and safety system unavailabilities. In the present case study, the frequency of occurrence of LOCA is calculated based on the piping inventory in the plant and their corresponding failure frequency. Generic data sources have been used for this study. Whereas, Class IV power supply failure has been estimated by using Bayesian analysis as explained earlier. The other initiating events frequencies have been calculated by performing system reliability analysis.

Similarly, safety system unavailabilities have been estimated from fault tree analysis. The different safety systems and their unavailabilities are shown in the Table 3.12. For the purpose of illustration, fault tree development of ECCS high-pressure injection is presented here.

3.9.1.8 Emergency Core Cooling System (ECCS)

In this case study, we develop fault tree of Emergency Core Cooling System (ECCS) of advanced reactors. While developing fault tree for any complex system, one should know the details about the system description, failure criteria, boundary conditions, support systems, logics, instrumentation and control, common cause failures, human reliability analysis, and failure modes of components. Once the fault trees are developed in evaluating the fault tree, one should generate the MCS and finally finding out the failure probability from the MCS. For finding out the

Table 3.12 List of safety system unavailabilities

S.No.	Safety system	Unavailability
1	Reactor protective system	1×10^{-6}
2	Emergency core cooling system-HPIS	2.95×10^{-5}
3	Emergency core cooling system-LPIS	9.98×10^{-5}
4	ECCS recirculation system	1.64×10^{-4}

system failure probability, one should have information about the failure probabilities at the component level. This is further explained in different sections (failure models). In this section, only up to fault tree development and MCS generation are explained and the details are given below.

3.9.1.9 System Description

ECCS is used as a decay heat removal system in the case of loss-of-coolant accident. It operates in three stages: high-pressure injection mode, low-pressure injection mode, and recirculation mode. In this example, for the purpose of illustration, ECCS in high-pressure injection mode is considered.

ECCS high-pressure injection consists of four accumulators, where each accumulator holds 60 m^3 of cold water and 10 m^3 of nitrogen at 5 MPa (50 kg/cm^2) pressure. Each of the accumulators is connected through a check valve, two numbers of motor-operated valves, and a rupture disk in each injection line to an emergency core cooling header (ECCH). Each ECCH is connected to all the channels. Each channel consists of a perforated water tube, through which cold water ejects directly on the fuel pins. The perforations in the water tubes are arranged in a manner to ensure adequate cooling of fuel pins during LOCA as shown in Fig. 3.28.

System Failure Criteria

- Water flow from any four accumulators is needed to meet cooling requirement. This implies the failure criterion, in this case, is failure of water flow from all the four accumulators.

System Boundary

- Accumulator is considered as boundary at inlet side and ECCS header at the outlet side.

Assumptions

- The valves above the accumulators are not modeled in the analysis. During LOCA condition, nitrogen supply will be blocked from the nitrogen cylinders and all the valves will be in closed position. Even if these valves get opened, it will pressurize the system. It is a failsafe situation.

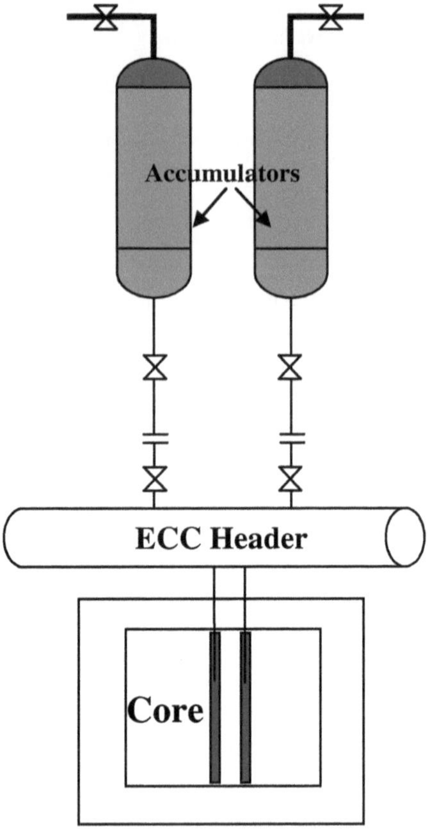

Fig. 3.28 Schematic of emergency core cooling system

Common Cause Failure (CCF)

- Common cause failure is considered for check valves and rupture disk in this analysis. A beta factor of 0.1 is considered in the analysis

Fault tree

- The fault tree for this system is shown in Fig. 3.29. There are four accumulator channels. Each channel consists of motor-operated valves, rupture disk, check valve, and accumulator. All the components are in series configuration. No redundancy exists at the component level.

Minimal Cut Sets (MCS) and System Unavailability

- Minimal cut sets of the system are generated. First-order cut sets are shown below.
  - HPIS-CHK-VLV-CCF
  - HPIS-RD-CCF

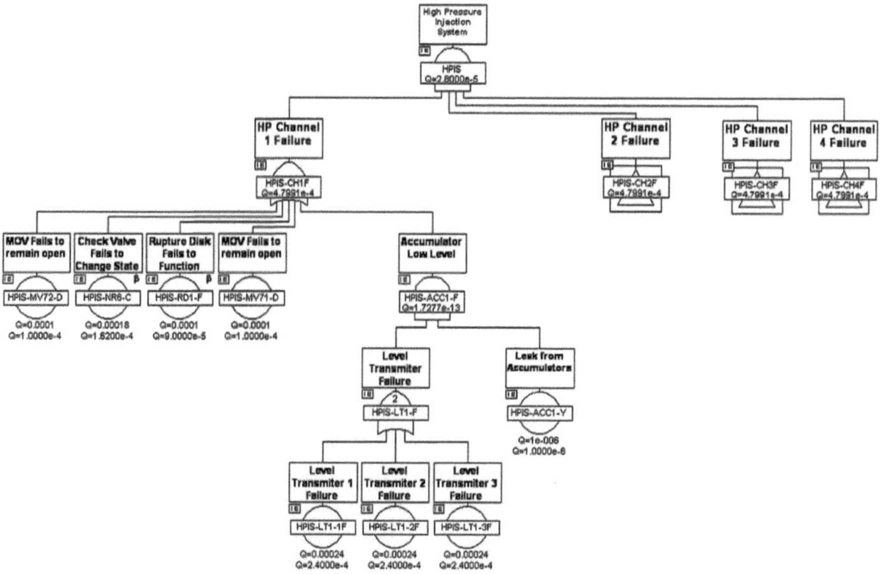

Fig. 3.29 Fault tree of ECCS-HPIS

- Common cause failure of check valve and rupture disk are contributing more toward the system unavailability and is estimated to be 2.95×10^{-5}

3.9.1.10 Quantification of Dominating Accident Sequence Frequency

Once the system unavailabilities are estimated, one can quantify the dominant accident sequence frequencies. For the case of large LOCA, the dominant accident sequences along with their frequencies are shown in the Table 3.13.

3.9.2 Level 2 PSA

The Level 2 PSA involves thermal hydraulic analysis of the core, evaluation of the inventory of fission products in the core and its release to the containment, thermal hydraulic analysis of the containment, and transport of radionuclides within the

Table 3.13 Dominant accident sequences for large LOCA

S. No.	Dominating accident sequences	Frequency
1	LLOCA-ECCS HP-ECCS LP-MODR	4.23E-15
2	LLOCA-ECCS LP- MODR	1.44E-10
3	LLOCA-RPS	6.30E-10

containment, assessment of leakage, and evaluation of the source term. For the purpose of illustration, the following accident sequence has been analyzed to assess the core state and the release of fission products from the reactor core to the environment

- Large LOCA with failure of reactor protection system (shut down systems 1 and 2)
- From Level 1 PSA, the frequency of occurrence of the accident sequence is estimated as 6.30×10^{-10}/year.

The above-mentioned accident sequence has been analyzed along with and without the operation of the designed containment engineered safety features.

3.9.2.1 Plant Damage State

The present accident sequence under consideration will fall under plant damage state class-1 due to which the radioactivity will be released into the containment without by passing the containment. Since, this event may lead to loss of core structural integrity this can be categorized as accident sequence category 1. This may cause simultaneous structural failure of more than one channel. Large fraction of fuel failures and significant metal water reaction are expected to be the consequences of this category.

3.9.2.2 Accident progression

In this task, plant-specific analyses of the progression of severe accidents are performed using appropriate analytical tools and computer codes. For the purpose of illustration, occurrence of Large LOCA with absence of shutdown system 1 and 2 is considered. The reactor behavior in case of LOCA scenario depends on the type of reactor whether reactivity increases or decreases due to void formation. In case of positive void reactivity reactors, void formation causes increase in reactor power, whereas in the case of negative void reactivity reactors void formation causes reduction in reactor power.

In case of positive void reactivity reactors, LOCA causes insertion of positive reactivity due to coolant channel voiding which results in increase in reactor power.

- The rate of addition of this reactivity depends on the thermal hydraulic behavior of coolant, size, and location of break.
- Following LOCA if both shutdown systems fail, power excursion continues due to positive void reactivity.
- Rapid power excursion causes accumulation of fission energy in fuel pellets leading to disintegration of fuel pellets, pressure tubes, and also in very short time fuel clad melting occurs in high-power channels which lead to disintegration of fuel pin and subsequent failure of coolant and calandria tube resulting in steam discharge from coolant channel to moderator.

- This causes rapid displacement of moderator resulting in large negative reactivity insertion which will terminate the power excursion instantaneously.

In case of negative void reactivity reactors, following LOCA the reactor, power drops due to increase in steam void in the reactor core.

- The power starts increasing following a void quench from accumulator injection. The rise in power causes the hot channels to get damaged due to overheating. The change in system pressure and clad temperature in one of the hot channels are shown in the Figs. 3.30 and 3.31.

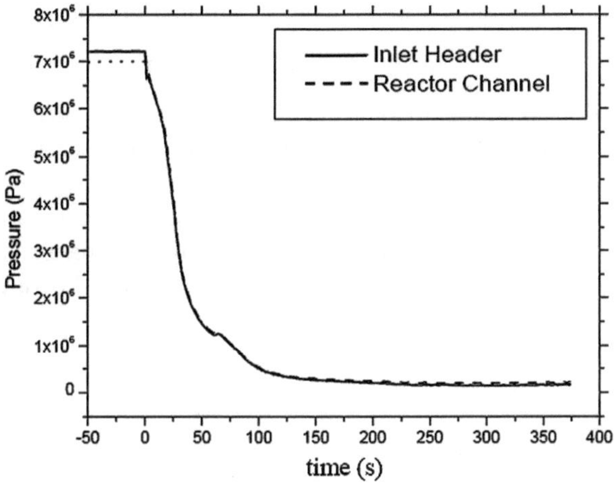

Fig. 3.30 System pressure

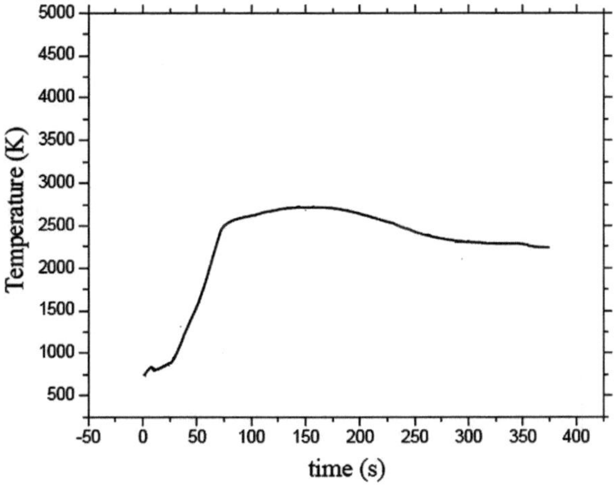

Fig. 3.31 Clad temperature in hot channel

3.9.2.3 Containment Analysis

The first step of containment analysis is to collect information on the containment design considerations and containment ESFs. These include

- Design temperature, design pressure, containment penetrations, and bypass pathways etc.
- Containment ESFs include:
  - Reactor Building Coolers (RBC),
  - Containment Isolation System (CIS),
  - Primary Containment Filtration and Pump Back System (PCFPB),
  - Secondary Containment Recirculation and Purge System (SCRP),
  - Primary Containment Controlled Discharge System (PCCD).

  The arrangement of containment ESFs in a typical NPP is shown in Fig. 3.32.
- Containment event trees have to be developed based on the availability of the containment ESFs. A typical containment event tree in which containment isolation is successful is shown in Fig. 3.33.
- In the next step, one has to estimate the failure probabilities of the containment ESFs based on the system reliability estimation methods like fault tree method. The failure probabilities of the ESFs are shown in Table 3.14.

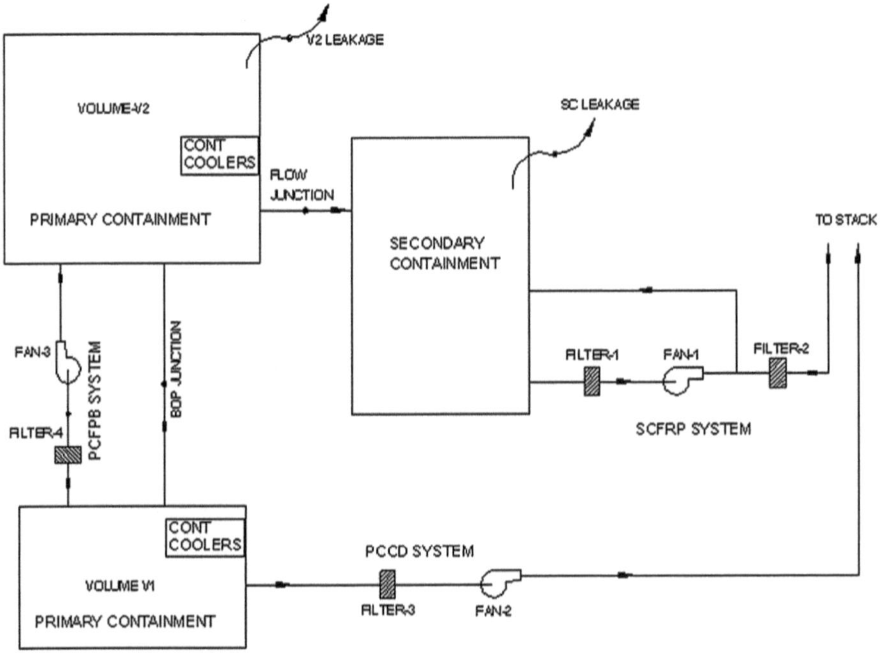

Fig. 3.32 Arrangement of containment ESFs

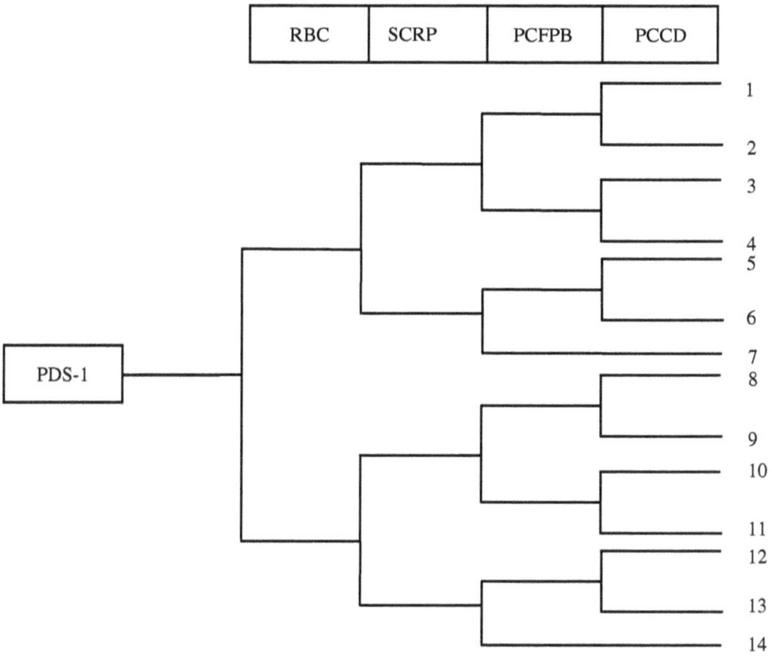

Fig. 3.33 Containment event tree for a particular PDS

Table 3.14 Failure probabilities of containment ESFs

System	Failure probability
RBC	5.1×10^{-4}
SCRP	7.2×10^{-3}
PCFPB	8.1×10^{-3}
PCCD	5.2×10^{-3}

- Thermal hydraulic calculation with and without building coolers of the containment show their influence in peak pressure and temperature.
  - With Containment coolers

    Containment Peak Pressure −0.251 MPa
    Containment Peak Temperature −406.5 K
  - Without Containment coolers

    Containment Peak Pressure −0.232 MPa
    Containment Peak Temperature −397.9 K
- In the next step, leakages through the containment has to be estimated based on the leak models for event under consideration as a function of the containment pressure and temperature.

- In the present case, based on the actual observations, a conservative value of 0.3 % containment volume/h at a given peak pressure has been assumed for analysis of leakage of fission products from primary containment to secondary containment.

3.9.2.4 Source Term Analysis

- For finding out the quantity of radioactive material released to the atmosphere, the core inventory should be known in hand, i.e., how much radioactive material (fission product inventory) is available in the core with different forms of radionuclides. The fission product inventory used for the present analysis is shown in the Table 3.15.
- Once any accident takes place, some percentage of radionuclides will be released either to the containment or directly to the atmosphere depending on the type of initiating event. The amount of fission products released to the containment is shown in the Table 3.16.
- Once the radionuclides are released to the containment, some percentage will be retained in the containment and remaining will be released to the atmosphere depending on the leakage paths and the availability of ESFs. The status of Fps in the containment is shown in the Table 3.17.

Table 3.15 Fission product inventory used for analysis

S. No.	Nuclide	Total inventory (kg)
1	Kr	3.42
2	Xe	164.45
6	Cs	127.49
9	I	4.26
11	Te	0.0348
12	Sb	0.537
13	Ag	3.71
14	Mo	30.78
15	Ru	80.36

Table 3.16 Fission product release to containment

Fission product	Core inventory (Kg)	Release from fuel (Kg)	Retention (Kg)	Release to containment (Kg)
Xe	164.45	11.227	7.901	3326×10^{-3}
Cs	127.49	10.24	9.636	604.8×10^{-3}
I	4.26	0.265	0.2503	14.68×10^{-3}
Kr	3.42	0.234	0.1567	77.29×10^{-3}

Table 3.17 Status of Cs and Iodine in containment when all the containment ESFs are functioning

Fission product	Quantity released (Kg)	Amount deposited (Kg)	Retention in filters (Kg)	Release to environment (Kg)
Cs	604.8×10^{-3}	602.76×10^{-3}	1.8944×10^{-3}	0.1456×10^{-3}
I	14.68×10^{-3}	8.10×10^{-3}	6.5708×10^{-3}	0.0092×10^{-3}

3.9.3 Level 3 PSA

- From the Level 2 analysis, the amount of fission products releases to the atmosphere is shown in the Table 3.18. The total amount is released in 72 h and is shown in Figs. 3.34, 3.35, 3.36, and 3.37 for I, Cs, Xe, and Kr. This will act as a input for Level 3 analysis.
- In the next step, based on the atmospheric conditions the released radionuclides (plume) will disperse in the atmosphere.

Table 3.18 Amount of fission products released to the environment in 72 h

Fission product	Stack release (Bq)	Ground release (Bq)	Total release (Bq)
I	3.0151×10^{12}	4.1001×10^{13}	4.4016×10^{13}
Cs	9.1575×10^{9}	4.7374×10^{11}	4.8290×10^{11}
Kr	3.3309×10^{13}	1.4998×10^{12}	3.4809×10^{13}
Xe	5.8253×10^{17}	2.9024×10^{16}	6.1156×10^{17}

Fig. 3.34 Release of CS

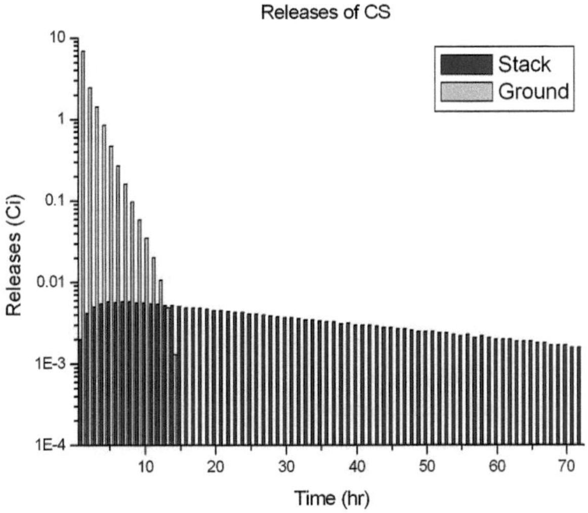

Fig. 3.35 Release of I

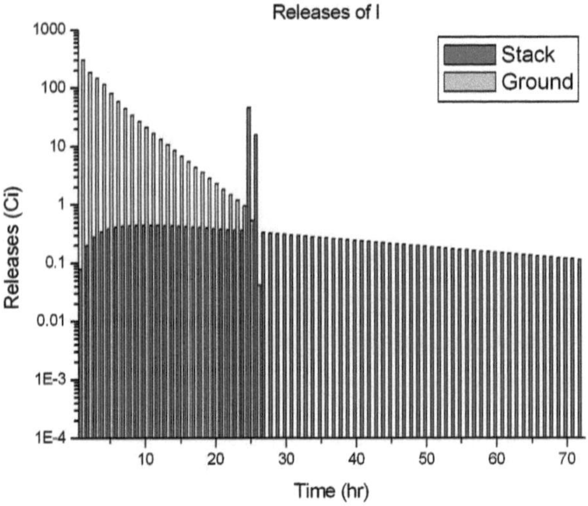

Fig. 3.36 Release of Kr

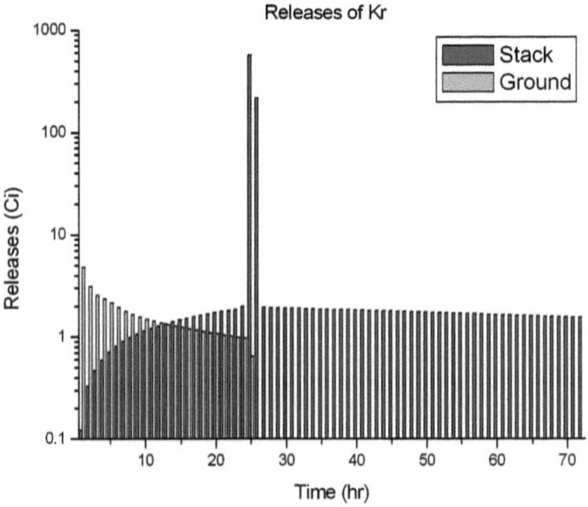

- The evaluations were carried out under the following assumptions.
  - The following meteorological parameters have been used in the analysis.

    wind speed
    stability category
    precipitation rate
    wind direction

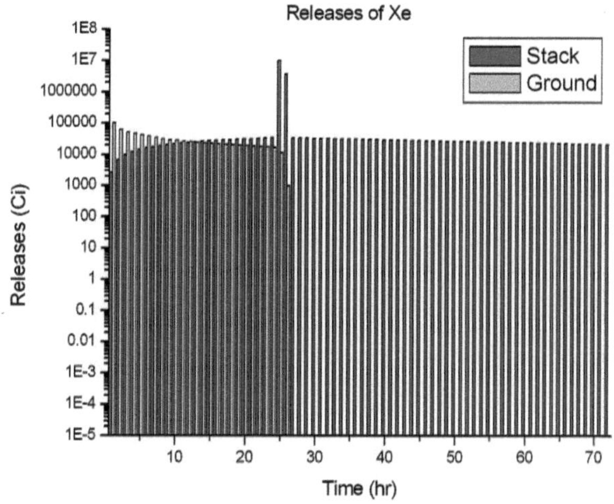

Fig. 3.37 Release of CS

- Stack height assumed is 100 m and ground release is assumed to be at 10 m.
- The release is for duration of 72 h. The dose is evaluated assuming an exposure time of 72 h. The individual is exposed all the time during these 72 h.
- Hourly changes of wind direction are taken into account while evaluating the radionuclide concentration and dose at various points.
- The dose-risk coefficients correspond to those of ICRP-60.
- The release is considered in six phases. The cumulative release in every 12 h is combined and is assumed to be release in the first hour of the phase.

• As explained, there are different pathways that the humans will be exposed to the radioactive material.

  - Cloud shine
  - Ground shine
  - Inhalation
  - Skin
  - Resuspension
  - Ingestion

• The dose rate will be calculated accordingly. For the present case, the thyroid dose for the person standing at 0.5 km from the point of release is shown in the Fig. 3.38.
• Finally, the complementary cumulative frequency distribution is shown in the Fig. 3.39. This shows the graph between thyroid dose vs frequency of exceedance.

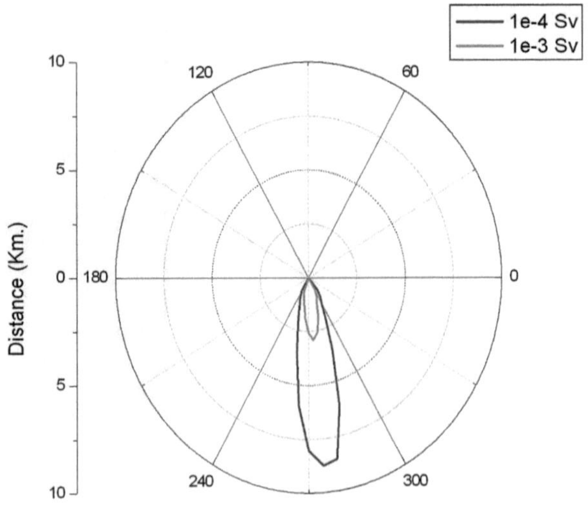

Fig. 3.38 Isodose curve for thyroid dose

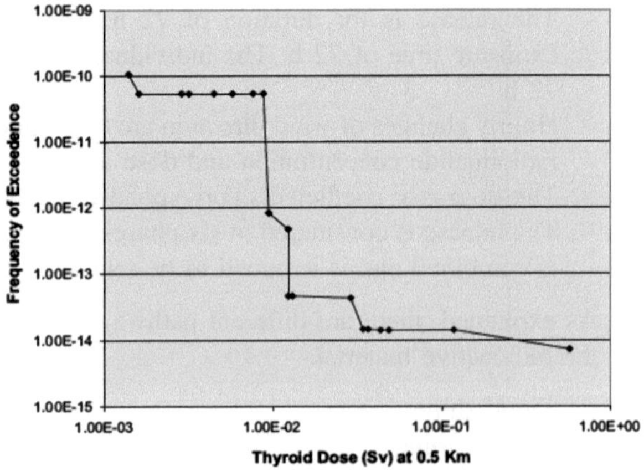

Fig. 3.39 CCFD for thyroid dose

References

1. ParthaSarathy KS (2011) Fukushima nuclear accident: what happened. Science Reporter, pp 8–16
2. Bajaj SS, Gore AR (2006) The Indian PHWR. Nucl Eng Des 236:701–722
3. IAEA Safety Series No. 50-P-4 (1992) Procedure for conducting probabilistic safety assessment of nuclear power plants (level 1). International Atomic Energy Agency, Vienna (http://gnssn.iaea.org/Superseded%20Safety%20Standards/Safety_Series_050-P-4_1992.pdf)
4. IAEA Safety Series No. 50-P-12 (1996) Procedure for conducting probabilistic safety assessment of nuclear power plants (level 3). International Atomic Energy Agency, Vienna (http://gnssn.iaea.org/Superseded%20Safety%20Standards/Safety_Series_050-P-12_1996.pdf)

5. IAEA Safety Standard Series No. NS-R-1 (1992) Safety of nuclear power plants: design requirements. International Atomic Energy Agency, Vienna (http://www-pub.iaea.org/MTCD/publications/PDF/Pub1099_scr.pdf)
6. AERB/NF/SM/O-2-R4 (2005) Radiation protection for nuclear facilities. AERB Safety manual, Atomic Energy Regulatory Board, Mumbai, India
7. Regulatory Document RD-337 (2008) Design of new nuclear power plants. Canadian Nuclear Safety Commission
8. Bedford T, Cooke R (2001) Probabilistic risk analysis: foundations and methods. Cambridge University Press, London
9. IAEA-TECDOC-593 (1991) Case study on the use of PSA methods: station blackout risk at millstone unit 3. International Atomic Energy Agency, Vienna (http://www.iaea.org/inis/collection/NCLCollectionStore/_Public/22/066/22066745.pdf)
10. NUREG/CR-2815 (1985). Probabilistic safety analysis procedures guide. USNRC
11. Fullwood RR (2000) Probabilistic safety assessment in the chemical and nuclear industries. Butterworth Heinemann, Oxford
12. IAEA Safety Series No. 50-P-8 (1995) Procedure for conducting probabilistic safety assessment of nuclear power plants (level 2). International Atomic Energy Agency, Vienna (https://gnssn.iaea.org/Superseded%20Safety%20Standards/Safety_Series_050-P-8_1995.pdf)
13. AERB/NF/SM-1-R3 (2002) Probabilistic safety assessment for nuclear power plants and research reactors. AERB Safety manual, Atomic Energy Regulatory Board, Mumbai, India
14. Hari Prasad M et al (2011) Level 1, 2 and 3 PSA for AHWR. Nucl Eng Des 241:3256–3269
15. NASA (2002) Fault tree hand book with aerospace applications. NASA, Washington DC
16. NUREG/CR-0492 (1981) Fault tree hand book. USNRC
17. WASH-1400 (1975) Reactor safety study: an assessment of accident risks in US commercial nuclear power plants. USNRC
18. Fleming KN (1975) A reliability model for common mode failure in redundant safety systems. General Atomic Report GA-A13284
19. NUREG/CR-4780, EPRI NP-5613 (1988) Procedures for treating common cause failures in safety and reliability studies, vols 1 and 2. U.S. Nuclear Regulatory Commission and Electric Power Research Institute
20. Fleming KN, Kalinowski AM (1983) An extension of the beta factor method to systems with high level of redundancy. Pickard, Lowe and Garric, Inc., PLG-0289
21. Mosleh A, Siu NO (1987) A multi parameter event based common cause failure model. In: Proceeding of ninth international conference on structural mechanics in reactor technology, Lausanne, Switzerland
22. IAEA Safety Standards Series No. SSG-3 (2010) Development and application of level 1 probabilistic safety assessment for nuclear power plants. International Atomic Energy Agency, Vienna (http://www-pub.iaea.org/MTCD/publications/PDF/Pub1430_web.pdf)
23. IAEA-TECDOC-478 (1988) Component reliability data for use in probabilistic safety assessment. International Atomic Energy Agency, Vienna (http://www-pub.iaea.org/MTCD/publications/PDF/te_478_web.pdf)
24. IAEA Safety Series No: 50-P-10 (1995) Human reliability analysis in PSA for nuclear power plants. IAEA, Vienna (https://gnssn.iaea.org/Superseded%20Safety%20Standards/Safety_Series_050-P-10_1995.pdf)
25. Apostolakis G (1999) The distinction between aleatory and epistemic uncertainties is important: an example from the inclusion of aging effects into PSA. In: Proceedings of PSA '99, international topical meeting on probabilistic safety assessment, American Nuclear Society, Washington DC, 22–26 August, pp 135–142
26. IAEA-TECDOC-737 (1994) Advances in reliability analysis and probabilistic safety assessment for nuclear power reactors. International Atomic Energy Agency, Vienna (http://www-pub.iaea.org/MTCD/ Publications/PDF/te_737_web.pdf)
27. Karanki DR, Kushwaha HS, Verma AK, Srividya A (2007) Quantification of epistemic and aleatory uncertainties in level 1 probabilistic safety assessment studies. Reliab Eng Syst Saf 92:947–956

28. Modarres M (1985) Statistical uncertainty analysis in reactor risk estimation. Nucl Eng Des 85:385–399
29. NUREG/CR-3385 (1983). Measures of risk importance and their applications, USNRC
30. Borgonovo E, Apostolakis GE (2001) A New importance measure for risk-informed decision making. Reliab Eng Syst Saf 72:193–212
31. Martin James E (2006) Physics for radiation protection: a hand book, 2nd edn. WILEY-VCH, Wein Heim

Chapter 4
Seismic PSA of Nuclear Power Plants

4.1 Introduction

Seismic Probabilistic Safety Assessment (Seismic PSA) [1] is an external event PSA. The objective of seismic PSA for the plants is to estimate the risk of the plant due to seismic events. Hence, in this analysis safety of the plant is numerically evaluated against the postulated earthquakes and appropriate measures will be taken to enhance the safety of the plant. In general, the plant consists of both process systems and emergency safety systems. Here the process systems represent the continuously operating systems; whereas, the safety systems are normally standby systems and they come into operation on demand. The failure of operating component/system during an earthquake may lead to change in the state of the plant. Here failure of a component can be either structural failure or subcomponent failure caused by excessive inertial stresses due to the dynamic loading (earthquake).

The failure of the operating component/system due to seismic event may lead to some internal initiating event which eventually calls for shutting down the reactor. Depending on the status of the other safety systems different scenarios may follow. These scenarios can be well represented with event trees as discussed in the earlier chapters. In the present case, these are termed as seismic event trees compare to the internal event PSA. Similarly, the safety system unavailability is modeled with seismic fault trees. In this analysis, the major difference with the internal events is the representation of failure modes of components of the systems. Here most of the component failures are considered in the form of structural failures such as support failure, anchorage failure, etc. These failures will eventually hamper the functioning of the components/systems.

In the seismic PSA, seismic event is the external initiating event, which in turn initiates the other internal initiating events as shown in Fig. 4.1 [1]. As it was mentioned earlier, in this analysis risk due to the seismic events is estimated. But in evaluating risk, all the three levels of PSA have to be performed. However, if our interest is only regarding the core damage scenario then one can perform the level 1

A.K. Verma et al., *Risk Management of Non-Renewable Energy Systems*,
Springer Series in Reliability Engineering, DOI 10.1007/978-3-319-16062-7_4

Fig. 4.1 Seismic with internal initiating events

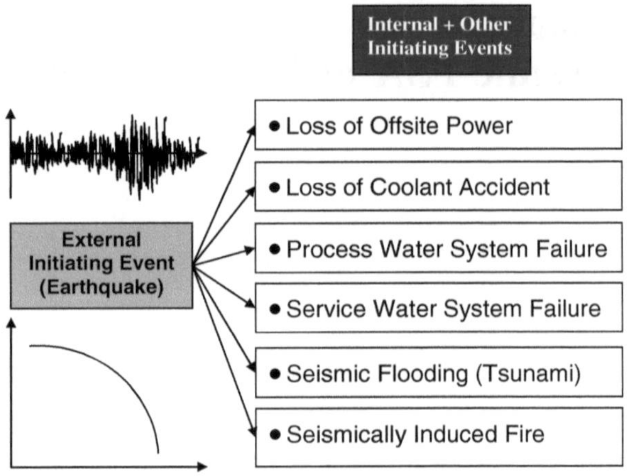

PSA analysis for external event and can estimate the core damage frequency due to the seismic events. In this case, risk will be defined in terms of CDF.

If one recollects the level 1 internal event PSA, the following steps have to be performed to estimate the CDF.

- Identification of Initiating Events
- Identification of Safety Systems
- Development of Event Trees
- Component data modeling
- Estimation of IE frequency and Safety system unavailability
- Identification of dominant accident sequences from event trees
- Estimation of accident sequence frequency
- Estimation of Core Damage Frequency

Similarly in the seismic PSA analysis, all the above-mentioned steps will be followed. However, the analysis part will be different. In this analysis, seismic event is the initiating event, but one has to identify the other internal initiating events that will be initiated due to seismic event. In this analysis, it is not needed to analyze all the IEs which have been analyzed as in the case of Level 1 PSA for internal events. Very few internal initiating events may have to be analyzed in the present analysis. In estimating the IE frequency (earthquake frequency), one has to depend on the seismic history of the site being considered. This analysis is altogether different from the internal event analysis. Similarly, in the case with the estimation of safety system unavailability, one should perform the response analysis and thereby estimate the failure probability of the systems. Hence, Seismic PSA evaluation process [1, 2] can be broadly divided into the following steps.

(i) Seismic Hazard Analysis
(ii) Seismic Fragility Analysis
(iii) Accident Sequence Analysis

In the seismic hazard analysis, the relationship between ground motion acceleration levels and their frequencies at a given site is evaluated using earthquake database, which gives seismic hazard curves. In the fragility analysis, capacities of components and structures against a seismic acceleration are probabilistically evaluated for various failure modes due to seismic events. In evaluating seismic fragilities, dynamic responses of structures and components are evaluated using input ground motion and response analysis model. In the accident sequence analysis, seismic event trees will be developed, dominating accident sequences will be identified, and each accident sequence frequency will be estimated based on the estimated failure probabilities of components/structures as well as estimated frequency of earthquakes to get the core damage frequency. The flowchart of seismic PSA is shown in Fig. 4.2 [3, 4].

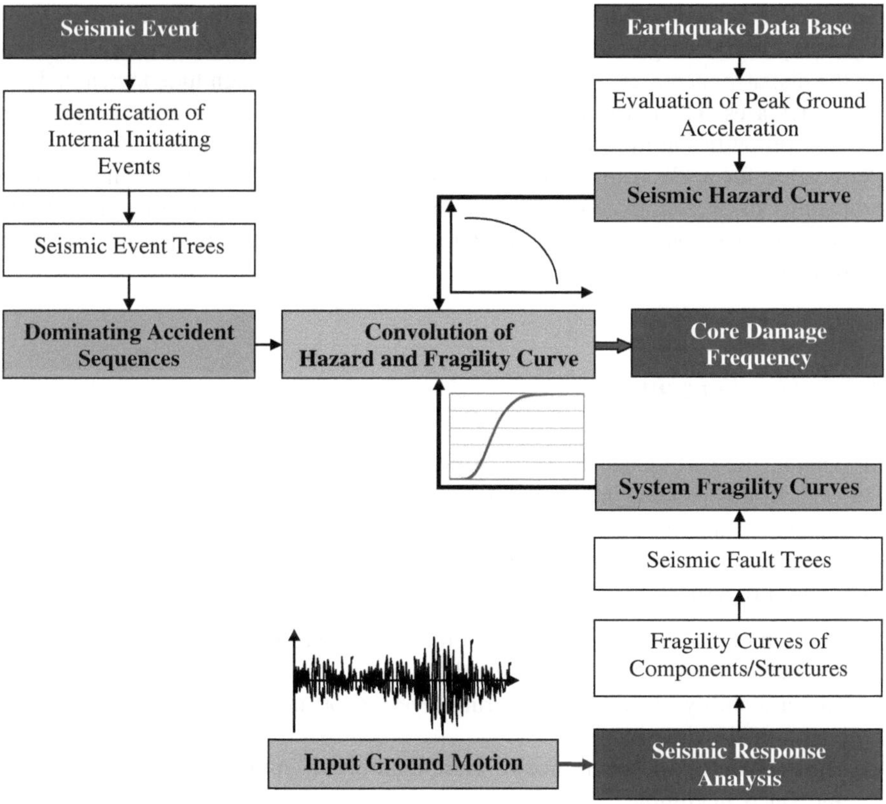

Fig. 4.2 Flow chart of seismic PSA

4.2 Probabilistic Seismic Hazard Analysis

The probabilistic seismic hazard analysis [5–8] is performed to estimate the annual frequency of occurrence of a hazard parameter such as the Peak Ground Acceleration (PGA), which characterizes the ground motion at a nuclear power plant site. The following input parameters are required for carrying out the hazard analysis:

- Potential sources of seismic activity,
- Seismic history of the region,
- Fault length, distance from source to the site
- Minimum and maximum magnitudes on each fault source
- Rate of occurrence of earthquakes from these sources,
- Attenuation of earthquake ground motion from source to the site.

Here basic idea in hazard analysis is that if an earthquake occurs somewhere on the fault source, it will produce some kind of ground motion at the site of interest depending on the magnitude of the earthquake and distance from source to the site. As the distance increases, PGA value reduces; and as the magnitude increases PGA value increases. The PGA at the site also depends on the path of wave propagation and soil properties at the site, etc. [9]. Hence, the aim of this analysis is to find the PGA value for a given magnitude and distance from different faults and find the probability of exceeding certain level of PGA at the site. Before proceeding into the hazard analysis, one should understand the process of earthquakes and their occurrences.

4.2.1 The Poisson Process

In general, the earthquake occurrence process is modeled with a Poisson process [10, 11]. In this process, number of occurrences of a particular event is considered as a random variable. This process will give a probability value for the number of occurrences of events over a given period of time. Poisson processes possess the following properties:

a. The number of occurrences in one time interval are independent of the number that occur in any other time interval.
b. The probability of occurrence during a very short interval is proportional to the length of the time interval.
c. The probability of occurrence of more than one event during a very short time interval is negligible.
d. The rate of occurrence of events is considered constant in a homogeneous Poisson model.

From the above properties, it is clear that the events of a Poisson process occur randomly and the process is a memory less process, i.e., the occurrence of an event does not account for the preceding events time of occurrence, size of the event, and location of occurrence.

Assume that N is the random variable representing number of earthquake events that can occur during a given period of time, and it follows Poisson process with a constant arrival rate of "v" per year. Then the probability of "n" earthquakes occurs in a time period of "t" in a given area can be well represented with a Poisson distribution as follows:

$$P_N(n) = P(N = n) = \frac{e^{-\upsilon t}(\upsilon t)^n}{n!} \quad n = 0, 1, 2, \ldots \tag{4.1}$$

Here N is a discrete random variable and the Poisson process is a discrete distribution.

Example 4.1 Consider that the earthquake occurrence follows a Poisson process with arrival rate of 1.05/year. Find the probability of three earthquakes occur during a period of 5 years. Plot the probability distribution for the same data.

Solution:

From the definition of Poisson process, the probability occurrence of "n" earthquakes occurring in a time period of "t" can be given as

$$P_N(n) = P(N = n) = \frac{e^{-\upsilon t}(\upsilon t)^n}{n!}$$

In the above equation, substitute $n = 3$, $t = 5$ years, and $v = 1.05$/year

Hence,

$$P(N = 3) = \frac{e^{-1.05\times5}(1.05 \times 5)^3}{3!} = 0.127$$

For finding out the distribution of the number of earthquakes occur, one has to find the probability values for different n values. This is tabulated in Table 4.1 and is shown in Fig. 4.3.

Table 4.1 Probability distribution for different n values

n	$P(N = n)$	n	$P(N = n)$
0	0.005248	6	0.152608
1	0.027549	7	0.114456
2	0.072317	8	0.075112
3	0.126555	9	0.043815
4	0.166104	10	0.023003
5	0.174409	11	0.010979

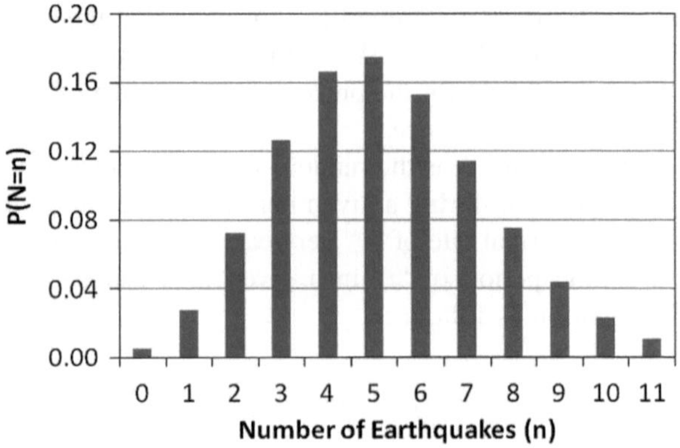

Fig. 4.3 Probability distribution of number of earthquakes

If one is interested in the earthquakes whose magnitude is greater than certain level, then the special events are Poisson arrivals with an average annual arrival rate "*pv*" (*p* is the probability that the magnitude is greater than certain level). Similarly, peak ground acceleration (PGA) is the ground motion parameter interested at a given site; if one is interested in the earthquakes, whose PGA is greater than certain level, the annual arrival rate of these earthquakes can be given as "*pv*" where *p* is the probability that the PGA is greater than certain level. Thus the number of times, *N*, that the PGA at the site will exceed "*z*" in an interval of time "*t*" is given as follows [11]:

$$P_N(n) = P(N = n|Z \geq z) = \frac{e^{-p\upsilon t}(p\upsilon\, t)^n}{n!} \quad n = 0,\ 1,\ 2,\ \ldots \tag{4.2}$$

Here *Z* represents the peak ground acceleration and is considered as a random variable. Now for the discrete probability distributions, the probability of occurrence of at least one event in a period of time t is given by

$$\begin{aligned} P(N \geq 1) &= P(N = 1) + P(N = 2) + \cdots + P(N = \infty) \\ &= 1 - P(N = 0) \end{aligned} \tag{4.3}$$

The probability distribution of *Z*, the PGA over an interval of time "*t*" is

$$\begin{aligned} P[Z > z] &= 1 - e^{-p\upsilon t} \\ P(Z \leq z) &= 1 - P[Z > z] \\ &= e^{-p\upsilon t} = P[N = 0|Z > z] \end{aligned} \tag{4.4}$$

Hence, the probability of exceedance of a ground motion level, z, in an exposure time or design time period, t, at a site is related to the annual frequency (or rate) of ground motion exceedance at the site, $v(z) = pv$, by:

$$P[Z > z] = 1 - e^{-v(z)t} \Rightarrow F_z = e^{-v(z)t} \tag{4.5}$$

The annual frequency of exceedance, $v(z)$, is derived in the following subsections. From the Eq. 4.5, one can find out the return period of a particular PGA at the site as follows: [11].

$$\begin{aligned}
&P[Z > z] = 1 - e^{-v(z)t} \\
&e^{-v(z)t} = 1 - P[Z > z] \\
&\ln\left[e^{-v(z)t}\right] = \ln[1 - P(Z > z)] \\
&-v(z)t = \ln[1 - P(Z > z)] \\
&v(z) = -\frac{\ln[1 - P(Z > z)]}{t} \Rightarrow \frac{1}{v(z)} = \frac{-t}{\ln[1 - P(Z > z)]} \\
&\therefore \text{Return Period} = \frac{-t}{\ln[1 - P(Z > z)]}
\end{aligned} \tag{4.6}$$

Return period refers to the time by which a particular PGA at the site will be returned. There are different steps in carrying out the probabilistic seismic hazard analysis (PSHA) that are explained in the following subsections.

Example 4.2 Find the return period of 0.1 g peak ground acceleration at a given site. The probability of exceedance of 0.1 g in 1 year is given as 1.51×10^{-4}.

Solution:
Consider Z is the random variable representing the peak ground acceleration that is expected at a given site. The probability of exceedance of a certain level of PGA at a site can be represented as $P(Z > z)$. The return period of a certain level of PGA can be given as

$$\text{Return Period} = \frac{-t}{\ln[1 - P(Z > z)]}$$

In the return period, equation substitute t = 1 year and $P(Z > z) = 1.51 \times 10^{-4}$

$$T_R = \frac{-1}{\ln[1 - 1.51 \times 10^{-4}]} = 6622\,\text{years}$$

Hence, the return period of 0.1 g at a site is obtained as 6622 years. It means if an earthquake which can produce 0.1 g PGA at the site is occurred today, then it will reoccur after 6622 years. This is only an average value. It does not mean that it will reoccur only after 6622 years but it can occur in between also because this is only estimation.

4.2.2 Identification of Fault Sources

The first and foremost step of this analysis is identification of potential sources [12, 13] (faults) of seismic activity around the site. These sources are treated as uniform earthquake potential sources, that is, the chance of occurrence of a given size of earthquake is same throughout the fault source. Depending on the geometry of the fault, there are different types of fault sources namely point source, line source, areal sources, etc., as shown in Fig. 4.4. For each fault, one has to identify the fault characteristics such as fault length, fault orientation, and shortest distance from the site to the fault. Here fault orientation is described by its strike and dip. The strike of a fault is defined as the horizontal line produced by the intersection of the fault plane and a hypothetical horizontal plane as shown in Fig. 4.5. The dip angle is described with downward slope of the fault plane, which is the angle between horizontal plane and the fault plane. The dip angle of 90° represents a vertical fault. In general for all the faults, it is very difficult to get the information

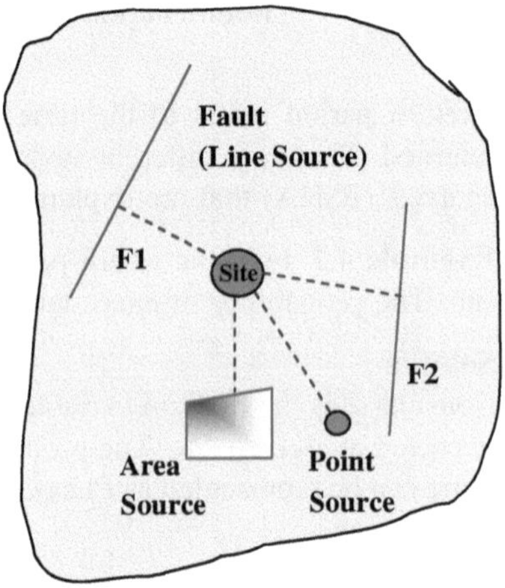

Fig. 4.4 Different fault sources around the site

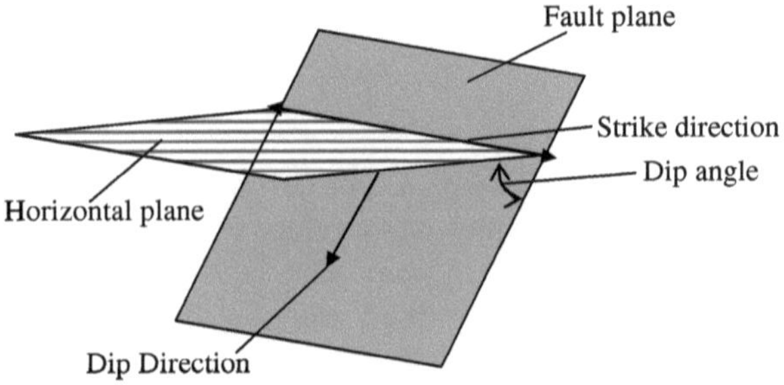

Fig. 4.5 Geometric representation of fault plane orientation

about the fault orientation, i.e., dip angle and depth of fault in dip direction. Strike direction can be obtained based on the fault projection on to the surface that can be verified during field checks.

4.2.3 Recurrence Relationship

Once the fault sources are identified around the site, then one has to determine the probability of occurrence of a particular magnitude on each fault. This is dealt with recurrence relationship. A recurrence relationship indicates the relation between earthquake magnitude and the frequency of occurrence of particular earthquake magnitude. Normally, frequency of occurrence is considered per year basis. Recurrence relationships can be developed for the region under consideration or for each fault source as well.

Recurrence relationship is usually derived from the earthquake data history of the particular region of interest. Hence to do the analysis, one has to collect the earthquake data around the site. Earthquake data can be both historical data (where one does not have any instrumented data) as well as instrumental data. Mostly the historical data will be the maximum observed earthquake magnitude over a period of time. Since we did not have instruments to collect the information, we cannot get the data on day-to-day basis. In general, this data is collected in a radius of 300 km around site. As such there is no upper limit on the consideration of this radius. This can be extended more than 300 km also.

These relationships are generally represented as straight line plots obtained from regression analysis. The recurrence relationship is generally represented by Gutenberg & Richter's Law as given below.

$$\log_{10} N_m = a - bm \quad (4.7)$$

If a graph is plotted between "$\log_{10}N_m$" as the ordinate and "m" as the abscissa, it will represent a straight line as shown in Fig. 4.6. Where N_m is the total number of

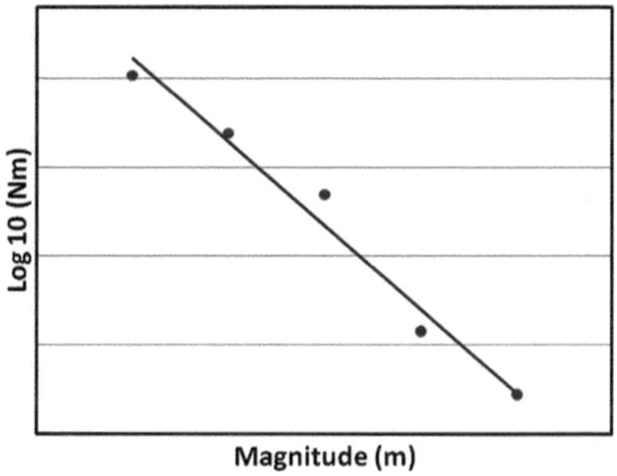

Fig. 4.6 Recurrence relationship—Graph between Magnitude versus $\text{Log}_{10}(N_m)$

earthquakes of a given magnitude, m, or larger that are expected to occur during a period of one year. "*a*" is intersection point on the ordinate and it represents the log of the total number of earthquakes of magnitude zero or greater expected to occur in a given region and "*b*" is the slope of the curve which describes the chance of occurrence of large and small earthquakes. It gives the information about the maximum magnitude of earthquake that can occur in that region. As the *b* value decreases, slope also decreases and the number of larger magnitude earthquakes increases compared to smaller magnitudes. Similarly as the *b* value increases, slope of the curve also increases and the number of larger magnitude earthquakes decreases compared to smaller magnitude earthquakes.

The parameters ("*a*" and "*b*") of the recurrence relationship can be derived from the earthquake occurrence records of that region. Assume that a_0 is the annual rate of occurrence of earthquakes in a particular region whose magnitude is greater than or equal to 0. Then, the probability of earthquakes whose magnitude is greater than "*m*" is given by

$$\begin{aligned}
P(M > m) &= \frac{N_m}{a_0} \\
\text{But} \quad \log_{10}(N_m) &= a - bm \\
N_m &= 10^{a-bm} \\
P(M > m) &= \frac{10^{a-bm}}{a_0} \\
&= 1 - P(M \leq m) \\
&= 1 - F_M(m) \\
\therefore F_M(m) &= 1 - \frac{10^{a-bm}}{a_0},
\end{aligned} \tag{4.8}$$

where $F_M(m)$ is the cumulative probability of earthquakes whose magnitude is $\leq m$. From the definition of probability density function

$$\begin{aligned}
f_M(m) &= \frac{\mathrm{d}F_M(m)}{\mathrm{d}m} \\
&= \frac{\mathrm{d}}{\mathrm{d}m}\left[1 - \frac{10^{a-bm}}{a_0}\right] \\
&= -\frac{1}{a_0} 10^{a-bm} \log 10(-b) \\
f(m) &= \frac{b}{a_0} 10^{a-bm} \quad 0 < m < \infty
\end{aligned} \tag{4.9}$$

In general for the analysis purposes, it is common practice to assume a threshold magnitude which is greater than zero (normally around 3.0). This is due to the reason that magnitude below a certain level the earthquakes will not produce significant ground motion at the site or will not have sufficient data. Also based on the earthquake data in the region an upper limit also can be imposed for that region. This is because the chance of occurrence of a magnitude greater than the maximum magnitude is very negligible. Now, one can derive the probability density function for the magnitude; the earthquake magnitude varies between certain limits such as between a lower limit of m_0 and an upper limit of m_u as follows:

Annual occurrence of earthquakes whose magnitude is greater than m_0 is given by

$$\begin{aligned} \log_{10} N_{m_0} &= a - bm_0 \\ N_{m_0} &= 10^{a-bm_0} \end{aligned} \tag{4.10}$$

Similarly, annual occurrence of earthquakes whose magnitude is greater than m_u is given by

$$\begin{aligned} \log_{10} N_{m_u} &= a - bm_u \\ N_{m_u} &= 10^{a-bm_u} \end{aligned} \tag{4.11}$$

Hence, annual occurrence of earthquakes between these limits can be given as

$$N_{m_u} - N_{m_0} = 10^{a-bm_u} - 10^{a-bm_0} \tag{4.12}$$

Now considering the lower limit, the probability of occurrence of magnitude less than certain value "m" can be given as ratio between annual occurrence of earthquakes occurring between m_0 and m and annual occurrence of earthquakes between m_0 and m_u. This is given as:

$$\begin{aligned} P(M \le m) &= \frac{N_{m_0} - N_m}{N_{m_0} - N_{m_u}} \\ \text{But}\quad N_m &= 10^{a-bm} \\ N_{m_0} &= 10^{a-bm_0} \\ N_{m_u} &= 10^{a-bm_u} \\ P(M \le m) &= \frac{10^{a-bm_0} - 10^{a-bm}}{10^{a-bm_0} - 10^{a-bm_u}} \\ P(M \le m) &= \frac{10^{-bm_0} - 10^{-bm}}{10^{-bm_0} - 10^{-bm_u}} = F_M(m) \end{aligned} \tag{4.13}$$

$$\begin{aligned} P(M > m) &= 1 - P(M \le m) = 1 - \frac{10^{-bm_0} - 10^{-bm}}{10^{-bm_0} - 10^{-bm_u}} \\ P(M > m) &= \frac{10^{-bm} - 10^{-bm_u}}{10^{-bm_0} - 10^{-bm_u}} \end{aligned} \tag{4.14}$$

From the definition of probability density function

$$
\begin{aligned}
f_M(m) &= \frac{dF_M(m)}{dm} \\
&= \frac{d}{dm}\left[\frac{10^{-bm_0} - 10^{-bm}}{10^{-bm_0} - 10^{-bm_u}}\right] \\
&= -\frac{1}{10^{-bm_0} - 10^{-bm_u}} 10^{-bm}(\log_e 10)(-b) \qquad (4.15) \\
f_M(m) &= \frac{\beta 10^{-bm}}{10^{-bm_0} - 10^{-bm_u}} = \frac{\beta 10^{-bm}}{10^{-bm_0}(1 - 10^{-b(m_u - m_0)})} \\
f_M(m) &= \frac{\beta 10^{-b(m-m_0)}}{1 - 10^{-b(m_u - m_0)}} \qquad m_0 \le M \le m_u \quad \beta = b \log_e 10
\end{aligned}
$$

Now annual occurrence of earthquakes whose magnitude is greater than m can be defined as:

$$
\begin{aligned}
\alpha(m) &= \alpha(m_0)P(M > m) \\
\alpha(m) &= \alpha(m_0)\frac{10^{-bm} - 10^{-bm_u}}{10^{-bm_0} - 10^{-bm_u}}, \qquad (4.16) \\
\text{where} \quad \alpha(m_0) &= 10^{a - bm_0}
\end{aligned}
$$

where $\alpha(m_0)$ is the annual occurrence of earthquakes whose magnitude is greater than or equal to m_0. Here $\alpha(m)$ can be termed as mean annual rate of exceedance of magnitude greater than or equal to m.

Example 4.3 Consider the Gutenberg–Richter recurrence relationship, in which parameters of the relationship are given as "*a*" = 2.3 and "*b*" = 0.95. Find the probability density function for the magnitude whose lower and upper limits are given as m_0 = 3.0 and m_u = 6.0. Find the probability of occurrence of magnitude between 3.5 and 4.0.

Solution:
The probability density function for magnitude *m* whose lower and upper limits are given as m_0 and m_u can be given as follows (from Eq. 4.15).

$$
f_M(m) = \frac{\beta 10^{a-bm}}{10^{a-bm_0} - 10^{a-bm_u}} \qquad m_0 \le M \le m_u
$$

Substituting $a = 2.3$, b = 0.95, $m_0 = 3.0$, and $m_u = 6.0$ in the above equation results into

$$
\begin{aligned}
\beta &= b \log_e 10 = 0.95 \times 2.3026 = 2.1875 \\
f_M(m) &= \frac{2.1875 \times 10^{2.3-0.95m}}{10^{2.3-0.95\times3.0} - 10^{2.3-0.95\times6.0}} \qquad 3.0 \le M \le 6.0 \\
f_M(m) &= \frac{1}{0.28144} \times 2.1875 \times 10^{2.3-0.95m} = 7.773 \times 10^{2.3-0.95m}
\end{aligned}
$$

This is graphically represented as shown in Fig. 4.7.

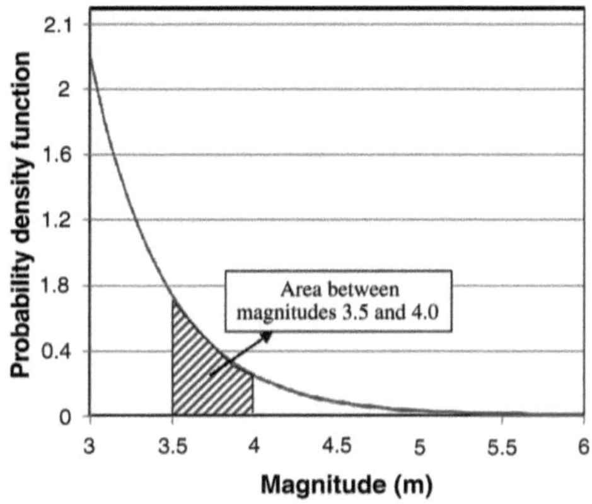

Fig. 4.7 Probability density function of magnitude "*M*"

Now the probability of occurrence of magnitude between 3.5 and 4.0 can be given as area under the probability density function between 3.5 and 4.0. This is shown in Fig. 4.7. This can be obtained as follows:

$$f_M(m) = \frac{\beta 10^{a-bm}}{10^{a-bm_0} - 10^{a-bm_u}} \qquad m_0 \le M \le m_u$$

$$f_M(m) = \frac{\mathrm{d}F}{\mathrm{d}m} \Rightarrow dF = f_M(m)\,\mathrm{d}m$$

$$\therefore F(m + \Delta m) - F(m) = f_M(m)\,\Delta m$$

$$\text{Here} \quad m = 3.5, \;\; \Delta m = 0.5 \Rightarrow m + \Delta m = 4.0$$

$$F(4.0) - F(3.5) = 0.5 f_M(3.5) = 0.5 \times 0.7338 = 0.3669$$

Example 4.4 From Example 4.3, draw a graph showing magnitude versus annual rate of exceedance and find the annual rate of exceedance of magnitude 5.0. Assume that the maximum magnitude is 7.0, repeat the above calculations.

Solution:

The annual rate of exceedance can be given as

$$\alpha(m) = \alpha(m_0)\frac{10^{-bm} - 10^{-bm_u}}{10^{-bm_0} - 10^{-bm_u}}$$

$$\text{where} \quad \alpha(m_0) = 10^{a-bm_0}$$

Substituting $a = 2.3$, $b = 0.95$, $m_0 = 3.0$, and $m_u = 6.0$ in the above equation results into

$$\alpha(m_0) = 10^{2.3-0.95\times 3.0} = 0.2818$$

$$\alpha(5) = 0.2818 \times \frac{10^{-0.95\times 5.0} - 10^{-0.95\times 6.0}}{10^{-0.95\times 3.0} - 10^{-0.95\times 6.0}} = 3.15 \times 10^{-3} \qquad m_u = 6.0$$

$$\alpha(5) = 0.2818 \times \frac{10^{-0.95\times 5.0} - 10^{-0.95\times 7.0}}{10^{-0.95\times 3.0} - 10^{-0.95\times 7.0}} = 3.5 \times 10^{-3} \qquad m_u = 7.0$$

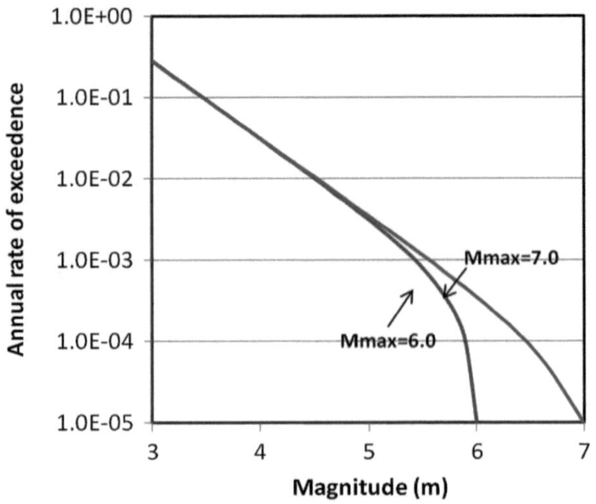

Fig. 4.8 Graph between magnitude versus annual rate of exceedance

Analysis of Data Completeness

As mentioned in the previous section on recurrence relationship, for finding out the parameters (a and b) of Gutenberg–Richter's law, one should consider the earthquake data in that region. The earthquake database can be divided into two parts, one is the historical data where one can get only the extreme values in a particular time period and the other part is instrumental data where there is a recording of the earthquakes. To estimate the parameters exactly, one should consider both the parts. In any data analysis, the correctness of the analysis depends on the degree of completeness of the data and also the homogeneity of each data point. Here homogeneity means each and every data point should be in the same scale, i.e., all the earthquake data points should be in terms of either intensity or magnitude. Nowadays, all the points are considered in magnitude scale, any intensity value is converted to magnitude using the following formula.

$$M = \frac{2}{3}I + 1 \tag{4.17}$$

where I is the intensity value and M is the moment magnitude. In general, whatever the historical data that is collected one gets mostly intensity values only. Now regarding completeness of data, here completeness means whether the data collected includes all the earthquakes occurred during certain period of time or not. If the data is complete, all the earthquakes occurred during certain period is considered in the analysis.

Now the question comes how to verify whether the data collected is complete or not? Or what is the period for which the data collected is complete? In order to check this, one has to perform data completeness analysis. As mentioned earlier, the occurrence of earthquakes follows Poisson arrival process, the assumption of

Poisson process is that the arrival rate of earthquakes is constant. Here one can utilize the concepts of sample mean and sample variance. In this analysis, the data is grouped into several magnitude classes over certain period of time. If the collected data is complete, the arrival rate of earthquakes will be constant irrespective of the time period. Let $x_1, x_2, x_3, \ldots x_n$ are the number of events per unit time interval. Then an unbiased estimate of the mean rate per unit time interval of this sample is [14–16]

$$\lambda = \frac{\sum_{i=1}^{n} x_i}{n} \tag{4.18}$$

and its variance is

$$\sigma_{\text{mean}}^2 = \frac{\sigma^2}{n} = \frac{\lambda}{n} \tag{4.19}$$

(Since, variance and mean are same for the Poisson process). Where n is the number of unit time intervals. Taking the unit time interval to be 1 year gives [17]

$$\sigma_{\text{mean}} = \frac{\sqrt{\lambda}}{\sqrt{T}} \tag{4.20}$$

as the standard deviation of the estimate of the mean, where T is the sample length. Thus, assuming stationary process, one can expect that σ behaves as $1/\sqrt{T}$ in the subinterval of the sample length, in which the mean rate of occurrence in a magnitude class is constant. In other words, when λ is constant, then the standard deviation σ_{mean} varies as $1/\sqrt{T}$ where T is the time interval of the sample. The above process is explained with an example as below.

Now consider the earthquake data [18] as shown in Table 4.2. For this data, one can perform data completeness test and can verify whether the data is complete for the entire period in which it is collected or not.

From the above data, it is clear that there are two parts in the data. One is that instrumental data from 1968 to 1997, and the other part is from 1800 to 1964. Now this data is sampled with time period increasing from 6 years to 198 years (Table 4.3). Consider the data in each time interval as one sample, as the sample size increases the variance gets reduced. As mentioned in the above, the standard deviation of the sample can be given as (Table 4.4)

Table 4.2 Earthquake data

Time period	$3 < m < 3.9$	$4 < m < 4.9$	$m > 5$
1800–1964	1	17	10
1968–1973	49	28	2
1974–1979	76	7	1
1980–1985	30	6	0
1986–1991	16	2	0
1992–1997	23	5	0

Table 4.3 Earthquake distribution by time and magnitude

Time period	Time interval (T)	$3 < m < 3.9$		$4 < m < 4.9$		$m > 5$	
		N	N/T	N	N/T	N	N/T
1992–1997	6	23	3.8333	5	0.8333	0	0
1986–1997	12	39	3.25	7	0.5833	0	0
1980–1997	18	69	3.8333	13	0.7222	0	0
1974–1997	24	145	6.0417	20	0.8333	1	0.0417
1968–1997	30	194	6.4667	48	1.6000	3	0.1000
1800–1997	198	195	0.9848	65	0.3282	13	0.0657

Table 4.4 Calculation of standard deviation

Time period	Time interval (T)	$1/\sqrt{T}$	$3 < m < 3.9$	$4 < m < 4.9$	$m > 5$
			$\sqrt{(N/T)}/\sqrt{T}$	$\sqrt{(N/T)}/\sqrt{T}$	$\sqrt{(N/T)}/\sqrt{T}$
1992–1997	6	0.4082	0.799305	0.372678	0
1986–1997	12	0.2887	0.520416	0.220479	0
1980–1997	18	0.2357	0.461479	0.200308	0
1974–1997	24	0.2041	0.501733	0.186339	0.041667
1968–1997	30	0.1826	0.46428	0.23094	0.057735
1800–1997	198	0.0711	0.070526	0.040718	0.018210

$$\sigma_{\text{mean}} = \frac{\sqrt{\lambda}}{\sqrt{T}} \qquad \text{where} \quad \lambda = \frac{N}{T} \tag{4.21}$$

where N is the total number of earthquakes in a time interval T.

From the above calculations, a graph with time versus standard deviation is plotted and is shown in Fig. 4.9.

For each magnitude interval shown in Fig. 4.9, the plotted points are supposed to define a straight line relation, as long as the dataset for that magnitude interval is complete. For a given seismic region, the slope of the lines for all magnitude intervals should be the same. For the sets $3 < M < 3.9$, the data appear complete for the past 60 years (1938–1997) (since the data points are lying on the line till 60 years), and for $4 < M < 4.9$ the data seems complete for last 60 years (1938–1997). There are less numbers of reported events for the magnitude greater than 5.0 compared to the other magnitude intervals. However, it is reasonable to assume that this set is complete for last 198 years. From the above analysis, it is revealed that the dataset is not complete for the interval 1800 through 1997.

The parameters of the Gutenberg-Richter's relationship (a and b values) are generally computed for the whole dataset without testing the completeness of the data which gives errors in the estimation of the parameters. From the above analysis, it is clear that the data is not complete for the entire period, i.e., 198 years. Hence, in the analysis it is very much needed to collect only that data which is

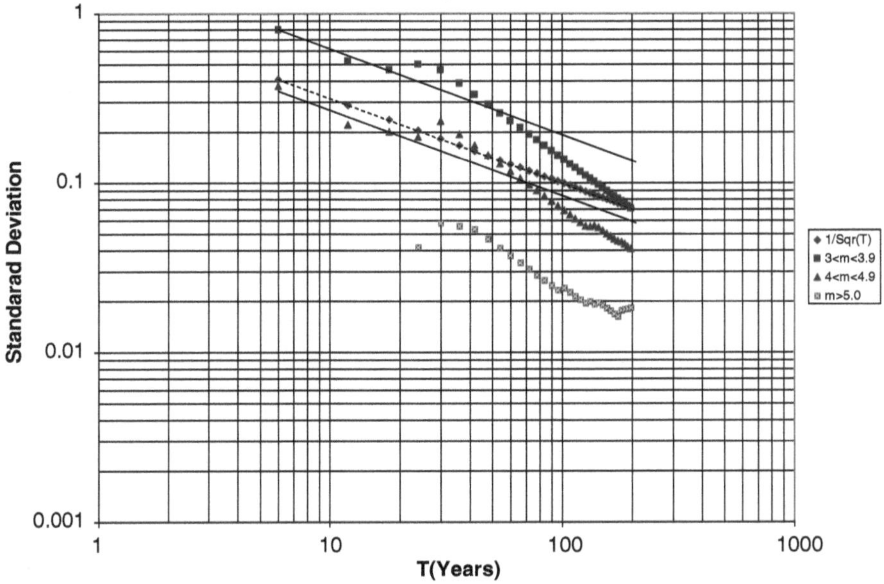

Fig. 4.9 Completeness test of earthquake data

completed for over a certain period of time. Table 4.5 shows the magnitudes and the time period for which the data is complete.

The data in each interval for time period and magnitude range is normalized by respective periods and then the cumulative number of events are taken for computation of "a" and "b" values (Table 4.5).

A plot between logarithm of the cumulative number of earthquakes per year for $M > m$ and magnitude is shown in Fig. 4.10. A straight line is fitted for the above data by using least square analysis and the fitted line is given as below

$$\begin{aligned} Log_{10}(N_m) &= 3.537 - 0.942\,m \quad m > 3 \\ \text{Here} \quad a &= 3.537 \\ b &= 0.942 \end{aligned}$$

Table 4.5 Recurrence rate for the region

Magnitude	Completed data in years	Events per year	Cumulative value (N)	$\text{Log}_{10}(N)$
3.0	1938–1997	2.1666	4.11565	0.614439
3.5	1938–1997	1.0666	1.94899	0.289810
4.0	1938–1997	0.7333	0.88232	−0.054370
4.5	1938–1997	0.0833	0.14899	−0.826840
5.0	1800–1997	0.0656	0.06565	−1.182720

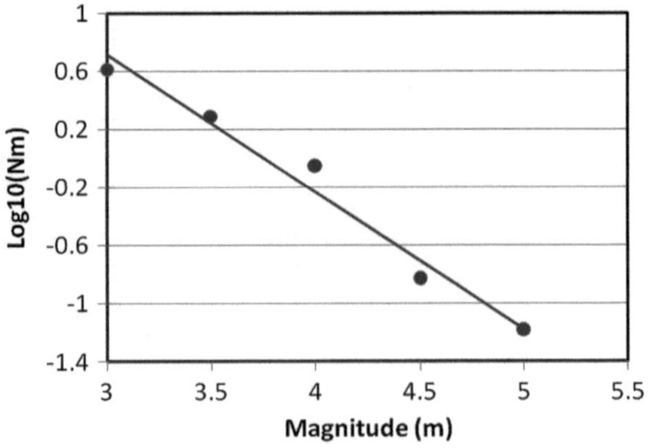

Fig. 4.10 Regional recurrence relationship

4.2.4 Source-to-Site Distance

The earthquake can occur anywhere on the fault. The point from where the rupture of the fault, first motion of an earthquake, and its elastic waves originate is called the focus of an earthquake. The rupture model of an earthquake is shown in Fig. 4.11.

Because these earthquakes are assumed to occur anywhere within the fault source, distances from all possible locations within that source to the site must be considered in the analysis instead of considering only a largest possible earthquake at the shortest distance from source to the site. Thus in this analysis, all the possible earthquake size at all the distances from source to site are taken into account. There is always a probability involved in occurrence of a particular magnitude at a particular distance; hence, associated source-to-site distance probabilities also needs to be considered in the analysis. In finding out these probability values, one should

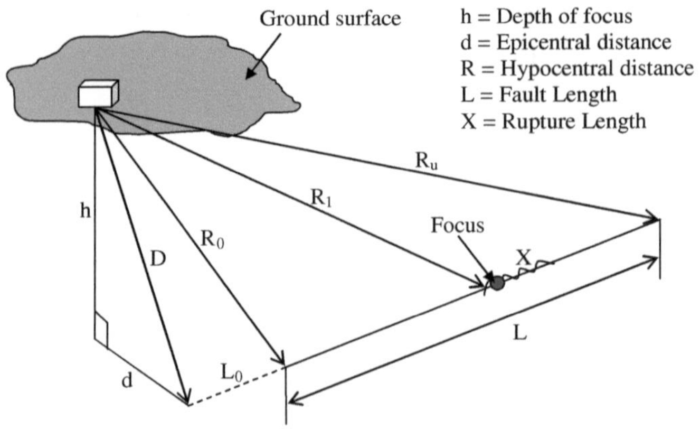

Fig. 4.11 Fault rupture model

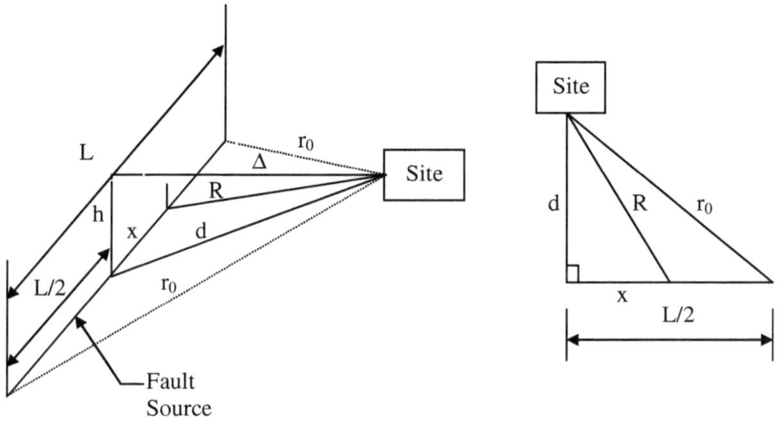

Fig. 4.12 Symmetrical fault with respect to site

develop probability distribution for source-to-site distances. These are explained for different cases as follows.

Case 1

In this case, fault source is considered as a line source (Fig. 4.12). The site is assumed to lie at perpendicular distance, Δ, from a line on the surface vertically above the fault at the focal depth, *h*. The earthquakes are expected to occur on this fault with different focal points [12]. Assume the fault is of length *L*, and the site is located symmetrically with respect to the length of the fault.

$$P(R \leq r|m) = \frac{x}{L/2} d \leq R \leq r_0$$

$$x = \sqrt{r^2 - d^2}$$

$$d = \sqrt{h^2 + \Delta^2}$$

$$P(R \leq r|m) = \frac{\sqrt{r^2 - d^2}}{L/2} = F_R(r) \tag{4.22}$$

$$f_R(r) = \frac{\mathrm{d}F_R(r)}{\mathrm{d}r} = \frac{\mathrm{d}}{\mathrm{d}r}\left[\frac{2\sqrt{r^2 - d^2}}{L}\right]$$

$$f_R(r) = \frac{2r}{L\sqrt{r^2 - d^2}} \quad d \leq R \leq r_0$$

Example 4.5 Find the probability distribution for source-to-site distance for line source model as shown in Fig. 4.12. Assume the length of the fault as 100 km, depth of focus as 10 km, and the epicentral distance as 50 km.

Solution:

The input parameters are given as

L = 100 km $L/2$ = 50 km h = 10 km Δ = 50 km

$$f_R(r) = \frac{2r}{L\sqrt{r^2 - d^2}} \quad d \le R \le r_0$$
$$F_R(r) = \frac{\sqrt{r^2 - d^2}}{L/2}$$
$$d = \sqrt{h^2 + \Delta^2} = \sqrt{10^2 + 50^2} = 50.99\,\text{km}$$
$$r_0 = \sqrt{d^2 + (L/2)^2} = \sqrt{50.99^2 + 50^2} = 71.414\,\text{km}.$$

Since it is a continuous distribution, it is not possible to find the probability of earthquake occurrence exactly at a particular point. Instead, one can find out the probability value for an interval. Here R can vary in between d and r_0. One can divide this range into subintervals and can find probability of occurrence of earthquakes in each interval. Later, the probability density function (pdf) is computed based on the following formula or else this can be calculated directly from the pdf equation as mentioned above.

$$f_R(r) = \frac{dF}{dr} = \operatorname*{Lim}_{\Delta r \to 0} \frac{\Delta F}{\Delta r} = \frac{F(r + \Delta r) - F(r)}{\Delta r}$$
$$r = 50.99$$
$$\Delta r = 1.0212 \Rightarrow r + \Delta r = 50.99 + 1.0212 = 52.0112$$
$$f_R(50.99) = \frac{F(52.0122) - F(50.99)}{1.0212} = 0.2009.$$

The calculation is shown in Table 4.6 and the both cumulative distribution function and probability density function are shown in Figs. 4.13 and 4.14, respectively.

Case 2: Line source is extended only in one direction (Fig. 4.15)

$$\begin{aligned} P(R \le r/m) &= \frac{x}{L} \qquad r_{01} < R < r_{02} \\ x &= \sqrt{r^2 - d^2} - L_0 \\ r_{01} &= \sqrt{d^2 + L_0^2} \\ r_{02} &= \sqrt{d^2 + (L + L_0)^2} \\ P(R \le r/m) &= \frac{\sqrt{r^2 - d^2} - L_0}{L} \end{aligned} \tag{4.23}$$

Table 4.6 Distribution function for hypocentral distance

S.no	R	$F_R(r)$	$f_R(r)$	S.no	R	$F_R(r)$	$f_R(r)$
1	50.99	0.0000	0.2009	12	62.2232	0.7132	0.0343
2	52.0112	0.2051	0.0846	13	63.2444	0.7483	0.0333
3	53.0324	0.2915	0.0659	14	64.2656	0.7823	0.0324
4	54.0536	0.3588	0.0563	15	65.2868	0.8154	0.0316
5	55.0748	0.4163	0.0503	16	66.308	0.8478	0.0309
6	56.096	0.4677	0.0461	17	67.3292	0.8794	0.0303
7	57.1172	0.5147	0.0430	18	68.3504	0.9103	0.0298
8	58.1384	0.5586	0.0405	19	69.3716	0.9407	0.0292
9	59.1596	0.6000	0.0385	20	70.3928	0.9706	0.0288
10	60.1808	0.6393	0.0369	21	71.414	1.0000	0.0286
11	61.202	0.6770	0.0355				

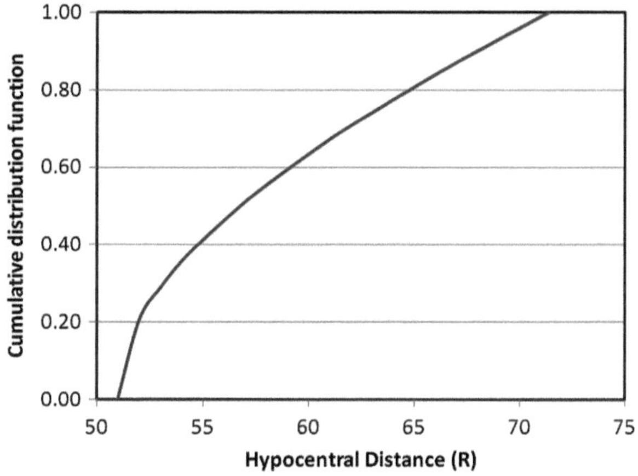

Fig. 4.13 Cumulative distribution function of hypocentral distance

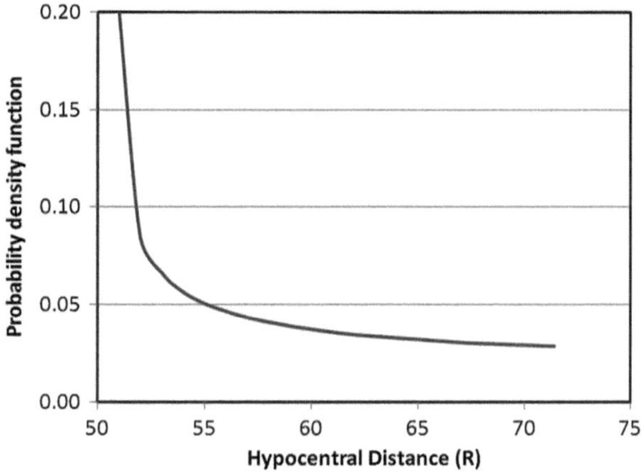

Fig. 4.14 Probability density function of hypocentral distance

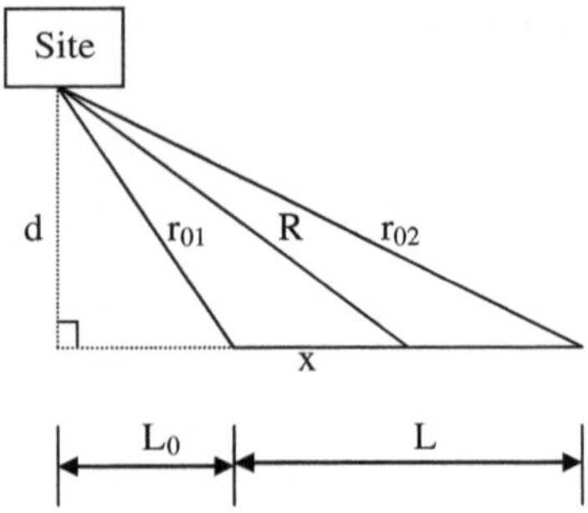

Fig. 4.15 Fault with eccentricity

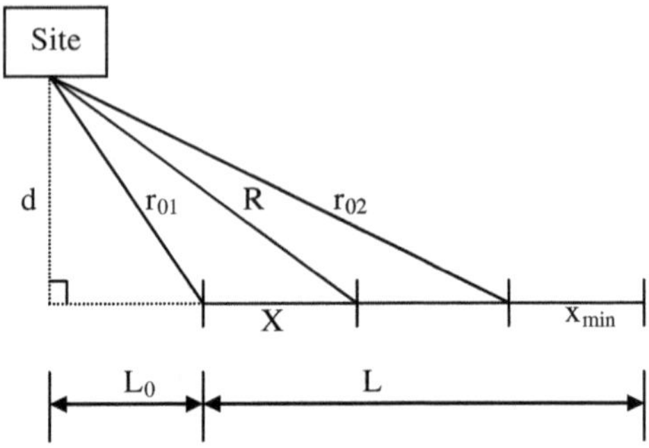

Fig. 4.16 Fault with rupture length

Case 3: Fault with rupture length (Fig. 4.16)

$$
\begin{aligned}
P(R \le r/m) &= \frac{x}{L_{\mathrm{eff}}} \qquad r_{01} < R < r_{02} \\
L_{\mathrm{eff}} &= L - X_{\min} \\
x &= \sqrt{r^2 - d^2} - L_0 \\
r_{01} &= \sqrt{d^2 + L_0^2} \\
r_{02} &= \sqrt{d^2 + (L + L_0 - X_{\min})^2} \\
P(R \le r/m) &= \frac{\sqrt{r^2 - d^2} - L_0}{L - X_{\min}},
\end{aligned}
\tag{4.24}
$$

where $X_{\min}$ = Rupture length
L = Fault length

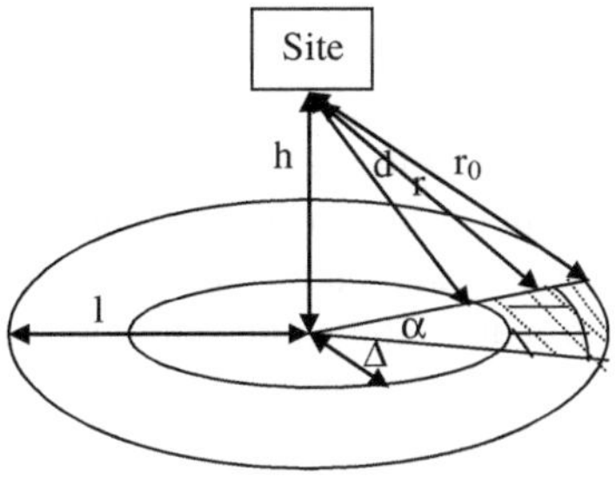

Fig. 4.17 Areal source model

Case 4: Areal source (Fig. 4.17) [2]

$$\begin{aligned} F_R(r) &= P[R \le r] \\ &= \frac{\alpha}{2\pi}\frac{\pi\left(r^2 - h^2 - \Delta^2\right)}{\pi\left(r_0^2 - h^2 - \Delta^2\right)} \\ &= \frac{\alpha}{2\pi}\frac{r^2 - d^2}{r_0^2 - d^2} \\ \therefore f_R(r) &= \frac{dF_R(r)}{dr} = \frac{\alpha}{2\pi}\frac{2r}{r_0^2 - d^2}, \end{aligned} \tag{4.25}$$

where

$$\begin{aligned} d^2 &= h^2 + \Delta^2 \\ r_0^2 &= l^2 + h^2 \end{aligned}$$

Case 5: Point Source

In this case, cluster of earthquakes are concentrated in a small region (Fig. 4.18) and the probability of occurrence of an earthquake in that region will be equal to one given that an earthquake of size "*m*" is going to occur in that region. Hence, for the case of point source

$$P(R = r) = 1 \text{ and } P(R \neq r) = 0 \tag{4.26}$$

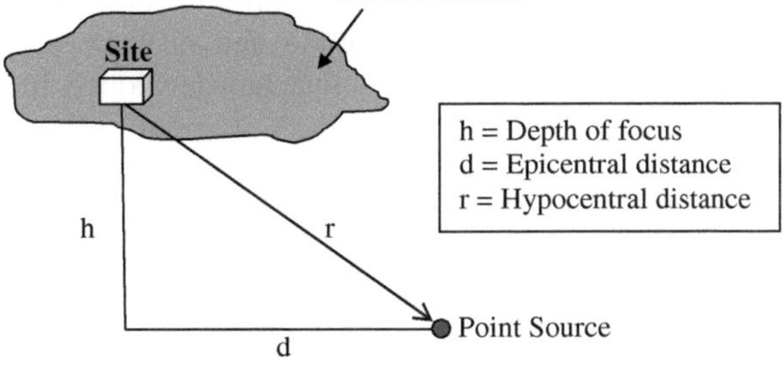

Fig. 4.18 Point source model

As mentioned in the above cases, the fault sources may not be always line source, point source, and areal sources. Some fault sources may be of complex geometry in such cases it is not possible to derive the probability density function for the source-to-site distance analytically. Hence, in those cases one can utilize the numerical methods to obtain the probability values. For example, one can divide the irregular fault sources into a large number of discrete elements of equal area and can find out the probability values by using simple probability concepts.

4.2.5 Attenuation Relationships

In hazard analysis, the ultimate aim is to find the ground motion parameter, i.e., peak ground acceleration (PGA) that is expected at the given site. Here the PGA depends on the earthquake magnitude and distance from source to site. This can be obtained through attenuation relationships. These are empirical relationships that are developed by regression analysis from the observed strong motion records in an area of interest. These are developed region wise and take into account the earthquake source, path of wave propagation, local site conditions, etc. These relationships usually relate ground motion parameter (PGA) and magnitude, distance, and some other parameters as described above. This can be written as a function as follows:

$$Y = f(M, R, P), \tag{4.27}$$

where Y is the ground motion parameter of interest,
M the magnitude of the earthquake,
R a measure of the distance from the source to the site being considered and the
P describes other parameters (earthquake source, wave propagation path, and/or local site conditions).

These relationships are bound to change as and when more strong motion records are available. There is lot of uncertainty involved in predicting the ground motion parameter despite the availability of huge data. The uncertainty is due to variability in the rupture mechanism, path of wave propagation, and local site conditions. Hence, the ground motion parameter will not be a deterministic value and it follows some kind of distribution with some mean and standard deviation. The relation between PGA versus magnitude and distance is shown in Fig. 4.19. Some of the attenuation relationships that are developed in the past are given as follows [11, 19, 20]:

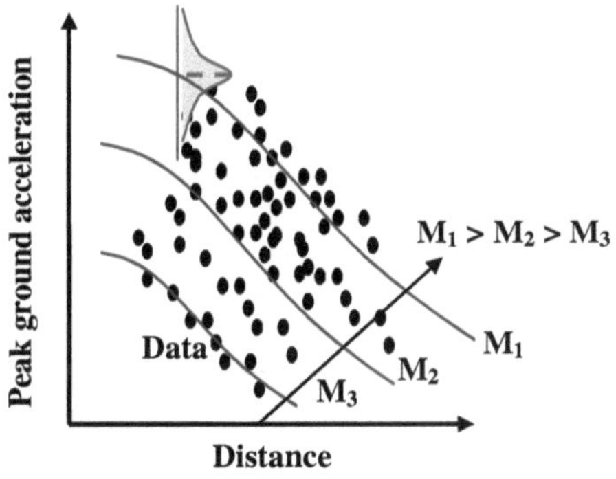

Fig. 4.19 Relationship between PGA versus magnitude and distance

Mc-Guire (1978)

$$Pgha(\mathrm{g}) = 0.0306\exp(0.89M)(R)^{-1.17} \tag{4.28}$$

R = Hypocentral distance
M = Magnitude

Campbell (1981)

$$\begin{aligned} &\ln \mathrm{PHA}(\mathrm{g}) = -4.141 + 0.868M - 1.09\ln\left[R + 0.0606\mathrm{e}^{0.7M}\right] \\ &\sigma_{\ln\ \mathrm{PHA}} = 0.37 \end{aligned} \tag{4.29}$$

Atkinson and Boore

$$\begin{aligned} &\ln(\mathrm{PGA}) = c_1 + c_2(M-6) + c_3(M-6)^2 - \ln(R) - c_4R \\ &c_1 = 1.841 \qquad c_2 = 0.686 \qquad c_3 = -0.123 \qquad c_4 = 0.00311 \end{aligned} \tag{4.30}$$

The standard error for ln(PGA) is 0.62.

4.2.6 Conditional Probability of Exceedence

As can be seen from Fig. 4.19, the ground motion parameter has got lot of uncertainty for a given magnitude and distance. The distribution of ground motion parameter (Peak Ground Acceleration) can be represented with Lognormal distribution whose density function can be given as follows [19]:

$$fz(z) = \frac{1}{\sigma\sqrt{2\pi}\,z}\exp\left[-\frac{1}{2}\left(\frac{\ln z - \mu_{\ln z}}{\sigma_{\ln z}}\right)^2\right] \quad 0 \le Z \le \infty \tag{4.31}$$

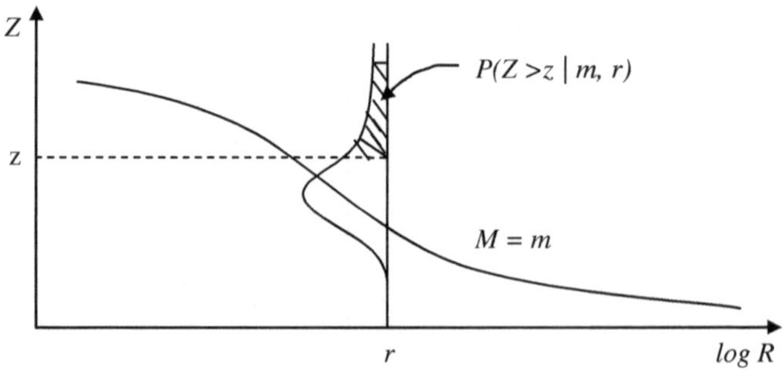

Fig. 4.20 Probability density function of ground motion parameter for a given magnitude and distance

Here Z is the random variable representing the ground motion parameter of interest (peak ground acceleration). Figure 4.20 shows the probability density function of PGA for a given magnitude and distance. The cumulative probability distribution for the lognormal distribution can be given by

$$F = P(Z \leq z) = \Phi\left[\frac{\ln z - \mu_{\ln z}}{\sigma_{\ln z}}\right] \tag{4.32}$$

And the probability of exceeding a certain level of PGA can be given as follows:

$$P(Z > z | m_i, r_j) = 1 - P(Z \leq z) = 1 - \Phi\left[\frac{\ln z - \mu_{\ln z}}{\sigma_{\ln z}}\right] \tag{4.33}$$

The above relation is valid in the range of 0 to infinity (∞). However, an earthquake of a given magnitude cannot produce infinity PGA value or in other words a smaller magnitude earthquake can produce only a smaller level of ground motion or smaller PGA at the site that too depending on the source distance, but it cannot produce very large PGA at the site. Hence, it is reasonable to have limits on the PGA values for a given magnitude and distance. In this case, one can derive truncated probability density function considering the lower and upper limits for the PGA values. This is explained as follows:

Consider X follows normal distribution and the range of X varies from $-\alpha$ to $+\alpha$. The probability density function for X can be given as

$$f_X(x) = \frac{1}{\sigma\sqrt{2\pi}} \exp\left[-\frac{1}{2}\left(\frac{x-\mu}{\sigma}\right)^2\right] \quad -\infty < X < +\infty \tag{4.34}$$

$$\text{Let, } u = \frac{x-\mu}{\sigma}$$

$$\therefore f(u) = \frac{1}{\sqrt{2\pi}} \exp\left[-\frac{u^2}{2}\right]$$

When X gets truncated at 3σ level, the expression gets changed and can be given as follows

$$f^1(u) = Cf(u) \qquad \mu - 3\sigma < x < \mu + 3\sigma \text{ and } -3 < u < +3$$

$$C\int_{-3}^{3} f(u)\mathrm{d}u = 1 \Rightarrow C = \frac{1}{\int_{-3}^{3} f(u)\mathrm{d}u}$$

$$\text{But,} \quad \Phi(u) = \int_{-\infty}^{u} f(u)\mathrm{d}u$$

$$\therefore \int_{-3}^{3} f(u)\mathrm{d}u = \int_{-\infty}^{\infty} f(u)\mathrm{d}u - \int_{-\infty}^{-3} f(u)\mathrm{d}u - \int_{3}^{\infty} f(u)\mathrm{d}u$$

$$= 1 - \Phi(-3) - \int_{-\infty}^{-3} f(u)\mathrm{d}u = 1 - 2\Phi(-3)$$

$$\therefore C = \frac{1}{1 - 2\Phi(-3)}$$

Hence,

$$f^1(u) = \frac{1}{1 - 2\Phi(-3)} \frac{1}{\sqrt{2\pi}} \exp\left[-\frac{u^2}{2}\right] \qquad -3 < u < +3$$

$$\therefore \Phi^1 = \int_{-3}^{u} f^1(u)du = \frac{1}{1 - 2\Phi(-3)}\left[\int_{-\infty}^{u} f(u)du - \int_{-\infty}^{-3} f(u)du\right] \tag{4.35}$$

$$\Phi^1 = \frac{\Phi(u) - \Phi(-3)}{1 - 2\Phi(-3)}$$

From the above, when the lognormal distribution for the ground motion level gets truncated at 3σ level, then

$$P(Z > z|m_i, r_j) = 1 - \Phi^1\left[\frac{\ln z - \mu_{\ln z}}{\sigma_{\ln z}}\right] \tag{4.36}$$

From the above relation, one can conclude that the PGA value at the site for a given magnitude and distance will have certain range in between the 3σ limits.

Example 4.6 Find the PGA for a magnitude of 6.0 and a hypocentral distance of 10 km using Campbell attenuation relationship. Develop probability density function and find the probability of exceedance of 0.1 g PGA at the site by considering the 3σ limits.

Solution:
Campbell attenuation relationship is given as

$$\ln \mathrm{PHA}(\mathrm{g}) = -4.141 + 0.868M - 1.09 \ln\left[R + 0.0606\mathrm{e}^{0.7M}\right]$$
$$\sigma_{\ln\,\mathrm{PHA}} = 0.37$$

By substituting M = 6.0 and R = 10 km in the above equation results into

$$\ln \mathrm{PHA}(\mathrm{g}) = -4.141 + 0.868 \times 6.0 - 1.09 \times \ln\left[10 + 0.0606\mathrm{e}^{0.7\times 6.0}\right]$$
$$\ln \mathrm{PHA}(\mathrm{g}) = -1.81277 \Rightarrow \mathrm{PHA}(\mathrm{g}) = \mathrm{e}^{-1.81277} = 0.1632\,g$$

Hence if an earthquake of magnitude 6.0 is occurred at a distance of 10 km from the site, it will produce a peak ground horizontal acceleration (PHA) of 0.1632 g. For constructing the probability density function, one needs the mean and standard deviation for normal parameters of PGA. It can be given as

$$\mu_{\ln\,\mathrm{PHA}} = -1.81277$$
$$\sigma_{\ln\,\mathrm{PHA}} = 0.37$$

This can be substituted in pdf function as given below

$$fz(z) = \frac{1}{0.37\sqrt{2\pi}\,z}\exp\left[-\frac{1}{2}\left(\frac{\ln z + 1.81277}{0.37}\right)^2\right] \qquad 0 \le Z \le \infty$$

The above density function is shown in Fig. 4.21.

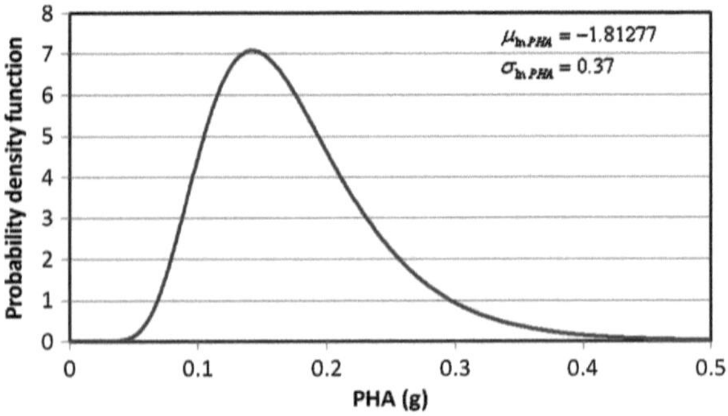

Fig. 4.21 Probability density function of PHA

Probability of exceedance can be given as follows:

$$P(Z > z|m_i, r_j) = 1 - P(Z \leq z) = 1 - \Phi\left[\frac{\ln z - \mu_{\ln z}}{\sigma_{\ln z}}\right]$$

$$P(Z > 0.1g|6, 10) = 1 - \Phi\left[\frac{\ln 0.1 + 1.81277}{0.37}\right] = 0.9072$$

If we consider the truncated lognormal distribution then

$$P(Z > z|m_i, r_j) = 1 - \Phi^1\left[\frac{\ln z - \mu_{\ln z}}{\sigma_{\ln z}}\right] = 1 - \Phi^1[u],$$

$$\text{where} \quad \Phi^1(u) = \frac{\Phi(u) - \Phi(-3)}{1 - 2\Phi(-3)}$$

$$u = \frac{\ln z - \mu_{\ln z}}{\sigma_{\ln z}} = \frac{\ln 0.1 + 1.81277}{0.37} = -1.32382$$

$$\Phi(u) = \Phi(-1.32382) = P(Z \leq -1.32382) = 0.09278$$

$$\Phi(-3) = 0.00135$$

$$\Phi^1(u) = \frac{\Phi(u) - \Phi(-3)}{1 - 2\Phi(-3)} = \frac{0.09278 - 0.00135}{1 - 2 \times 0.00135} = 0.091678$$

$$P(Z > z|m_i, r_j) = 1 - \Phi^1[u] = 1 - 0.091678 = 0.90832$$

The 3σ limits of ln(PHA) can be given as follows:

$$\text{Lower Limit} = \mu_{\ln \text{PHA}} - 3\sigma_{\ln \text{PHA}} = -1.81277 - 3 \times 0.37 = -2.9228$$

$$\text{Upper Limit} = \mu_{\ln \text{PHA}} - 3\sigma_{\ln \text{PHA}} = -1.81277 + 3 \times 0.37 = -0.7028$$

The lower and upper limits of PHA can be obtained as

$$\text{Lower Limit} = e^{-2.9228} = 0.0538\,g$$

$$\text{Upper Limit} = e^{-0.7028} = 0.4952\,g$$

From the above, it can be seen that a magnitude of 6.0 at a distance of 10 km can produce a minimum PHA of 0.0538 g and a maximum PHA of 0.4952 g at the site.

4.2.7 Determining the Hazard at the Site

This is the final step of hazard analysis in which the hazard at a given site is defined in terms of the probability of exceedance of a particular level of peak ground acceleration due to the postulated earthquakes occurring at different distances considering all the fault sources around the site. Hence, the expected number of exceedances of ground motion level "z" during a period of one year is given as [11]

$$\nu(z) = \sum_{i=1}^{N} \upsilon_i(m_0) \int_{m_0}^{m_u} \int_{d}^{r_0} f_M(m) f_R(r)\, P[Z > z/m \geq m_0, r]\, \mathrm{d}r\, \mathrm{d}m$$

or

$$\nu(z) = \sum_{i=1}^{N} \upsilon_i(m_0) \left[\sum_{M} \sum_{R} f_M(m) f_R(r)\, P[Z > z/m \geq m_0, r]\, P(Z > z/m_i, r_j) \Delta R \Delta m \right] \quad (4.37)$$

in which

$\upsilon(m_0)$ = the annual frequency of occurrence of earthquakes on seismic source n whose magnitudes are greater than m_0 and below the maximum event size, m_u.
$f_R(r)$ = the probability of an earthquake of magnitude m_i on source n occurring at a certain distance r_j from the site
$f_M(m_i)$ = the occurrence probability of an earthquake of magnitude m_i on source n
$P(Z > z| m_i, r_j)$ = the probability that ground motion level z will be exceeded, given n earthquake of magnitude m_i at distance of r_j from the site.

A typical hazard curve is shown in Fig. 4.22.

Example 4.7 Consider a site surrounding two sources of earthquakes are located (2 Line Sources) as shown in Fig. 4.23. The input information is provided for each fault in Table 4.7. Develop the hazard curve for the site under consideration.

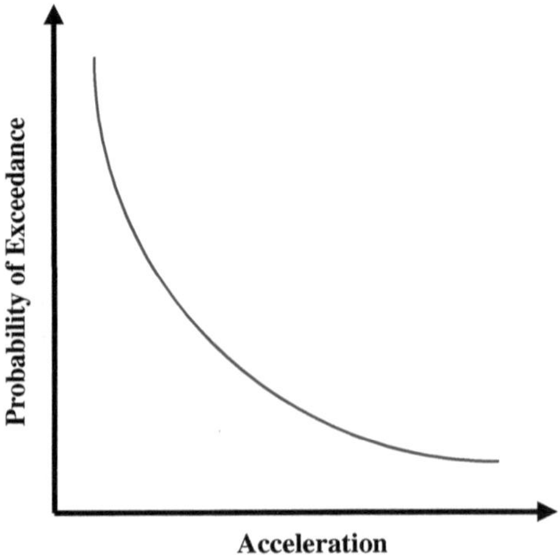

Fig. 4.22 Hazard curve of a typical site

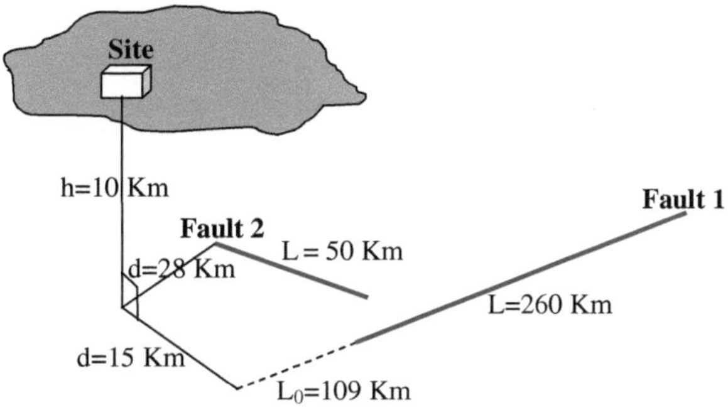

Fig. 4.23 Faults around the site

Table 4.7 Characteristics of the faults

Source	m_0	m_U	a	b	h (km)	d (km)	L (km)	L_0 (km)
Fault 1	4.0	6.0	3.892	1.294	10	15	260	109
Fault 2	4.0	5.0	3.176	1.294	10	28	50	0

Solution:

The annual rate of exceedance of a certain level of PGA at the site can be given as

$$v(z) = \sum_{i=1}^{N} \upsilon_i(m_0) \int_{m_0}^{m_u} \int_{d}^{r_0} f_M(m) f_R(r)\, P[Z > z/m \geq m_0, r]\, \mathrm{d}r\, \mathrm{d}m$$

or

$$v(z) = \sum_{i=1}^{N} \upsilon_i(m_0) \left[\sum_{M} \sum_{R} f_M(m) f_R(r)\, P[Z > z/m \geq m_0, r]\, P(Z > z/m_i, r_j) \Delta R \Delta m \right]$$

In this problem, there are two fault sources around the site, i.e., $N = 2$. In the above equation, the variables that should be determined are

- $v_i(m_0)$
- $P(M = m_i)$
- $P(R = r_j|m_i)$
- $P(Z > z|m_i, r_j)$

Analyzing the Fault 1

Now consider the fault 1 separately as shown in Fig. 4.24. From the figure, one can find out the hypocentral distance of the fault from the source.

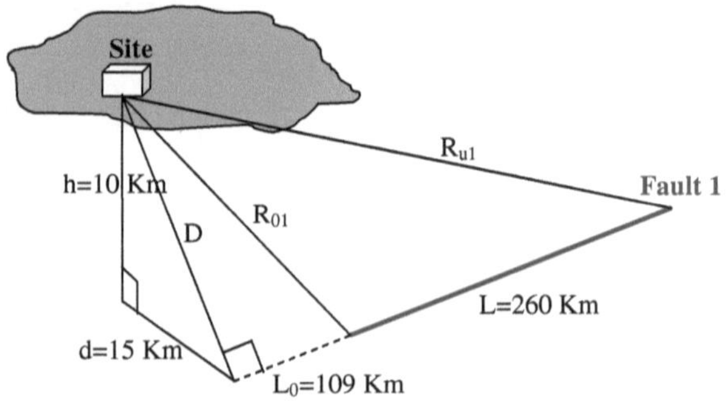

Fig. 4.24 Line source model-1

From the Figure

$$D = \sqrt{h^2 + d^2} = \sqrt{10^2 + 15^2} = 18.028\,\text{km}$$
$$R_{01} = \sqrt{D^2 + L_0^2} = \sqrt{18.02776^2 + 109^2} = 110.5\,\text{km}$$
$$R_{u1} = \sqrt{D^2 + (L_0 + L)^2} = \sqrt{18.02776^2 + (109 + 260)^2} = 369.4\,\text{km}$$

From the data

$m_0 = 4.0$ $a = 3.892$
$m_u = 6.0$ $b = 1.294$

- Hence, on fault 1, an earthquake can occur with a magnitude ranging from a minimum value of 4.0 to a maximum value of 6.0. And it can occur anywhere on the fault with hypocentral distance varies from 110.5 to 369.4 km.

Step 1. Calculation of $\nu_1(m_0)$

- This represents the annual rate of exceedance of magnitude m_0 on the fault 1. This can be calculated from the Gutenberg-Richter's relationship as given below

$$\upsilon_1(m_0) = 10^{a-bm_0}$$
$$\upsilon_1(4) = 10^{3.892-1.294\times4.0} = 0.052/\text{year}$$

Step 2. Calculating $f_M(m)\Delta m$

- The probability of occurrence of a particular magnitude on the fault can be calculated from its probability density function.
- The probability density function (pdf) of magnitude "m" whose lower and upper limits are given as m_0 and m_u can be given as follows:

$$f_M(m) = \frac{\beta 10^{a-bm}}{10^{a-bm_0} - 10^{a-bm_u}} \quad m_0 \le M \le m_u$$

- The pdf can be calculated for different magnitude values as follows

$$\beta = b\log_e(10) = 1.294 \times 2.3026 = 2.98$$

$$f_M(4.0) = \frac{2.98 \times 10^{-1.294\times4.0}}{10^{-1.294\times4.0} - 10^{-1.294\times6.0}} = 2.9873$$

- This calculation can be repeated for different values of magnitude and the results are shown in Table 4.8.
- The probability density function of magnitude is shown in Fig. 4.25.
- Since the above pdf is continuous in nature, it is not possible to find the value of $P(M = m)$ directly. But one can find the probability value in between some interval. Hence one can divide the range of magnitudes into small interval and can find the probability of occurrence of magnitude in that interval.

Table 4.8 Calculation of $f_M(m)$

S.no	m	$f_M(m)$
1	4	2.9873
2	4.2	1.6462
3	4.4	0.9071
4	4.6	0.4999
5	4.8	0.2755
6	5	0.1518
7	5.2	0.0837
8	5.4	0.0461
9	5.6	0.0254
10	5.8	0.0140
11	6.0	0.0077

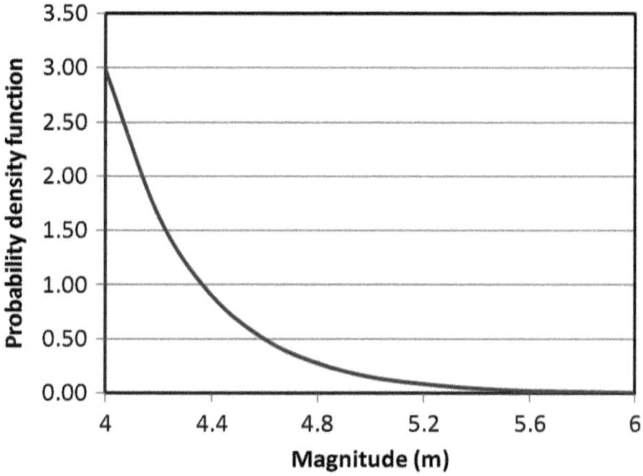

Fig. 4.25 Probability density function of magnitude

Table 4.9 Magnitude intervals

S.no	Interval	S.no	Interval
1	4.0–4.2	6	5.0–5.2
2	4.2–4.4	7	5.2–5.4
3	4.4–4.6	8	5.4–5.6
4	4.6–4.8	9	5.6–5.8
5	4.8–5.0	10	5.8–6.0

- In the present case, the lower and upper limit of magnitude is given as
  - $m_0 = 4.0$
  - $m_u = 6.0$
  - Range $= m_u - m_0 = 6.0 - 4.0 = 2.0$
- This range can be divided into 10 small and equal intervals. Then interval size becomes

$$\Delta m = \frac{(m_u - m_0)}{10} = \frac{(6.0 - 4.0)}{10} = 0.2$$

- Now the magnitude range can be divided into 10 small intervals as follows: (Table 4.9)
- From the definition of probability density function

$$f_M(m) = \frac{\mathrm{d}F}{\mathrm{d}m} = \operatorname*{Lim}_{\Delta r \to 0} \frac{\Delta F}{\Delta m} = \frac{F(m + \Delta m) - F(m)}{\Delta m}$$

$$f_M(m)\Delta m = F(m + \Delta m) - F(m) = P(m + \Delta m \leq M \leq m)$$

$$m = 4.0$$

$$\Delta m = 0.2 \Rightarrow m + \Delta m = 4.0 + 0.2 = 4.2$$

$$P(4.2 \leq M \leq 4.0) = F(4.2) - F(4.0)$$

$$F_M(m) = \frac{10^{-bm_0} - 10^{-bm}}{10^{-bm_0} - 10^{-bm_u}}$$

$$F_M(4.2) = \frac{10^{-1.294 \times 4.0} - 10^{-1.294 \times 4.2}}{10^{-1.294 \times 4.0} - 10^{-1.294 \times 6.0}} = \frac{6.668x10^{-6} - 3.675x10^{-6}}{6.668x10^{-6} - 1.722x10^{-8}} = 0.4501$$

$$F_M(4.0) = \frac{10^{-1.294 \times 4.0} - 10^{-1.294 \times 4.0}}{10^{-1.294 \times 4.0} - 10^{-1.294 \times 6.0}} = \frac{6.668x10^{-6} - 6.668x10^{-6}}{6.668x10^{-6} - 1.722x10^{-8}} = 0.0$$

$$P(4.2 \leq M \leq 4.0) = F(4.2) - F(4.0) = 0.4501 - 0.0 = 0.4501$$

- This calculation can be repeated for different intervals and the results are shown in Table 4.10.
- The cumulative distribution function is shown in Fig. 4.26.
- The cumulative distribution function can be calculated from the probability density function as follows.

Table 4.10 Cumulative distribution function values

S.no	Interval	m	$F_M(m)$ I	$m+\Delta m$	$F_M(m+\Delta m)$ II	$\Delta F_M(m)$ $[f_M(m)\ \Delta m]$ II–I
1	4.0–4.2	4	0.0000	4.2	0.4501	0.4501
2	4.2–4.4	4.2	0.4501	4.4	0.6981	0.2480
3	4.4–4.6	4.4	0.6981	4.6	0.8348	0.1367
4	4.6–4.8	4.6	0.8348	4.8	0.9101	0.0753
5	4.8–5.0	4.8	0.9101	5	0.9516	0.0415
6	5.0–5.2	5	0.9516	5.2	0.9745	0.0229
7	5.2–5.4	5.2	0.9745	5.4	0.9871	0.0126
8	5.4–5.6	5.4	0.9871	5.6	0.9941	0.0069
9	5.6–5.8	5.6	0.9941	5.8	0.9979	0.0038
10	5.8–6.0	5.8	0.9979	6.0	1.0000	0.0021

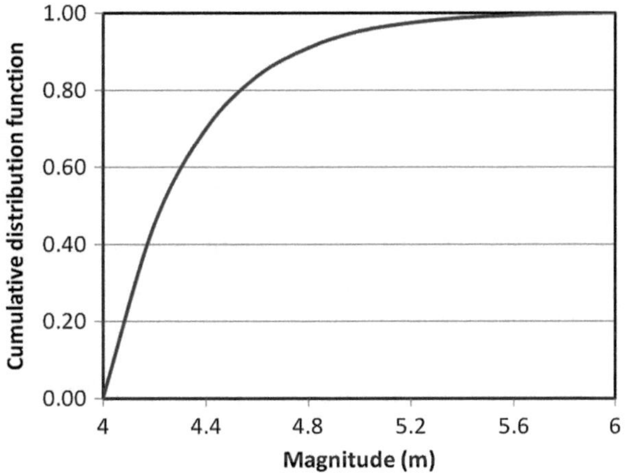

Fig. 4.26 Cumulative distribution function of magnitude

- From the definition of probability density function

$$f_M(m) = \frac{\mathrm{d}F}{\mathrm{d}m} = \underset{\Delta r \to 0}{Lim}\frac{\Delta F}{\Delta m} = \frac{F(m+\Delta m)-F(m)}{\Delta m}$$
$$f_M(m)\Delta m = F(m+\Delta m) - F(m) = P(m+\Delta m \leq M \leq m)$$
$$P(m+\Delta m \leq M \leq m) = f_M(m)\Delta m$$

- The above function can also be written as follows:

$$f_M(m) = \frac{F\left(m+\frac{\Delta m}{2}\right) - F\left(m-\frac{\Delta m}{2}\right)}{\Delta m}$$
$$f_M(m)\Delta m = F\left(m+\frac{\Delta m}{2}\right) - F\left(m-\frac{\Delta m}{2}\right) = P\left(m-\frac{\Delta m}{2} \leq M \leq m+\frac{\Delta m}{2}\right)$$
$$P(m_1 \leq M \leq m_2) = f_M(m)\Delta m$$

- Now consider the first interval, i.e., 4.0 − 4.2, where

$$\begin{aligned} m_1 &= 4.0 \\ m_2 &= 4.2 \\ \Delta m &= 4.2 - 4.0 = 0.2 \end{aligned}$$

$$\begin{aligned} &m_1 = m - \frac{\Delta m}{2} \Rightarrow m = m_1 + \frac{\Delta m}{2} \\ &m = 4.0 + \frac{0.2}{2} = 4.1 \\ &\therefore P(4.0 \le M \le 4.2) = f_M(4.1) \times 0.2 \end{aligned}$$

$$f_M(m) = \frac{\beta 10^{-bm}}{10^{-bm_0} - 10^{-bm_u}} \qquad m_0 \le M \le m_u$$

$$\begin{aligned} &\beta = b \log_e(10) = 1.294 \times 2.3026 = 2.98 \\ &f_M(4.1) = \frac{2.98 \times 10^{-1.294 \times 4.1}}{10^{-1.294 \times 4.0} - 10^{-1.294 \times 6.0}} = 2.2175 \\ &P(4.0 \le M \le 4.2) = f_M(4.1) \times 0.2 = 2.2175 \times 0.2 = 0.4435 \end{aligned}$$

- These results are compared with the previous one from Table 4.10. There is a small difference in the values. This can be eliminated if we choose very small interval size.

$$P(4.0 \le M \le 4.2) = \begin{cases} 0.4501 & \text{From Table 4.10} \\ 0.4435 & \text{Pr esent Value} \end{cases}$$

- This calculation can be repeated for different intervals and the results are shown in Table 4.11.
- The graph between magnitudes *Vs* $\Delta F(m)$ is shown in Fig. 4.27. This graph represents the probability of occurrence of a given earthquake in a particular interval.
- Now one can find out the annual rate of exceedance of different magnitudes on fault 1 as shown below

$$\begin{aligned} &\alpha(m) = \alpha(m_0) \frac{10^{-bm} - 10^{-bm_u}}{10^{-bm_0} - 10^{-bm_u}} \\ &\text{where} \quad \alpha(m_0) = 10^{a - bm_0}. \end{aligned}$$

Table 4.11 Cumulative distribution function values from pdf

S.no	Interval	m	$f_M(m)$	$\Delta F_M(m) = f_M(m)\ \Delta m$
1	4.0–4.2	4.1	2.2175	0.4435
2	4.2–4.4	4.3	1.2220	0.2444
3	4.4–4.6	4.5	0.6734	0.1347
4	4.6–4.8	4.7	0.3711	0.0742
5	4.8–5.0	4.9	0.2045	0.0409
6	5.0–5.2	5.1	0.1127	0.0225
7	5.2–5.4	5.3	0.0621	0.0124
8	5.4–5.6	5.5	0.0342	0.0068
9	5.6–5.8	5.7	0.0189	0.0038
10	5.8–6.0	5.9	0.0104	0.0021

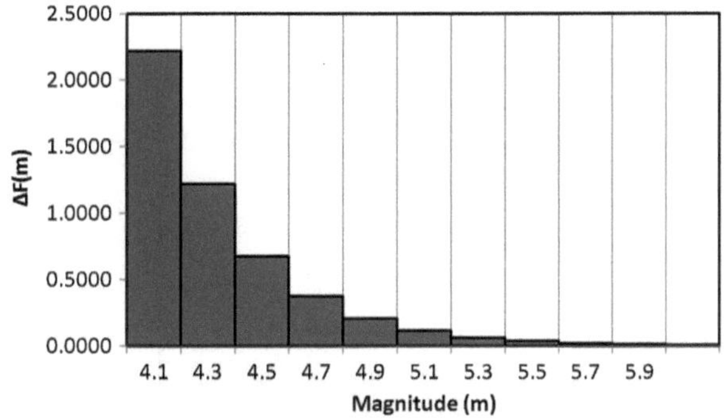

Fig. 4.27 Graph between magnitude versus $\Delta F(m)$

- Substituting $a = 3.892$, $b = 1.294$, $m_0 = 4.0$, and $m_u = 6.0$ in the above equation results into

$$\alpha(m_0) = 10^{3.892-1.294\times4.0} = 0.052$$

$$\alpha(4.4) = 0.2818 \times \frac{10^{-1.294\times4.4} - 10^{-1.294\times6.0}}{10^{-1.294\times4.0} - 10^{-1.294\times6.0}} = 0.3019$$

- Similarly the above calculations can be repeated for different values of magnitude and the results are shown in Table 4.12.
- The graph between magnitude and annual rate of exceedance is shown in Fig. 4.28.

Step 3: Calculating $f_R(r)\ \Delta r$

- In this step, we calculate the probability distribution for source-to-site distance. This can be obtained for two cases one without considering the concept of rupture length and another one with considering the rupture length concept.
- Without considering rupture length concept, the probability distribution for the hypocentral distance can be calculated as follows (Fig. 4.29).

Table 4.12 Annual rate of exceedance of a particular magnitude

S.no	m	$\mathbf{v_1}(m)$
1	4	5.2000E-02
2	4.2	2.8595E-02
3	4.4	1.5697E-02
4	4.6	8.5896E-03
5	4.8	4.6730E-03
6	5	2.5146E-03
7	5.2	1.3253E-03
8	5.4	6.6988E-04
9	5.6	3.0870E-04
10	5.8	1.0968E-04
11	6.0	0.0000E-00

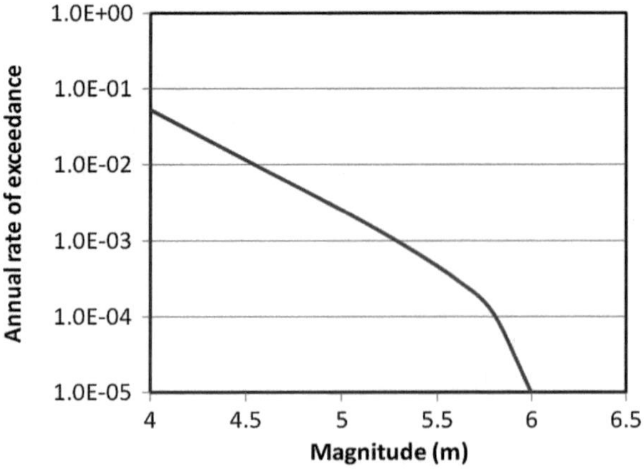

Fig. 4.28 Graph between magnitude versus annual rate of exceedance

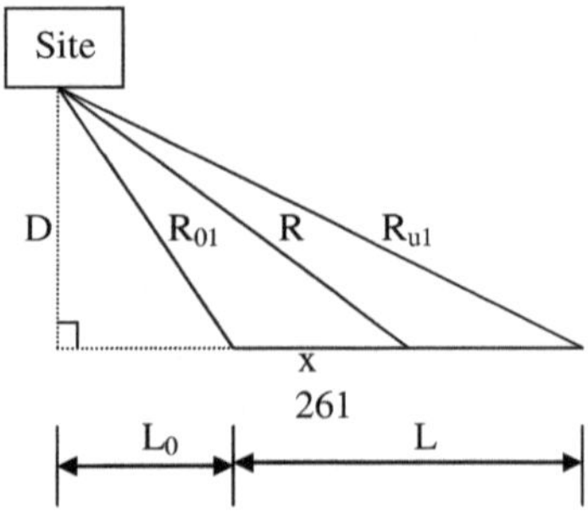

Fig. 4.29 Fault with eccentricity

$$P(R \leq r/m) = \frac{x}{L} \qquad r_{01} < R < r_{02}$$

$$P(R \leq r/m) = \frac{\sqrt{r^2 - D^2} - L_0}{L}$$

- In the above diagram

$$L_0 = 109\,\text{km} \quad L = 260\,\text{km}$$
$$D = 18.02776\,\text{km}$$
$$R_{01} = 110.4808\,\text{km} \quad R_{u1} = 369.4401\,\text{km}$$

- Now the earthquake can occur at any point on the fault with shortest distance of 110 km to longest distance of 369 km from the site.
- This distance is continuous in nature, but in the calculation one should find out what is the probability that earthquake will occur at a particular distance, i.e., $P(R = r)$; in the case of continuous distributions, it is not possible to find this value. But one can find out probability of occurrence in a particular interval like probability of occurrence of an earthquake in between 110 and 111 km, i.e., $P(110 \leq R \leq 111)$. This can be done by dividing the entire range into small intervals.
- The range of distance is

$$R_{01} = 110.5\,\text{km}$$
$$R_{u1} = 369.4\,\text{km}$$
$$\text{Range} = R_{u1} - R_{01} = 369.4 - 110.5 = 258.9\,\text{km}$$

- This range can be divided into 20 equal intervals and the interval size becomes 12.95 km and the intervals are shown in Table 4.13.

Table 4.13 Different intervals of hypocentral distance with lower and upper limits

S. no	Interval (km)		S. no	Interval (km)	
	Lower	Upper		Lower	Upper
1	110.5	123.445	11	239.95	252.895
2	123.445	136.39	12	252.895	265.84
3	136.39	149.335	13	265.84	278.785
4	149.335	162.28	14	278.785	291.73
5	162.28	175.225	15	291.73	304.675
6	175.225	188.17	16	304.675	317.62
7	188.17	201.115	17	317.62	330.565
8	201.115	214.06	18	330.565	343.51
9	214.06	227.005	19	343.51	356.455
10	227.005	239.95	20	356.455	369.4

$$\Delta R = \frac{258.9}{20} = 12.95\,\text{km}$$

- Now we can calculate the probability of occurrence of earthquake in each interval by using the following formula

$$P(r_l \le R \le r_u) = P(R \le r_u) - P(R \le r_l)$$

$$P(R \le r) = \frac{\sqrt{r^2 - D^2} - L_0}{L}$$

- For the first interval 110.5–123.445 km

$$r_l = 110.5\,\text{km} \qquad r_u = 123.445\,\text{km}$$

$$P(R \le r_l) = \frac{\sqrt{r_l^2 - D^2} - L_0}{L}$$

$$D = 18.028\,\text{km}\ L_0 = 109\,\text{km} \qquad L = 260\,\text{km}$$

$$P(R \le 110.5) = \frac{\sqrt{110.5^2 - 18.028^2} - 109}{260} = 0.0$$

$$P(R \le 123.445) = \frac{\sqrt{123.445^2 - 18.028^2} - 109}{260} = 5.0467 \times 10^{-2}$$

$$P(110.5 \le R \le 123.445) = P(R \le 123.445) - P(R \le 110.5) = 5.0467 \times 10^{-2}$$

- The above calculations can be repeated for the other intervals and the results are shown in Table 4.14.
- The cumulative distribution function is shown in Fig. 4.30 and also a graph between interval and the probability of occurrence is shown in Fig. 4.31.
- **Now consider the case of rupture length** (Fig. 4.32). Considering rupture length concept, the probability distribution for the hypocentral distance can be calculated as follows:

$$P(R \le r/m) = \frac{x}{L_{\text{eff}}} \qquad R_{01} < R < R_{02}$$

$$P(R \le r/m) = \frac{\sqrt{R^2 - D^2} - L_0}{L - X_{\min}}$$

where $X_{\min}$ = Rupture length

- In the above equation, one has to find out the rupture length for a given magnitude. It is calculated based on the following formula

Table 4.14 Cumulative distribution function values

S.no	Interval (km)		$P(R \leq r)$ I	$P(R \leq r + \mathrm{d}r)$ II	$f_R(r)\ \Delta r$ II–I
	r	$r + \Delta r$			
1	110.5	123.445	0.0000E+00	5.0467E-02	5.0467E-02
2	123.445	136.39	5.0467E-02	1.0074E-01	5.0276E-02
3	136.39	149.335	1.0074E-01	1.5093E-01	5.0191E-02
4	149.335	162.28	1.5093E-01	2.0106E-01	5.0126E-02
5	162.28	175.225	2.0106E-01	2.5114E-01	5.0075E-02
6	175.225	188.17	2.5114E-01	3.0117E-01	5.0036E-02
7	188.17	201.115	3.0117E-01	3.5117E-01	5.0004E-02
8	201.115	214.06	3.5117E-01	4.0115E-01	4.9977E-02
9	214.06	227.005	4.0115E-01	4.5111E-01	4.9956E-02
10	227.005	239.95	4.5111E-01	5.0105E-01	4.9938E-02
11	239.95	252.895	5.0105E-01	5.5097E-01	4.9922E-02
12	252.895	265.84	5.5097E-01	6.0088E-01	4.9909E-02
13	265.84	278.785	6.0088E-01	6.5077E-01	4.9898E-02
14	278.785	291.73	6.5077E-01	7.0066E-01	4.9888E-02
15	291.73	304.675	7.0066E-01	7.5054E-01	4.9880E-02
16	304.675	317.62	7.5054E-01	8.0042E-01	4.9872E-02
17	317.62	330.565	8.0042E-01	8.5028E-01	4.9866E-02
18	330.565	343.51	8.5028E-01	9.0014E-01	4.9860E-02
19	343.51	356.455	9.0014E-01	9.5000E-01	4.9855E-02
20	356.455	369.4	9.5000E-01	1.0000E + 00	5.0005E-02

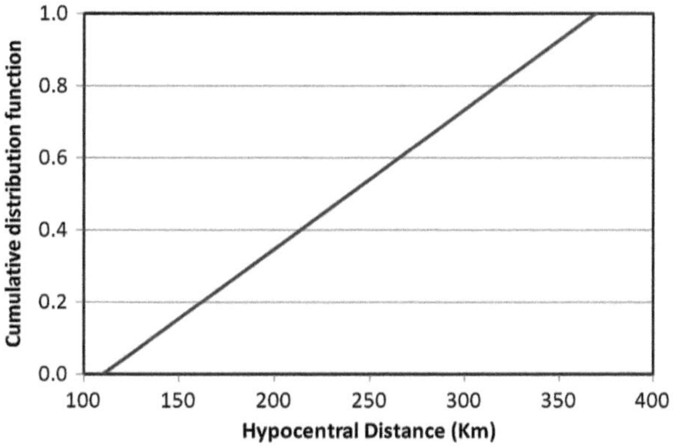

Fig. 4.30 Cumulative distribution function of hypocentral distance

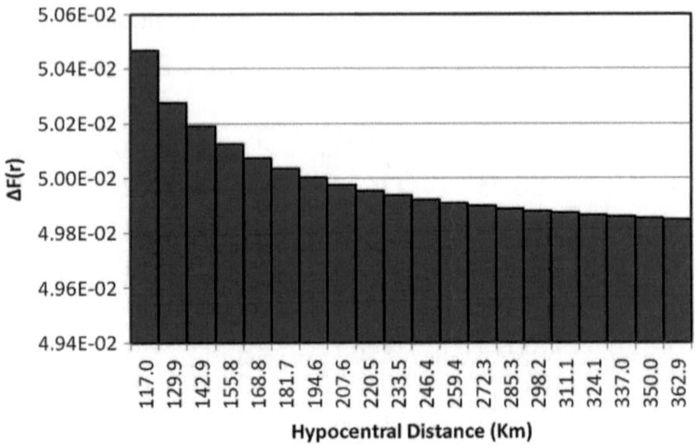

Fig. 4.31 Probability of occurrence in each interval

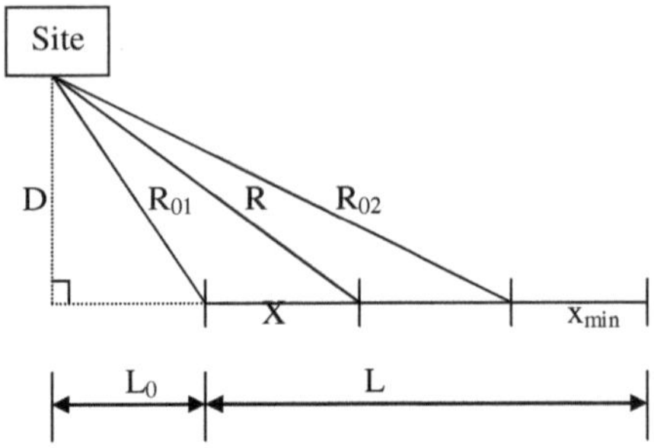

Fig. 4.32 Fault with rupture length

$$X_{\min} = \text{Crack Length} = \text{Min}[\exp(-4.654 + 1.189 m_i), \text{fault length}]$$

- Given the earthquake magnitude of 5.9, the rupture length for this corresponding magnitude will be

$$X_{\min} = \text{Min}[\exp(-4.654 + 1.189 \times 5.9), 260]$$
$$X_{\min} = \text{Min}[10.603, 260] = 10.603\ \text{km}$$

- Now this magnitude can occur anywhere on the fault with updated length of the fault and hypocentral distance as follows.

$$L_{\text{eff}} = L - X_{\min} = 260 - 10.603 = 249.397$$

$$R_{u1} = \sqrt{D^2 + \left(L_0 + L_{eff}\right)^2}$$

$$D = 18.028\ \text{km}\ L_0 = 109\ \text{km}$$

$$R_{u1} = \sqrt{18.028^2 + (109 + 249.397)^2} = 358.850\ \text{km}$$

- The probability distribution for distance from the site to the rupture on the source is computed conditionally on the earthquake magnitude because it is affected by the rupture size of the earthquake rupture.
- The range of distance is

$$R_{01} = 110.5\,\text{km}$$
$$R_{u1} = 358.850\,\text{km}$$
$$\text{Range} = R_{u1} - R_{01} = 358.85 - 110.5 = 248.35\,\text{km}$$

- This range can be divided into 20 equal intervals and the interval size becomes 12.42 km as given below.

$$\Delta R = \frac{248.35}{20} = 12.42\,\text{km}$$

- Now the first interval limits will be
  - Lower limit = 110.5 km
  - Upper limit = 110.5 + 12.42 = 122.92 km
- Now consider the first interval 110.5–122.92 km

$$r_l = 110.5\,\text{km} \qquad r_u = 122.92\,\text{km}$$
$$P(R \le r_l|m_i) = \frac{\sqrt{r_l^2 - D^2} - L_0}{L - X_{\min}}$$
$$D = 18.028\ \text{km}\,L_0 = 109\,\text{km}$$
$$L = 260\,\text{km} \qquad X_{\min} = 10.603\,\text{km}$$
$$P(R \le 110.5|5.9) = \frac{\sqrt{110.5^2 - 18.028^2} - 109}{260 - 10.603} = 0.0$$
$$P(R \le 122.92) = \frac{\sqrt{122.92^2 - 18.028^2} - 109}{260 - 10.603} = 5.0475 \times 10^{-2}$$
$$P(110.5 \le R \le 122.92) = P(R \le 122.92) - P(R \le 110.5) = 5.0475 \times 10^{-2}$$

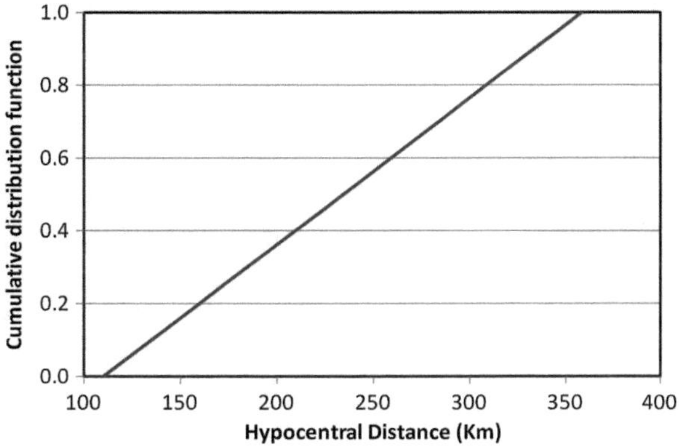

Fig. 4.33 Cumulative distribution function of hypocentral distance

- The above calculations can be repeated for the other intervals and the cumulative distribution function is shown in Fig. 4.33.

4. Attenuation relationship

- With the help of attenuation relationships, one can find the peak ground acceleration at the site due to a certain magnitude of earthquake occurring at a certain distance.
- There are different attenuation relationships available, care should be taken when applying a particular one for a particular region and site conditions.
- For the purpose of illustration in this example, Campbell attenuation relationship has been utilized for finding out the PGA at the site and is given below

$$\ln \mathrm{PHA(g)} = -4.141 + 0.868M - 1.09 \ln\left[R + 0.0606e^{0.7M}\right]$$
$$\sigma_{\ln \mathrm{PHA}} = 0.37$$

- Now consider an earthquake of magnitude 5.9 will occur at a hypocentral distance of 117 km. Then one can find the PGA at the site by substituting $M = 5.9$ and $R = 117$ km in the attenuation relationship as follows.

$$\ln \mathrm{PHA(g)} = -4.141 + 0.868 \times 5.9 - 1.09 \times \ln\left[117 + 0.0606e^{0.7\times 5.9}\right]$$
$$\ln \mathrm{PHA(g)} = -4.24512 \Rightarrow \mathrm{PHA(g)} = e^{-4.24512} = 0.0143\,g$$

- From the above equation, an earthquake of magnitude 5.9 occurring at a distance of 117 km will produce a PGA of 0.0143 g at the site. The variation of

Table 4.15 Variation of PGA with respect to distance and magnitude

Hypocentral distance R (km)	Peak ground acceleration (g)				
	$m = 4.0$	$m = 4.5$	$m = 5.0$	$M = 5.5$	$M = 5.9$
110	0.0030	0.0046	0.0071	0.0109	0.0153
130	0.0025	0.0039	0.0060	0.0091	0.0128
150	0.0022	0.0033	0.0051	0.0078	0.0110
170	0.0019	0.0029	0.0045	0.0069	0.0096
190	0.0017	0.0026	0.0040	0.0061	0.0086
210	0.0015	0.0023	0.0036	0.0055	0.0077
230	0.0014	0.0021	0.0032	0.0050	0.0070
250	0.0012	0.0019	0.0029	0.0045	0.0064
270	0.0011	0.0018	0.0027	0.0042	0.0059
290	0.0011	0.0016	0.0025	0.0039	0.0054
310	0.0010	0.0015	0.0023	0.0036	0.0051
330	0.0009	0.0014	0.0022	0.0034	0.0047
350	0.0009	0.0013	0.0020	0.0031	0.0044
370	0.0008	0.0012	0.0019	0.0030	0.0042

Fig. 4.34 Variation of PGA with distance and magnitude

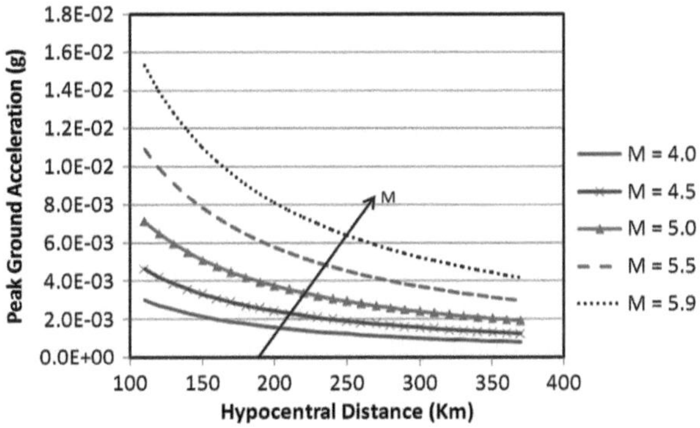

PGA with respect to magnitude and hypocentral distance is given in Table 4.15 and also shown in Fig. 4.34.

- The above calculated PGA is not a constant value but it is a random variable which follows lognormal distribution with the following parameters and the corresponding probability density function is shown in Fig. 4.35.

$$\mu_{\ln \text{PHA}} = -4.24512$$
$$\sigma_{\ln \text{PHA}} = 0.37$$

5. Computation of Conditional Probability of Exceedence

- Probability of exceedance for a given value of PGA can be computed using the following formula

Fig. 4.35 Probability density function of PGA for $m = 5.9$ and $R = 117$ km

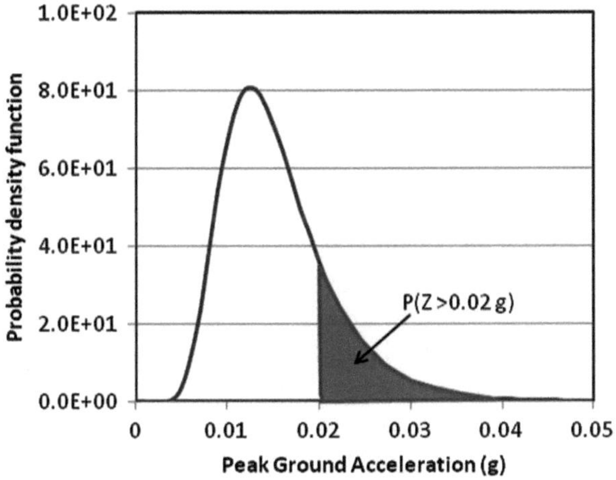

$$P(Z > z|m_i, r_j) = 1 - P(Z \leq z) = 1 - \Phi\left[\frac{\ln z - \mu_{\ln z}}{\sigma_{\ln z}}\right]$$

$$P(Z > 0.02g|5.9, 117) = 1 - \Phi\left[\frac{\ln 0.02 + 4.24512}{0.37}\right] = 0.184$$

- If we consider the truncated lognormal distribution, the 3σ limits of ln(PHA) can be given as follows:

$$\text{Lower Limit} = \mu_{\ln \text{PHA}} - 3\sigma_{\ln \text{PHA}} = -4.24512 - 3 \times 0.37 = -5.35512$$
$$\text{Upper Limit} = \mu_{\ln \text{PHA}} + 3\sigma_{\ln \text{PHA}} = -4.24512 + 3 \times 0.37 = -3.13512$$

- The lower and upper limits of PHA can be obtained as

$$\text{Lower Limit} = e^{-5.35512} = 0.004724\, g$$
$$\text{Upper Limit} = e^{-3.13512} = 0.043495\, g$$

- Hence PHA will vary in between 0.0047 and 0.043 g. With these limits, the probability of exceedance can be calculated as follows:

$$P(Z > z|m_i, r_j) = 1 - \Phi^1\left[\frac{\ln z - \mu_{\ln z}}{\sigma_{\ln z}}\right] = 1 - \Phi^1[u]$$

$$\text{where} \quad \Phi^1(u) = \frac{\Phi(u) - \Phi(-3)}{1 - 2\Phi(-3)}$$

$$u = \frac{\ln z - \mu_{\ln z}}{\sigma_{\ln z}} = \frac{\ln 0.02 + 4.24512}{0.37} = 0.9003$$

$$\Phi(u) = \Phi(0.9003) = P(Z \leq 0.9003) = 0.816$$

$$\Phi(-3) = 0.00135$$

$$\Phi^1(u) = \frac{\Phi(u) - \Phi(-3)}{1 - 2\Phi(-3)} = \frac{0.816 - 0.00135}{1 - 2 \times 0.00135} = 0.8168$$

$$P(Z > 0.02|5.9, 117) = 1 - \Phi^1[u] = 1 - 0.8168 = 0.183$$

- From the above equation, one can observe that the probability of exceedance in the case of without truncating the PHA limits is more when compared with truncating the PHA limits.
- This is due to the fact that any earthquake with certain magnitude at a particular distance cannot produce any amount of PHA at the site, it can produce in between its 3σ limits with 99.9 % confidence.
- By considering these limits, probability exceeding its $+3\sigma$ limit will be zero.

6. Estimation of hazard

- Here hazard is defined as the frequency of exceedance of a particular level of peak ground acceleration due to the postulated earthquakes occurring at different distances considering all the fault sources around the site.
- The frequency of exceedance of ground motion level "z" during a period of 1 year is given as

$$\nu(z) = \sum_{i=1}^{N} \upsilon_i(m_0) \int_{m_0}^{m_u} \int_{d}^{r_0} f_M(m) f_R(r)\, P[Z > z/m \geq m_0, r]\, \mathrm{d}r\, \mathrm{d}m$$

or

$$\nu(z) = \sum_{i=1}^{N} \upsilon_i(m_0) \left[\sum_{M} \sum_{R} f_M(m) f_R(r)\, P(Z > z/m_i, r_j) \Delta R \Delta m \right]$$

- In the present case, there are two faults around the site with different maximum magnitudes on each fault. Hence, the above expression can be represented as

$$\begin{aligned} \nu(z) &= \sum_{i=1}^{2} \upsilon_i(m_0) \left[\sum_{M} \sum_{R} f_M(m) f_R(r)\, P(Z > z/m_i, r_j) \Delta R \Delta m \right] \\ &= \upsilon_1(4) \left[\sum_{M=4.0}^{M=6.0} \sum_{R=110}^{369} f_M(m) f_R(r)\, P(Z > z/m_i, r_j) \Delta R \Delta m \right] \\ &\quad + \upsilon_2(4) \left[\sum_{M=4.0}^{M=5.0} \sum_{R=110}^{129} f_M(m) f_R(r)\, P(Z > z/m_i, r_j) \Delta R \Delta m \right] \end{aligned}$$

- For fault 1, the above expression can be written as

$$\nu_1(z) \quad = \upsilon_1(4) \left[\sum_{M=4.0}^{M=6.0} \sum_{R=110}^{369} f_M(m) f_R(r)\, P(Z > z/m_i, r_j) \Delta R \Delta m \right]$$

- From the previous steps, the above parameters can be rewritten as

$$\begin{aligned} &\upsilon_1(4) = 0.052\ /\text{year} \\ &\text{for}\, m = 5.9 \quad f_M(5.9)\Delta m = 2.10 \times 10^{-3} \\ &\text{for}\, r = 117 \quad f_R(117)\Delta r = 5.0467 \times 10^{-2} \\ &\text{for}\, m = 5.9 \quad r = 117\,\text{km} \quad z = 0.02g \\ &P(Z > 0.02g/5.9, 117) = 0.183 \end{aligned}$$

- Here one should understand that $f_M(5.9)\ \Delta m$ is nothing but the probability of earthquake occurring in the interval between 5.8 and 6.0 magnitude. The average of this interval is 5.9 which is taken as reference magnitude in the calculation of PHA at the site.

- Similarly, $f_R(117)\ \Delta r$ represents the probability of occurrence of a particular magnitude on the fault in between 110.5 and 123.445 km from the site. The average of this interval is 117 km which is taken as reference hypocentral distance in the calculation of PHA at the site.
- Now the frequency of exceedance of 0.02 g at the site can be calculated as

$$
\begin{aligned}
\therefore \upsilon_1(0.02\,\mathrm{g}) &= 0.052 \times 2.10 \times 10^{-3} \times 5.0467 \times 10^{-2} \times 0.183 \\
&= 1.00851 \times 10^{-6}
\end{aligned}
$$

- From the above equation, the frequency of exceedance of 0.02 g at the site given that an earthquake with a magnitude of 5.9 occurring at a distance of 117 km is obtained as 1.00851×10^{-6}.
- Similarly the above calculations can be repeated for different magnitudes occurring at different hypocentral distances from the site and frequency of exceedance can be calculated.
- Table 4.16 shows frequency of exceeding a particular level of PGA (0.02 g) at the site for different combinations of magnitudes and distances. This is shown in Fig. 4.36 which represents the magnitude–distance bins for PGA of 0.02 g.
- Summing up all these values will give frequency of exceeding a PGA of 0.02 g at the site due to first fault source. This is obtained as 3.0871×10^{-4}/year.
- Repeating the above calculations one can obtain the annual rate of exceedance for different PGA values at the site due to fault source 1. These are given in

Table 4.16 Annual rate of exceedance of 0.02 g at the site for different combinations of magnitudes and distances

Magnitude	Distance (km)					
	110.5	116.9725	129.9175	142.8625	155.8075	168.7525
4	0.00E+00	0.00E+00	0.00E+00	0.00E+00	0.00E+00	0.00E+00
4.1	0.00E+00	0.00E+00	0.00E+00	0.00E+00	0.00E+00	0.00E+00
4.3	0.00E+00	0.00E+00	0.00E+00	0.00E+00	0.00E+00	0.00E+00
4.5	0.00E+00	0.00E+00	0.00E+00	0.00E+00	0.00E+00	0.00E+00
4.7	0.00E+00	0.00E+00	0.00E+00	0.00E+00	0.00E+00	0.00E+00
4.9	0.00E+00	0.00E+00	0.00E+00	0.00E+00	0.00E+00	0.00E+00
5.1	2.23E-07	1.06E-07	0.00E+00	0.00E+00	0.00E+00	0.00E+00
5.3	5.31E-07	3.35E-07	1.19E-07	2.68E-08	0.00E+00	0.00E+00
5.5	8.76E-07	6.10E-07	2.87E-07	1.28E-07	5.02E-08	1.23E-08
5.7	1.16E-06	8.67E-07	4.75E-07	2.54E-07	1.32E-07	6.59E-08
5.9	1.27E-06	1.01E-06	6.29E-07	3.82E-07	2.27E-07	1.33E-07
6	1.69E-06	1.39E-06	9.09E-07	5.83E-07	3.67E-07	2.27E-07

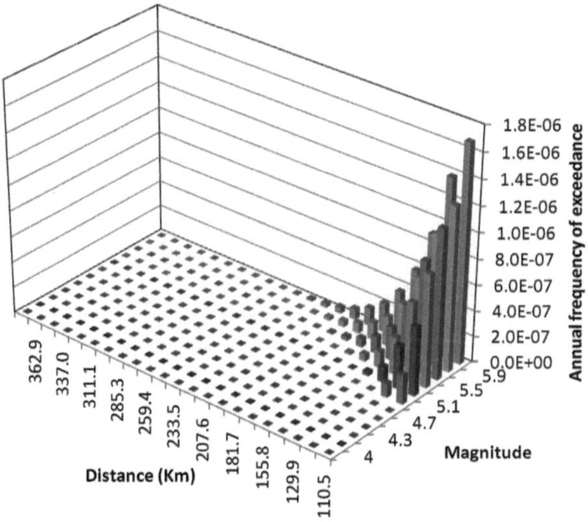

Fig. 4.36 Magnitude distance bins corresponding to annual frequency of exceedance of PGA of 0.02 g at the site due to 1st fault source

Table 4.17 Annual rate of exceedance with respect to PGA

PGA (g)	Annual rate of exceedance (/year)
0.005	3.7950E-03
0.010	3.2795E-04
0.015	6.3466E-05
0.020	1.6053E-05
0.025	4.5112E-06
0.030	1.3134E-06
0.035	3.7832E-07
0.040	9.9571E-08
0.045	1.9851E-08
0.050	7.8534E-10
0.055	0.0000E+00
0.060	0.0000E+00

Table 4.17 and a graph between PGA and Annual rate of exceedance is shown in Fig. 4.37.

Analyzing the Fault 2

Now consider the fault 2 separately as shown in Fig. 4.38. From the figure one can find out the hypocentral distance of the fault from the source.

From the Figure

$$R_{02} = \sqrt{h^2 + d^2} = \sqrt{10^2 + 28^2} = 29.73\,\text{km}$$

$$R_{u2} = \sqrt{R_{02}^2 + L^2} = \sqrt{18.02776^2 + 50^2} = 58.17\,\text{km}$$

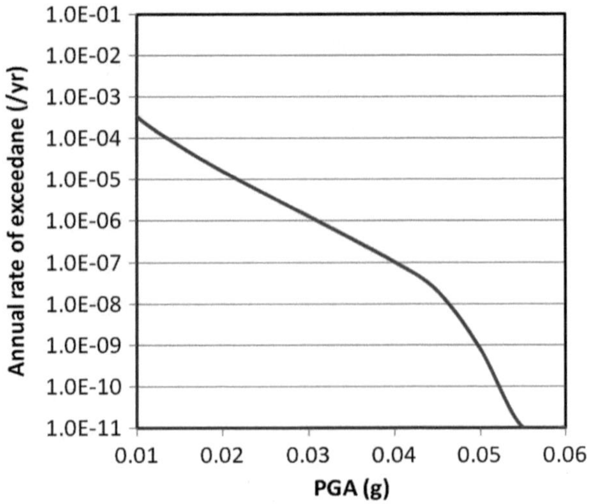

Fig. 4.37 PGA versus annual rate of exceedance due to fault 1

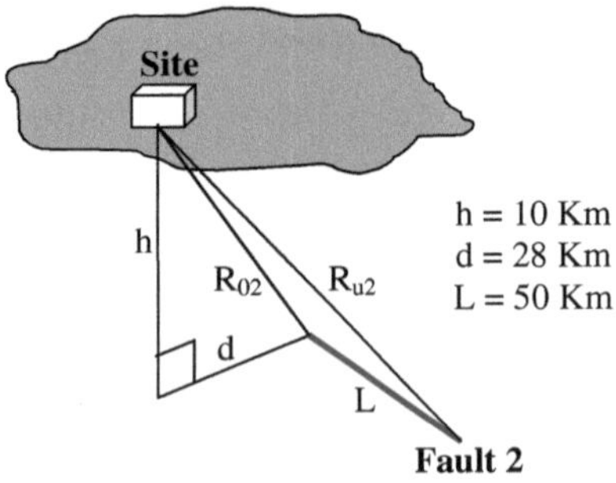

Fig. 4.38 2nd Fault source model

From the data

$m_0 = 4.0$ $a = 3.176$

$m_u = 5.0$ $b = 1.294$

- Hence, on fault 2 an earthquake can occur with a magnitude ranging from a minimum value of 4.0 to a maximum value of 5.0. And it can occur anywhere on the fault with hypocentral distance varies from 29.73 to 58.17 km.

Step 1. Calculation of $\nu_2(m_0)$

- This represents the annual rate of exceedance of magnitude m_0 on the fault 2. This can be calculated from the Gutenberg–Richter's relationship as given below

$$\upsilon_2(m_0) = 10^{a-bm_0}$$
$$\upsilon_2(4) = 10^{3.176-1.294\times4.0} = 0.01/\text{year}$$

Step 2. Calculating $f_M(m)\ \Delta m$

- The probability density function (pdf) of magnitude "m" whose lower and upper limits are given as m_O and m_u can be given as follows.

$$f_M(m) = \frac{\beta 10^{a-bm}}{10^{a-bm_0} - 10^{a-bm_u}} \qquad m_0 \le M \le m_u$$

- This is similar to the previous calculations and can be repeated for different values of magnitude and the results are shown in Table 4.18.
- The probability density function of magnitude is shown in Fig. 4.39.
- From the cumulative distribution function, one can calculate $f_M(m)\ \Delta m$ which represents the probability of occurrence of a magnitude in between some interval m and $m + \Delta m$. where Δm is interval size which is calculated as follows:
- In the present case, the lower and upper limit of magnitude is given as

Table 4.18 Calculation of $f_M(m)$

S.no	m	$f_M(m)$
1	4.0	3.1395
2	4.2	1.7301
3	4.4	0.9534
4	4.6	0.5254
5	4.8	0.2895
6	5.0	0.1595

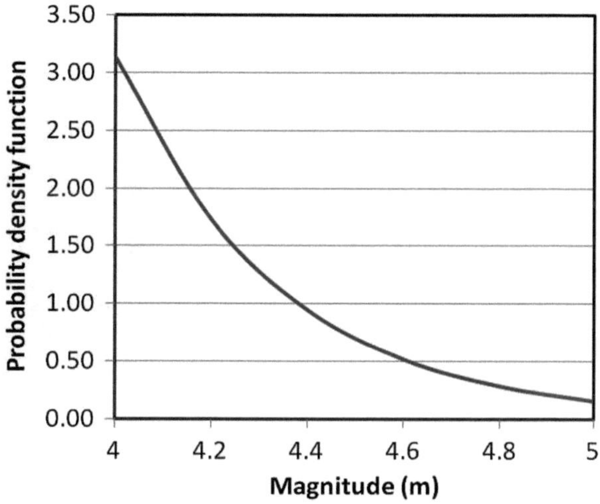

Fig. 4.39 Probability density function of magnitude

- $m_0 = 4.0$
- $m_u = 5.0$
- Range = $m_u – m_0 = 5.0 - 4.0 = 1.0$

- This range can be divided into five small and equal intervals. Then interval size becomes

$$\Delta m = \frac{(m_u - m_0)}{5} = \frac{(5.0 - 4.0)}{5} = 0.2$$

- Here interval size is kept as 0.2 similar to the fault source 1. Hence number of intervals has been reduced.
- Now the magnitude range can be divided into five small intervals and the $f_M(m)\ \Delta m$ can be calculated and the results are shown in Table 4.19.
- The graph between magnitude versus $f_M(m)\ \Delta m$ is shown in Fig. 4.40. This graph represents the probability of occurrence of a given earthquake in a particular interval.
- Now one can find out the annual rate of exceedance of different magnitudes on fault 2 by using the following formula

Table 4.19 Calculation of $f_M(m)$

S.no	Interval	$f_M(m)\ \Delta m$
1	4.0–4.2	0.4730
2	4.2–4.4	0.2606
3	4.4–4.6	0.1436
4	4.6–4.8	0.0791
5	4.8–5.0	0.0436

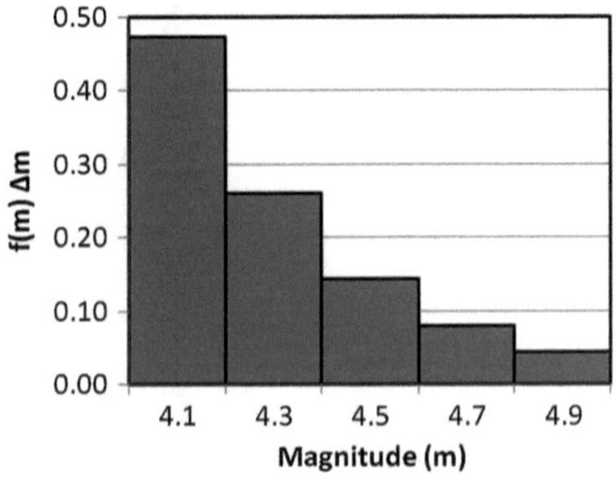

Fig. 4.40 Graph between magnitude versus $f_M(m)\ \Delta m$

$$\alpha(m) = \alpha(m_0)\frac{10^{-bm} - 10^{-bm_u}}{10^{-bm_0} - 10^{-bm_u}}$$
$$\text{where} \quad \alpha(m_0) = 10^{a-bm_0}$$

- Substituting $a = 3.176$, $b = 1.294$, $m_0 = 4.0$, and $m_u = 6.0$ in the above equation results into

$$\alpha(m_0) = 10^{3.176-1.294\times4.0} = 0.01$$
$$\alpha(4.4) = 0.2818 \times \frac{10^{-1.294\times4.4} - 10^{-1.294\times5.0}}{10^{-1.294\times4.0} - 10^{-1.294\times5.0}} =$$

- The calculations can be repeated for different values of magnitude and the results are shown in Table 4.20.
- The graph between magnitude versus annual rate of exceedance is shown in Fig. 4.41.

Table 4.20 Annual rate of exceedance of a particular magnitude

S.no	m	$v_2(m)$
1	4	0.0100
2	4.2	0.0053
3	4.4	0.0027
4	4.6	0.0012
5	4.8	0.0004
6	5	0.0000

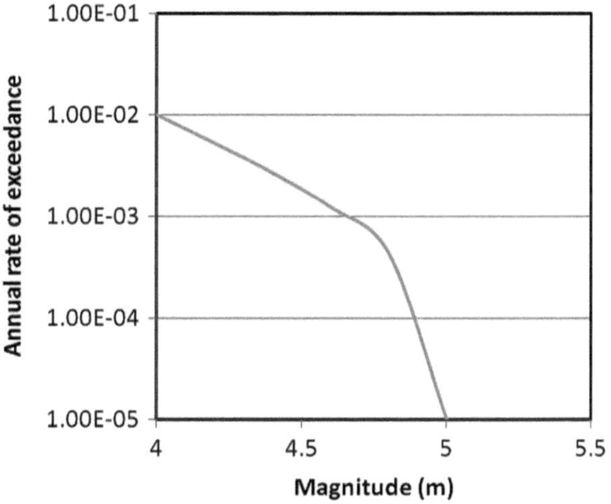

Fig. 4.41 Graph between magnitude versus annual rate of exceedance

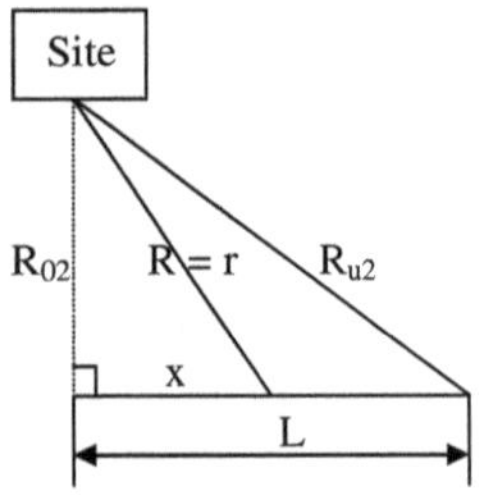

Fig. 4.42 Fault with eccentricity

Step 3: Calculating $f_R(r)\ \Delta r$

- In this step, we calculate the probability distribution for source-to-site distance.
- Without considering rupture length concept, the probability distribution for the hypocentral distance can be calculated as follows (Fig. 4.42).

$$P(R \le r/m) = \frac{x}{L} \qquad R_{02} < R < R_{u2}$$

$$P(R \le r/m) = \frac{\sqrt{r^2 - R_{02}^2}}{L}$$

$$f_R(r)\Delta r = P(R \le r + \Delta r) - P(R \le r)$$

- In the above diagram

$$L = 50\,\text{km}$$

$$R_{02} = 29.73\,\text{km} \qquad R_{u2} = 58.17\,\text{km}$$

- Now the earthquake can occur at any point on the fault with shortest distance of 29.73 km to longest distance of 58.17 km from the site.
- One can calculate $f_R(r)\ \Delta r$ by dividing the entire range into small intervals.
- The range of distance is

$$R_{02} = 29.73\,\text{km}$$

$$R_{u1} = 58.17\,\text{km}$$

$$\text{Range} = R_{u2} - R_{02} = 58.17 - 29.73 = 28.44\,\text{km}$$

- This range can be divided into 10 equal intervals and the interval size becomes,

$$\Delta R = \frac{28.44}{10} = 2.844\,\text{km}$$

Table 4.21 Cumulative distribution function values

S.no	Interval (km)		$P(R \leq r)$ I	$P(R \leq r + \mathrm{d}r)$ II	$f_R(r)\ \Delta r$ II–I
	r	$r + \Delta r$			
1	29.730	32.574	0.0000	0.2662	0.2662
2	32.574	35.418	0.2662	0.3850	0.1188
3	35.418	38.262	0.3850	0.4817	0.0967
4	38.262	41.106	0.4817	0.5677	0.0860
5	41.106	43.950	0.5677	0.6474	0.0796
6	43.950	46.794	0.6474	0.7227	0.0753
7	46.794	49.638	0.7227	0.7950	0.0723
8	49.638	52.482	0.7950	0.8650	0.0700
9	52.482	55.326	0.8650	0.9332	0.0682
10	55.326	58.170	0.9332	1.0000	0.0668

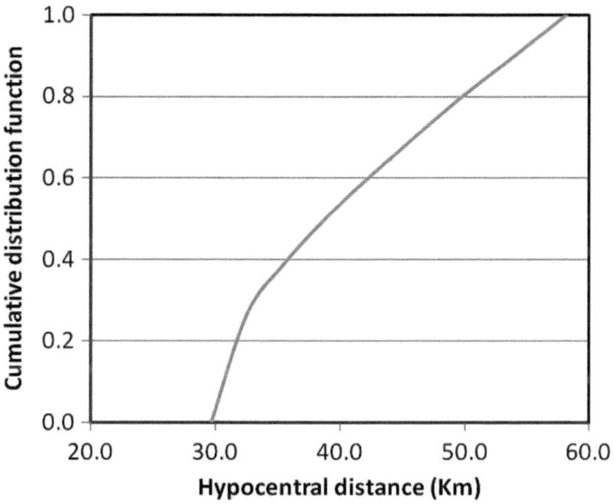

Fig. 4.43 Cumulative distribution function of hypocentral distance

- One can calculate $f_R(r)\ \Delta r$ interval wise and the results are shown in Table 4.21.
- The cumulative distribution function is shown in Fig. 4.43 and also a graph between interval Vs the probability of occurrence is shown in Fig. 4.44.

4. Attenuation relationship

- Campbell attenuation relationship has been utilized for finding out the PGA at the site and is given below

$$\ln \mathrm{PHA(g)} = -4.141 + 0.868M - 1.09 \ln\left[R + 0.0606\mathrm{e}^{0.7M}\right]$$
$$\sigma_{\ln\ \mathrm{PHA}} = 0.37$$

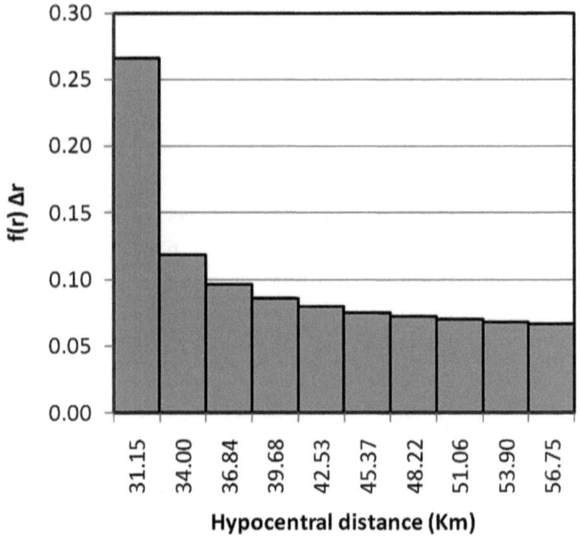

Fig. 4.44 Probability of occurrence in each interval

Table 4.22 Variation of PGA with respect to distance and magnitude

Hypocentral distance R (km)	Peak ground acceleration (g)		
	$m = 4.0$	$m = 4.5$	$m = 5.0$
29.730	0.0122	0.0186	0.0282
32.574	0.0111	0.0169	0.0257
35.418	0.0102	0.0155	0.0235
38.262	0.0094	0.0143	0.0217
41.106	0.0087	0.0133	0.0202
43.950	0.0081	0.0124	0.0188
46.794	0.0076	0.0116	0.0176
49.638	0.0071	0.0109	0.0166
52.482	0.0067	0.0102	0.0156
55.326	0.0063	0.0097	0.0148
58.170	0.0060	0.0092	0.0140

- The variation of PGA with respect to magnitude and hypocentral distance is given in Table 4.22 and is shown in Fig. 4.45.
- As was discussed previously, the calculated PGA for a given magnitude and hypocentral distance is not a constant value but it is a random variable which follows lognormal distribution with mean as the calculated value and standard deviation as mentioned above.

5. Computation of Conditional Probability of Exceedence

- Probability of exceedance for a given value of PGA can be computed using the truncated lognormal distribution. The 3σ limits of ln(PHA) can be given as follows for a mean value of -3.5696 ($m = 5.0$, $R = 29.73$ km, PGA = 0.0282 g):

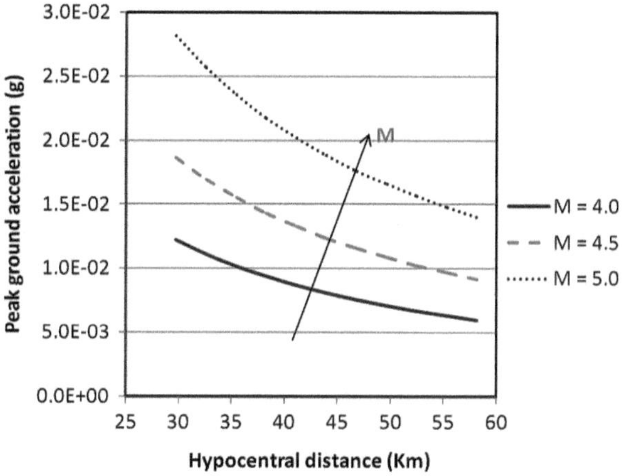

Fig. 4.45 Variation of PGA with distance and magnitude

$$\text{Lower Limit} = \mu_{\ln \text{PHA}} - 3\sigma_{\ln \text{PHA}} = -3.5696 - 3 \times 0.37 = -4.6796$$
$$\text{Upper Limit} = \mu_{\ln \text{PHA}} + 3\sigma_{\ln \text{PHA}} = -3.5696 + 3 \times 0.37 = -2.4596$$

- The lower and upper limits of PHA can be obtained as

$$\text{Lower Limit} = e^{-4.6796} = 0.0093\,\text{g}$$
$$\text{Upper Limit} = e^{-2.4596} = 0.0855\,\text{g}$$

- Hence PHA will vary in between 0.0093 and 0.0855 g. With these limits, the probability of exceedance can be calculated as follows:

$$P(Z > z|m_i, r_j) = 1 - \Phi^1\left[\frac{\ln z - \mu_{\ln z}}{\sigma_{\ln z}}\right] = 1 - \Phi^1[u]$$

$$\text{where} \quad \Phi^1(u) = \frac{\Phi(u) - \Phi(-3)}{1 - 2\Phi(-3)}$$

$$u = \frac{\ln z - \mu_{\ln z}}{\sigma_{\ln z}} = \frac{\ln 0.02 + 3.5696}{0.37} = -0.9255$$

$$\Phi(u) = \Phi(-0.9255) = P(Z \leq -0.9255) = 0.1774$$

$$\Phi(-3) = 0.00135$$

$$\Phi^1(u) = \frac{\Phi(u) - \Phi(-3)}{1 - 2\Phi(-3)} = \frac{0.1774 - 0.00135}{1 - 2 \times 0.00135} = 0.1765$$

$$P(Z > 0.02|5.0, 29.73) = 1 - \Phi^1[u] = 1 - 0.1765 = 0.8235$$

6. Estimation of hazard

- For fault 2, the frequency of exceedance can be can be written as

$$\nu_2(z) = \upsilon_2(4)\left[\sum_{M=4.0}^{M=5.0}\sum_{R=110}^{369} f_M(m)f_R(r)\,P(Z > z/m_i, r_j)\,\Delta R\Delta m\right]$$

- From the previous steps, the above parameters can be rewritten as

$$\begin{aligned} &\upsilon_{21}(4) = 0.01\ /yr \\ &\text{for } m = 5.0 \quad f_M(5.0)\Delta m = 0.0436 \\ &\text{for } r = 29.73 \quad f_R(29.73)\Delta r = 0.186 \\ &\text{for } m = 5.0 \quad r = 29.73\,\text{km} \quad z = 0.02g \\ &P(Z > 0.02g\ /5.0, 29.73) = 0.8235 \end{aligned}$$

- Now the frequency of exceedance of 0.02 g at the site can be calculated as

$$\therefore \upsilon_2(0.02g) = 0.01 \times 0.0436 \times 0.186 \times 0.8235 = 6.7 \times 10^{-5}/\text{year}$$

- From the above equation, the frequency of exceedance of 0.02 g at the site given that an earthquake with a magnitude of 5.0 occurring at a distance of 29.73 km is obtained as 6.7×10^{-5}/year.
- Similarly the above calculations can be repeated for different magnitudes occurring at different hypocentral distances from the site and frequency of exceedance can be calculated.
- Table 4.23 shows frequency of exceeding a particular level of PGA (0.02 g) at the site for different combinations of magnitudes and distances. This is also shown in Fig. 4.46 which represents the magnitude–distance bins for PGA of 0.02 g.
- Summing up all these values will give frequency of exceeding a PGA of 0.02 g at the site due to 2nd fault source (Fig. 4.47, Table 4.24). This is obtained as 2.57×10^{-3}/year.

Table 4.23 Annual rate of exceedance of 0.02 g at the site for different combinations of magnitudes and distances

Magnitude	Distance (km)					
	29.73	31.15	34.00	36.84	39.69	42.53
4	1.17E-04	9.15E-05	2.45E-05	1.17E-05	6.03E-06	3.13E-06
4.1	1.73E-04	1.39E-04	3.92E-05	1.99E-05	1.09E-05	6.07E-06
4.3	1.82E-04	1.54E-04	4.79E-05	2.68E-05	1.61E-05	9.98E-06
4.5	1.64E-04	1.44E-04	4.91E-05	2.99E-05	1.96E-05	1.32E-05
4.7	1.28E-04	1.17E-04	4.30E-05	2.83E-05	2.00E-05	1.45E-05
4.9	8.87E-05	8.38E-05	3.29E-05	2.31E-05	1.74E-05	1.35E-05
5	9.60E-05	9.19E-05	3.71E-05	2.68E-05	2.08E-05	1.65E-05

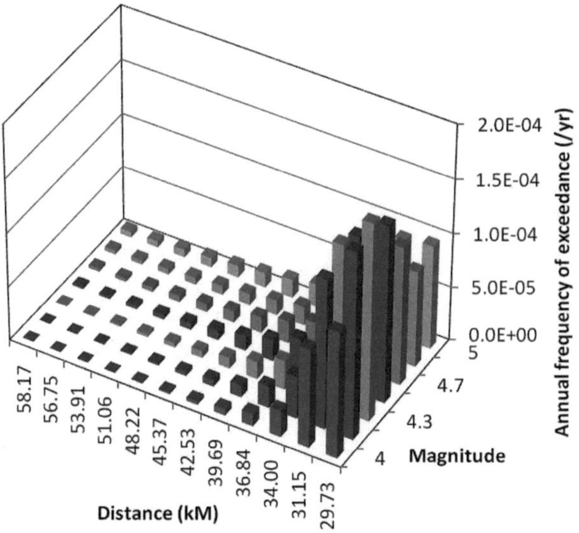

Fig. 4.46 Magnitude distance bins corresponding to annual frequency of exceedance of PGA of 0.02 g at the site due to 2nd fault source

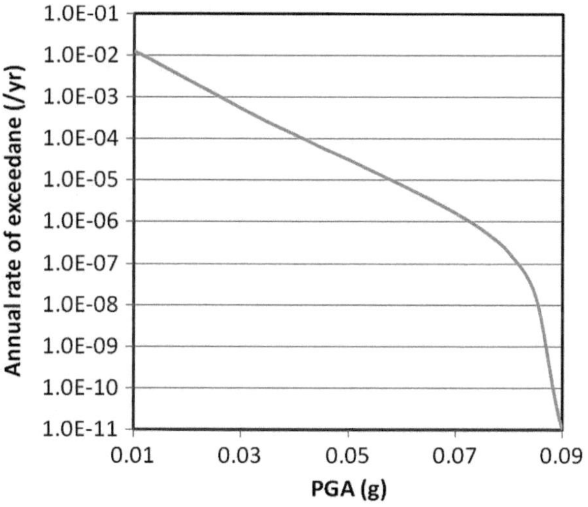

Fig. 4.47 Hazard curve of the site due to fault source 2

Total Hazard at the site

- This is the summation of hazard contribution from all the fault sources around the site. This is given in Table 4.25.
- Figure 4.48 shows the deaggregation of hazard with respect to each fault. Figure 4.49 shows the total hazard at the site.
- From Fig. 4.48, it is understood that the second fault will be contributing more compared to fault 1 toward final hazard at the site.
- Once the annual rate of exceedance is calculated, one can calculate probability of exceedance for a given period of time and return period of particular level of PGA as follows:

Table 4.24 Annual rate of exceedance with respect to PGA

PGA (g)	Annual rate of exceedance (/year)
0.005	1.9252E-02
0.010	1.2076E-02
0.015	5.7642E-03
0.020	2.5701E-03
0.025	1.1593E-03
0.030	5.3725E-04
0.035	2.5418E-04
0.040	1.2304E-04
0.045	6.0877E-05
0.050	3.0317E-05
0.055	1.4999E-05
0.060	7.3989E-06
0.065	3.4905E-06
0.070	1.5866E-06
0.075	6.1132E-07
0.080	1.6858E-07
0.085	1.5102E-08
0.090	0.0000E + 00

Table 4.25 Annual rate of exceedance with respect to PGA

PGA (g)	Annual rate of exceedance (/year)	Probability of exceedance			Return period (year)
		t = 1 year	t = 50 year	t = 100 year	
0.005	2.305E-02	2.278E-02	6.841E-01	9.002E-01	4.339E + 01
0.010	1.240E-02	1.233E-02	4.622E-01	7.107E-01	8.062E + 01
0.015	5.828E-03	5.811E-03	2.528E-01	4.416E-01	1.716E + 02
0.020	2.586E-03	2.583E-03	1.213E-01	2.279E-01	3.867E + 02
0.025	1.164E-03	1.163E-03	5.653E-02	1.099E-01	8.592E + 02
0.030	5.386E-04	5.384E-04	2.657E-02	5.243E-02	1.857E + 03
0.035	2.546E-04	2.545E-04	1.265E-02	2.513E-02	3.928E + 03
0.040	1.231E-04	1.231E-04	6.138E-03	1.224E-02	8.121E + 03
0.045	6.090E-05	6.089E-05	3.040E-03	6.071E-03	1.642E + 04
0.050	3.032E-05	3.032E-05	1.515E-03	3.027E-03	3.298E + 04
0.055	1.500E-05	1.500E-05	7.497E-04	1.499E-03	6.667E + 04
0.060	7.399E-06	7.399E-06	3.699E-04	7.396E-04	1.352E + 05
0.065	3.491E-06	3.490E-06	1.745E-04	3.490E-04	2.865E + 05
0.070	1.587E-06	1.587E-06	7.933E-05	1.586E-04	6.303E + 05
0.075	6.113E-07	6.113E-07	3.057E-05	6.113E-05	1.636E + 06
0.080	1.686E-07	1.686E-07	8.429E-06	1.686E-05	5.932E + 06
0.085	1.510E-08	1.510E-08	7.551E-07	1.510E-06	6.622E + 07

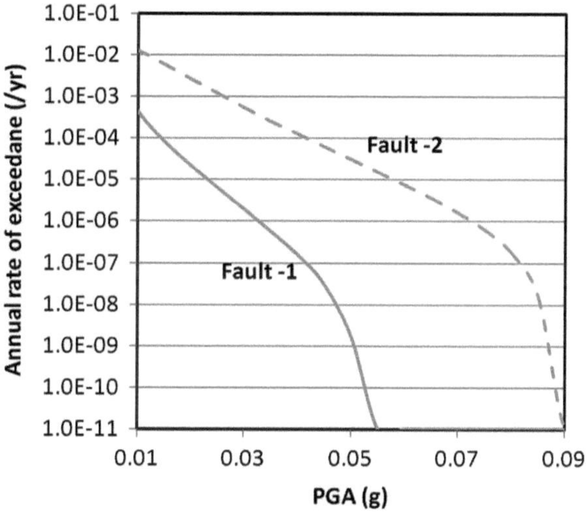

Fig. 4.48 Deaggregation of hazard curves with respect to faults

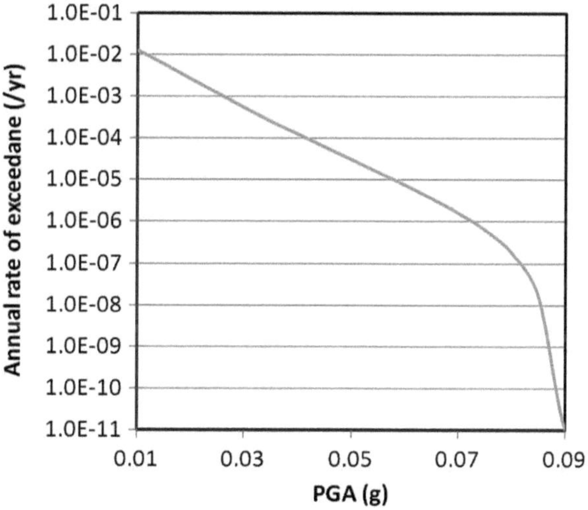

Fig. 4.49 Hazard curve at the site

$$P[Z > z] = 1 - \mathrm{e}^{-\nu(z)t}$$

$$\text{Return Period} = \frac{-t}{\ln[1 - P(Z > z)]},$$

- where "t" can be plant life time. In Table 4.25, probability of exceedance is given for 50 years as well as 100 years and also return period of each PGA level is provided (Fig. 4.50).

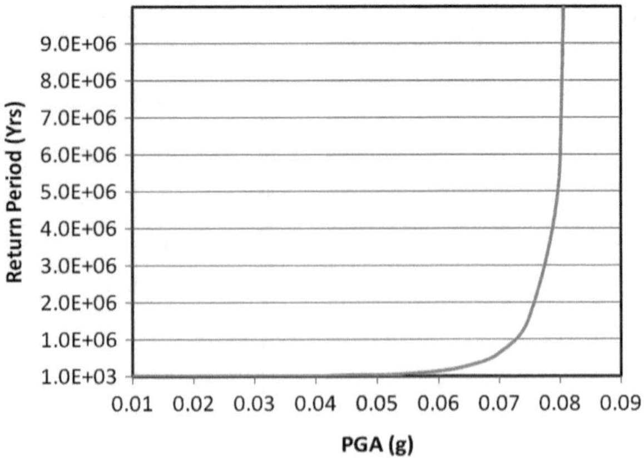

Fig. 4.50 Graph between PGA versus Return period

4.2.8 Logic Tree Methods

In the development of hazard curve for a site, there are several parameters whose uncertainty have been treated probabilistically such as earthquake occurrence process, earthquake magnitude, earthquake occurrence distance from the site, peak ground acceleration for a particular magnitude and distance. However, uncertainty can be existed in several other parameters such as Gutenberg-Richter's parameters ("*a*" and "*b*"), assigning of Maximum magnitude for each fault source, attenuation models, depth of focus. In the previous analysis, given the values of "*a*" and "*b*" for each fault, given the maximum magnitude that can occur on each fault and given the attenuation model the hazard curve has been developed. In order to consider the uncertainty in these parameters also one can utilize the concept of "logic trees".

In the logic tree approach, one can use different attenuation models and different values for parameters with some weighting factors assigned to each value depending on the applicability of the models that is interpreted as the relative likelihood of that model being correct. It consists of a series of nodes, representing points at which models are specified and branches that represent the different models specified at each node. The sum of the probabilities of all branches connected to a given node must be 1. The simple logic tree shown in Fig. 4.51 allows uncertainty in selection of models for attenuation, fault depth, and maximum magnitude to be considered.

In this logic tree, attenuation according to the models of McGuire and Boore et al. are considered equally likely to be correct, hence each is assigned a relative likelihood of 0.5. Proceeding to the next level of nodes, fault depth of 10 km is considered to be more likely to correct (0.6) than the 20 km (0.4). At the final level of nodes, different relative likelihoods are assigned to the maximum magnitude. This logic tree terminates with a total of $2 \times 2 \times 3 = 12$ (no. of attenuation models × no. of depths × no. of maximum magnitudes) branches. The relative likelihood of the combination of models and parameters implied by each terminal branch is given

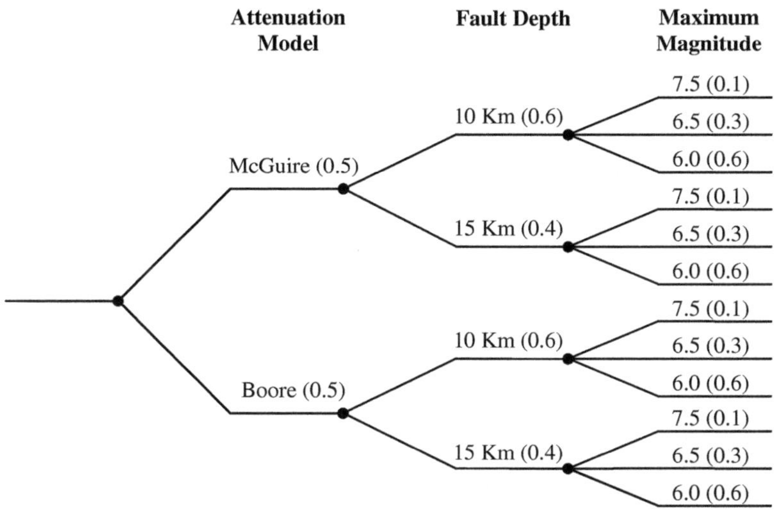

Fig. 4.51 Simple Logic Tree for incorporation of model uncertainty

by the product of the relative likelihood of the terminal branch and all prior branches leading to it. Hence, the relative likelihood of the combination of the McGuire's attenuation model, fault depth of 10 km and maximum magnitude of 7.5 is $0.1 \times 0.6 \times 0.3 = 0.018$. The sum of the relative likelihoods of the terminal branches, or of those at any prior level, is equal to 1.

To use the logic tree, a seismic hazard analysis is carried out for the combination of models and parameters associated with each terminal branch. The result of each analysis is weighted by the relative likelihood of its combination of branches, with the final result taken as the sum of the weighted individual results. It is easy to see that the required computational effort increases quickly with increasing number of nodes and branches. Nevertheless, the logic tree is very useful tool for the analysis of seismic hazards.

Example 4.8 Consider the data from Example 4.7. Develop the hazard curve at the site using logic tree method. Consider uncertainty only in the attenuation models with the following weighting factors, Campbell (0.4) and Boore (0.6).

Solution:
In this example, there are two attenuation models that can be applicable for the site with different weighting factors such as

Campbell-0.4
Boore-0.6

One can develop logic tree as shown in Fig. 4.52.

From Fig. 4.52, it is clear that there are only two nodes and our calculation should be repeated for two attenuation models one using Campbell model and other using Boore model (Fig. 4.53). In the previous Example 4.7, the hazard curve has

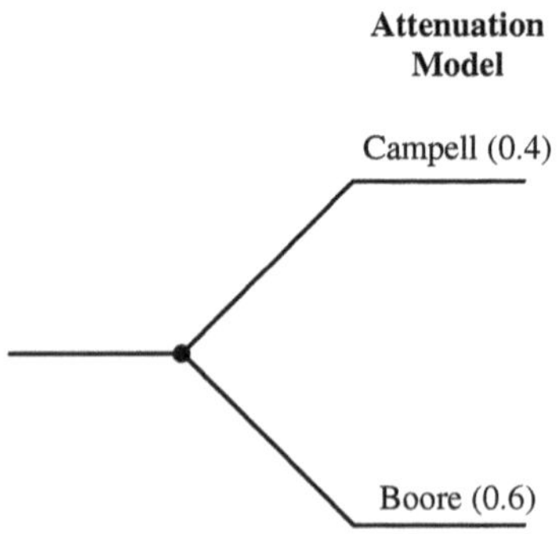

Fig. 4.52 Logic Tree for uncertainty in attenuation models

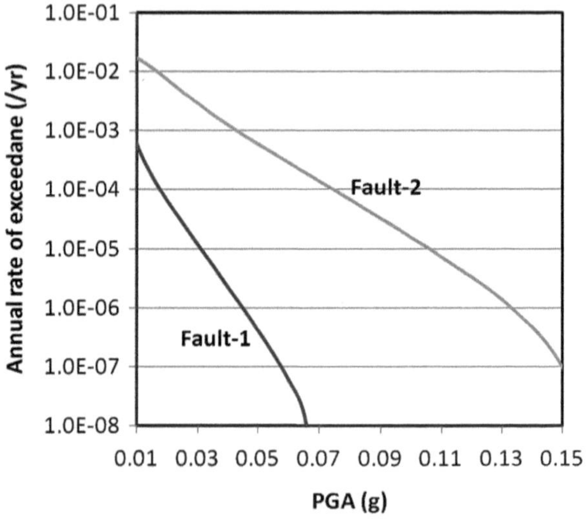

Fig. 4.53 Deaggregation of hazard at the site due to different fault by using Boore's attenuation model

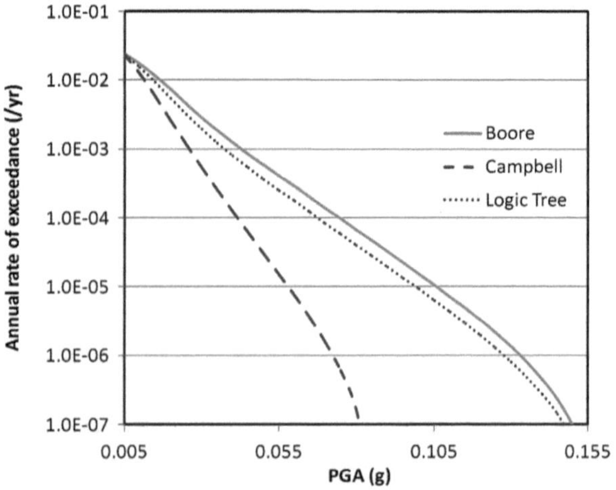

Fig. 4.54 Hazard at the site due to different attenuation models

been developed using Campbell attenuation model. Similarly by repeating the same procedure, one can develop the hazard curve using Boore attenuation model. Hazard curves due to different attenuation models and applying logic tree method are shown in Fig. 4.54.

4.3 Seismic Fragility Evaluation

The seismic fragility of a structure or equipment or component is defined as the conditional probability of its failure for a given level of seismic input parameter, typically the peak ground acceleration (PGA). In the fragility evaluation, the conditional probability of component failure is determined by considering the capacities of the components in various failure modes. Seismic-induced fragility data is generally unavailable for components and structures. Thus, fragility curves must be developed primarily from analysis combined heavily with engineering judgment supported by very limited test data. Such fragility curves will contain a great deal of uncertainty. The uncertainty can be of aleatory or epistemic in nature. The aleatory uncertainty can be represented with the help of probability distributions and the uncertainty in the parameters of the distributions is of epistemic in nature. The fragility curve for any component can be defined with the help of its median ground acceleration capacity, A_m, and the corresponding uncertainties β_R(aleatory) and β_U(epistemic). Hence, the probability of failure (P_f) at a non exceedance probability (Q) can be expressed as [21, 4, 22]:

$$p_f = \Phi\left[\frac{\ln\left(\frac{a}{A_m}\right) + \beta_u \Phi^{-1}(Q)}{\beta_R}\right] \tag{4.38}$$

If both the uncertainties are combined together then the probability of failure can be given as follows:

$$p_{f(a)} = \Phi\left[\frac{\ln\left(\frac{a}{A_m}\right)}{\beta_C}\right] \tag{4.39}$$

where

$$\beta_C = \sqrt{\beta_R^2 + \beta_U^2}$$

in which $\Phi(.)$ is the standard Gaussian cumulative function.

The seismic capacity in terms of peak ground acceleration of a given structure depends on a number of factors related to the basic ground motion, soil structure interaction, structure modeling, method of evaluation of dynamic response, and material properties. The seismic capacity of a given structure or component is given as

$$A_m = F\,A_{\text{design}} \tag{4.40}$$

A_{design} is the design basis PGA, usually it is the PGA value associated with Safe Shutdown Earthquake (SSE) and F is the factor of safety. In case of structures the factor of safety, F, can be modeled as a product of three random variables [4, 21]:

$$F = F_S \cdot F_\mu \cdot F_{RS} \tag{4.41}$$

where,

F_S: Strength factor.
F_μ: Inelastic-absorption factor.
F_{RS}: Structure response conservatism factor.

For equipment and other components, the factor of safety is made up of three parts consisting of a capacity factor F_C, a structure response factor F_{RS}, and an equipment response (relative to the structure) factor F_{RE}.

$$F = F_C \cdot F_{RE} \cdot F_{RS} \tag{4.42}$$

The details about these factors can be found in following references [21–24].

Example 4.9 Develop the fragility curve for a component whose median ground acceleration capacity and combined uncertainty are given as $A_m = 0.2$ g, $\beta_C = 0.55$.

Solution:
The fragility curve can be developed by using the following formula.

$$p_{f(a)} = \Phi\left[\frac{\ln\left(\frac{a}{A_m}\right)}{\beta_C}\right]$$

$$\text{where} \quad A_m = 0.2g$$
$$\beta_C = 0.55$$

- Assume that PGA is 0.1 g. Then probability of failure for this level of ground motion can be calculated as

$$P_f(0.1) = \Phi\left[\frac{\ln\left(\frac{0.1}{0.2}\right)}{0.55}\right] = \Phi[-1.26] = 0.1038$$

- Hence, the probability of failure is obtained as 0.1038. These calculations can be repeated for different PGA values and a graph between PGA and probability of

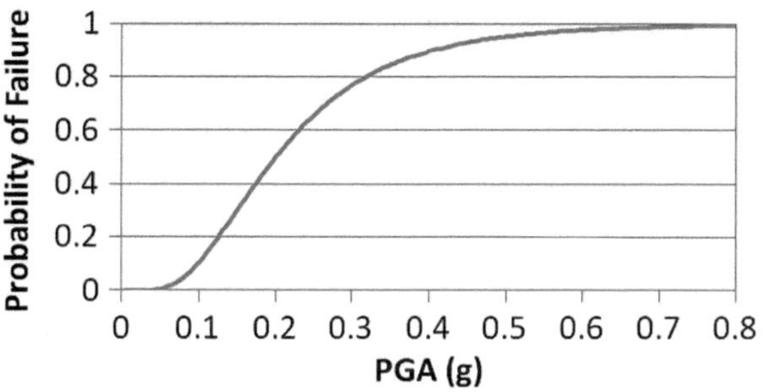

Fig. 4.55 Fragility Curve of a component

failure can be drawn which gives fragility curve of a given component (Fig. 4.55). From the curve, it can be seen that probability of failure of the component at 0.2 g will be 50 % because it is the median value.

4.4 Accident Sequence Analysis

Seismic events are treated as initiating events that can cause adverse impacts on support systems, frontline systems and structural integrity. The method of dealing with these various challenges is to use a single event tree for many of the consequential seismic events. The initiating event for the seismic event trees is the occurrence of seismic event and whenever seismic event occurs it will in turn initiate other internal events like loss of offsite power, LOCA, loss of process water systems, etc. Hence, event trees should be generated for all the initiating events and dominating accident sequences should be identified for the Core Damage Frequency (CDF) estimation.

4.4.1 Seismic Event Trees

As discussed above in seismic event trees, seismic event is the initiating event and the other internal initiating events are due to seismic event. As a case study, event tree for seismically induced Class-IV power supply failure is developed and is shown in Fig. 4.56 [1].

Upon failure of Class-IV power, reactor trips on “No PCP running”. This leads to the actuation of Reactor Protection System (RPS), initially with Shutdown System (SDS-1) and with SDS-2, if SDS-1 fails. Class-III power supply or Emergency Power Supply (EPS) is a system with four diesel generator (DG) sets. If Class-III is available, the mode of decay heat removal and long-term reactivity control will be with secondary steam relief system (SSR), auxiliary boiler feed water system (ABFWS) and shutdown cooling system (SDC). If there is a failure in

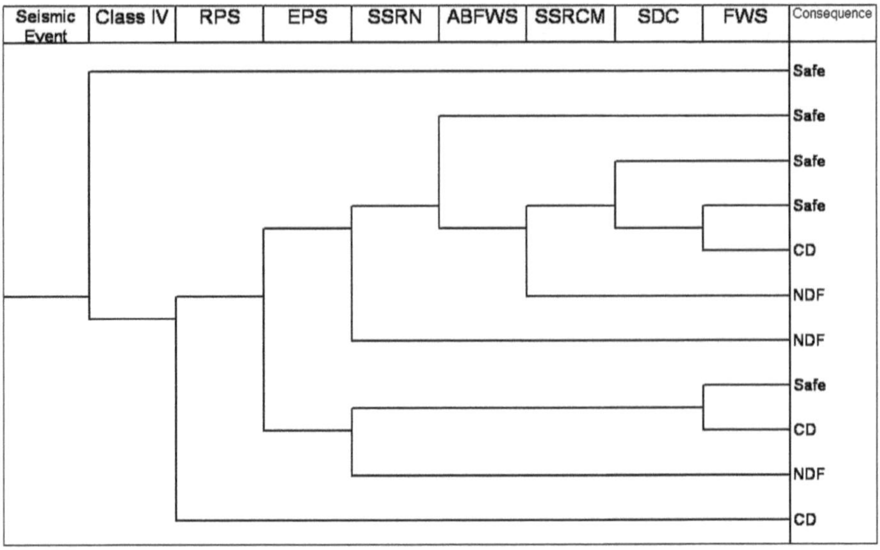

Fig. 4.56 Seismic Event tree of Class-IV failure

decay heat removal systems, core cooling will be achieved through valving in of fire water system (FWS). Class-IV failure followed with complete loss of Class-III failure leads to a station blackout scenario. During station blackout scenario, core cooling will be achieved through valving in of fire water system. The dominating accident sequences, in terms of consequences, are given below.

Seismic-Class IV-RPS
Seismic-Class IV-EPS-FWS
Seismic-Class IV-ABFWS-SDC-FWS

4.4.2 Seismic Fault Trees

In evaluating the accident sequence frequency from the seismic event trees, one should have the information about the initiating event frequency and the seismically induced failure probabilities of process systems and safety systems. Initiating event frequency (frequency of occurrence of seismic events) can be derived from hazard curve analysis as explained in the previous sections and the seismically induced failure probabilities of systems can be evaluated by developing seismic fault trees. Unlike the traditional fault trees, these fault trees will consist of component failures mainly from structures point of view. The fault trees are developed based on the assumption that components of a similar design, located at the same elevation and with the same orientation will fail in a given seismic event if one of these groups fails and are considered as a single component. One should also consider the failure probability contribution from the random failures of the components in the analysis.

Once the seismic fault trees are developed, next step is to develop component fragilities depending on their seismic capacities [21] as explained in the previous section. In finding out the seismic capacities of the components, one has to perform seismic response analysis. The system fragility curve can be generated from the component fragilities depending on the system configuration and its failure criteria. This can be well represented with the seismic fault trees. Initially component fragility curves can be derived based on their respective seismic capacities (A_m) and their uncertainty parameter (β_C). From the component fragility, one can obtain the conditional probability of failure of the component for a given level of earthquake (PGA). By considering one PGA level at a time and propagating the failure of the components according to seismic fault trees to the system level (top event) failure and by applying boolean algebraic concepts, one can obtain the conditional probability of failure of the system for that level of earthquake. Repeating the same procedure, one can obtain the conditional failure probability of the system for different levels of earthquake. When a graph between PGA and probability of failure is plotted it will give fragility curve for that particular system. Seismic fault trees of Class-IV power supply system, RPS (SDS-1 and SDS-2), Class-III power supply system, and fire water system (FWS) of a typical NPP are explained briefly in the following subsections.

Class-IV power supply system

Class-IV power supply is derived from 220 kV grid through start-up transformer and from the turbo generator 400 kV system through generator transformer and unit transformer. The most fragile components for the offsite grid and delivery system are ceramic insulators in the switchyard. The list of component is given in Table 4.26. The fragility of these components significantly dominates the failure of the system. The seismic fault tree of this system is shown in Fig. 4.57. The fragility curve of Class-IV system is shown in Fig. 4.58.

Reactor Protection System

There are two fast acting independent shutdown systems provided as part of the reactor protective system (RPS), viz., primary and secondary shutdown systems. The fault tree of the RPS is shown in Fig. 4.59. Primary shutdown system (SDS-1) is the primary means for reducing the reactor power from full power to subcritical state during operating condition as well as in accident conditions. The PSS consists of a set of mechanical rods and a driving mechanism at the top of the reactor.

Table 4.26 List of components of Class-IV power supply

S.no	Name	Description	A_m	β_C
1	Transformer	Failure of transformer	3.0 g	0.55
2	Ceramic insulators	Failure of ceramic insulators	0.2 g	0.55

Fig. 4.57 Seismic fault tree of Class-IV power supply

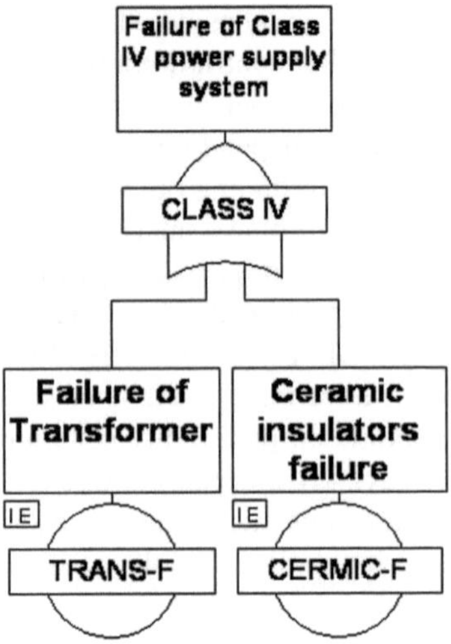

Fig. 4.58 Fragility Curve of Class-IV Power Supply System

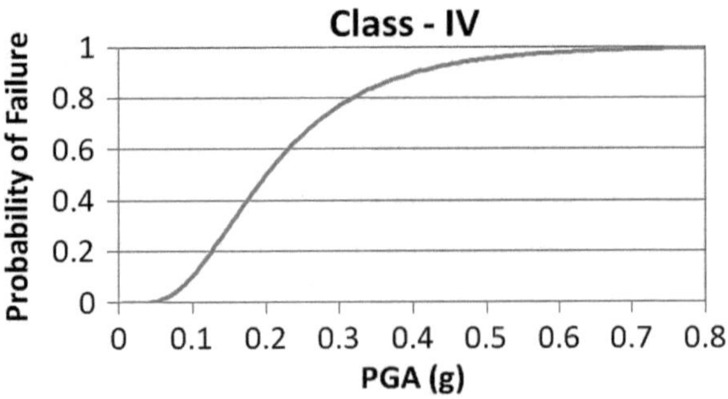

Fig. 4.59 Seismic fault tree of RPS

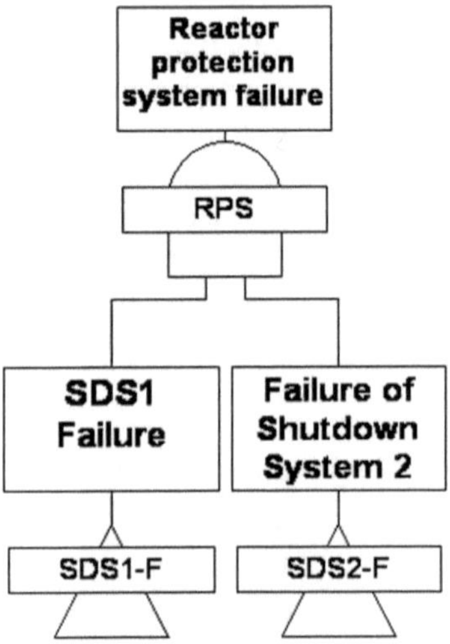

Table 4.27 List of components considered in fragility evaluation of SDS-1

S.no	Name	Description	A_m	β_C
1	Control rod drive mechanism	Failure of control rod drive mechanism	0.657 g	0.46
2	Shut off rod	Failure of shut off rods	0.591 g	0.36
3	Clutch	Failure of clutch	0.657 g	0.46

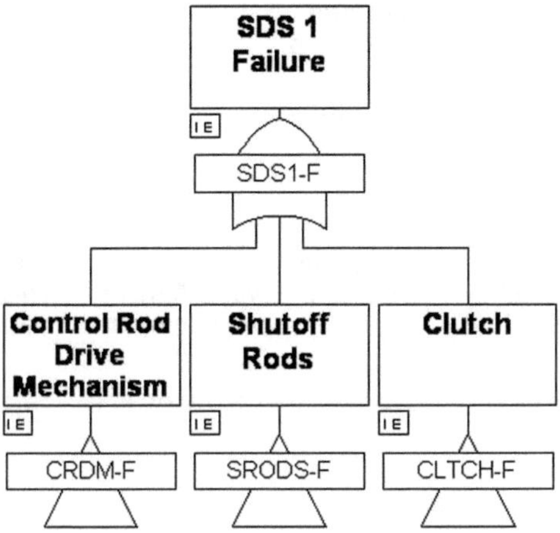

Fig. 4.60 Seismic fault tree of SDS-1

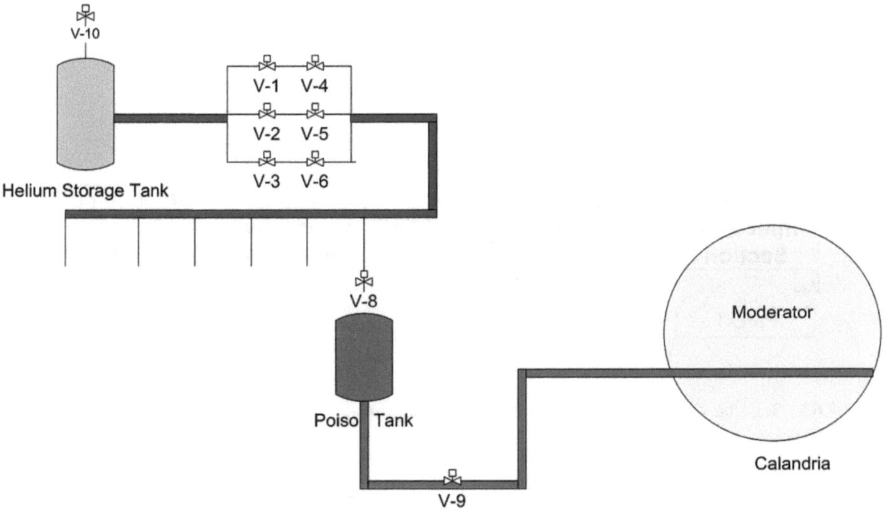

Fig. 4.61 Schematic of shutdown system -2

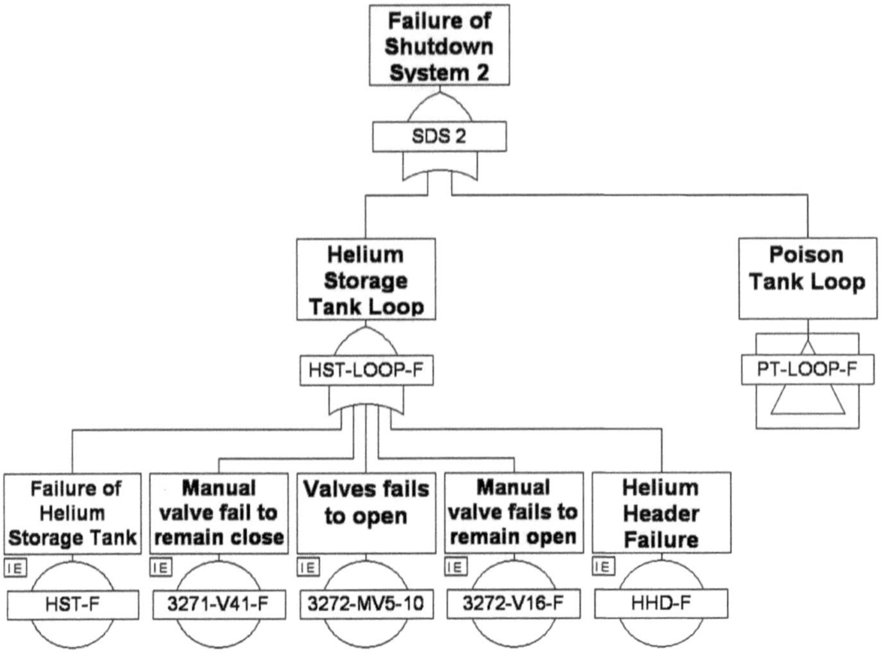

Fig. 4.62 Seismic fault tree of SDS-2

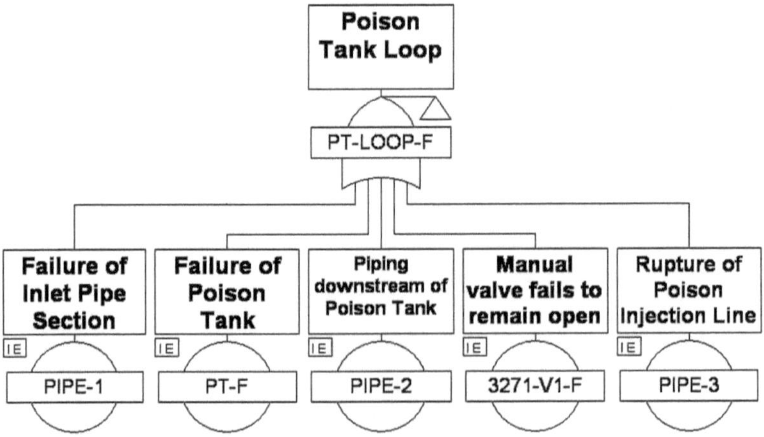

Fig. 4.63 Seismic fault tree of poison tank loop of SDS-2

The list of components considered in the fragility evaluation of the SDS1 is given in Table 4.27. The seismic fault tree of SDS-1 is shown in Fig. 4.60.

The secondary shutdown system (SDS-2) which serves as a stand-by provides fast injection of liquid poison in a set of vertical tubes located inside the core. The

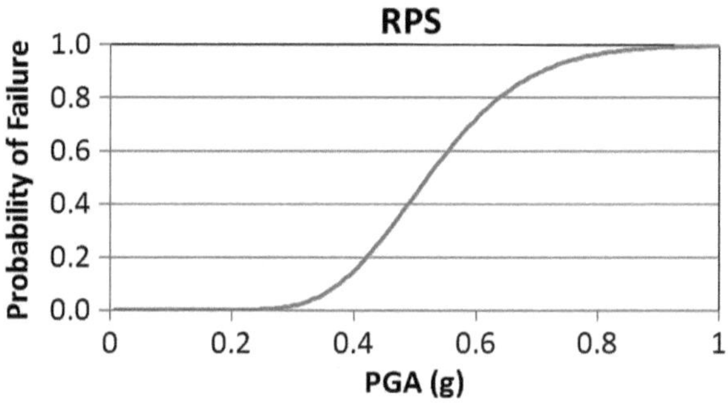

Fig. 4.64 Fragility of Reactor protection system (RPS)

schematic diagram of SDS-2 is shown in Fig. 4.61. The seismic fault trees of SDS-2 are shown in Figs. 4.62 and 4.63. The fragility of RPS is shown in Fig. 4.64.

Class-III power supply system

The system consists of four nos. of diesel generator (DG) sets each of 50 % capacity. This system derives its power from Class-IV system under normal operating conditions. When normal supply fails, DG sets will be able to meet the station emergency loads. Any two DG sets will meet emergency load of one unit. Each DG set is provided with compressed air starting system. Fuel oil is supplied from a day tank with sufficient capacity for 8 h of full load operation of the DG set. Automatic pressure lubrication is provided with engine-driven lube oil pump. Engine jacket cooling, lube oil cooling, and turbo charger cooling are by DM

Table 4.28 List of component of Class-III power supply system

S.no	Name	Description	A_m (g)	β_C
1	DG	Diesel generator	1.84	0.46
2	DT	Day tank	1.9	0.46
3	FOP	Fuel oil pipe	3.8	0.61
4	AC	Air compressor	1.84	0.46

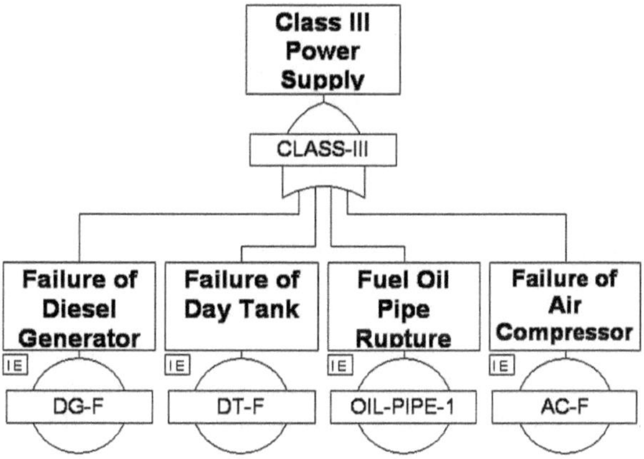

Fig. 4.65 Seismic fault tree of Class-III power supply

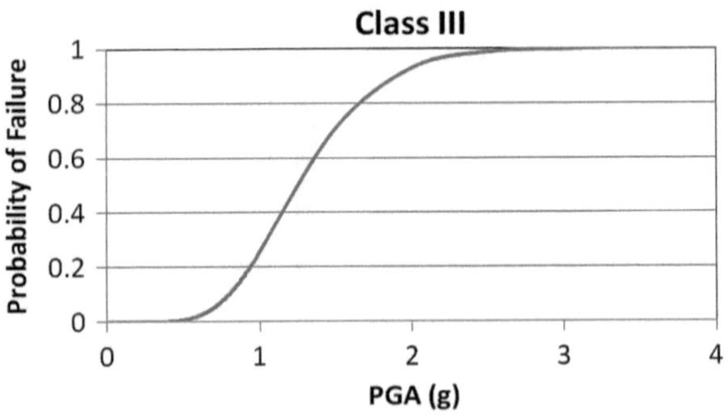

Fig. 4.66 Fragility of Class-III power supply system

water in a closed-loop system through jacket water heat exchanger. Secondary side of jacket water heat exchanger is provided with cooling water from nonactive process water. The list of components considered in the fragility evaluation of Class-III system is given in Table 4.28. The seismic fault tree of the same is shown in Fig. 4.65. The corresponding fragility curve is shown in Fig. 4.66.

Fire water system

The fire water system is very important for a Nuclear Power Plant from viewpoint of safety. The main objective of fire water system is to ensure availability of water for firefighting under normal operating conditions and anticipated operational occurrences. Depending up on the type of equipment to be protected, either hydrant system or automatic or nonautomatic high-velocity sprinkler system has been adopted. The system has been designed to be available under safe shutdown earthquake condition for firefighting of related equipment or system.

It has been proved by safety analysis that fire water system has a major role in case of emergency, hence, it is considered as one of the important safety system in nuclear power plants. Firewater is supplied to steam generators in case there is a failure in feed water system, i.e., feed water pump failure or in case of station black out condition. By the operation of SSR steam generator, pressure falls down and fire water is injected to steam generators, which prevents excessive hike in core temperature. In addition to the above important safety functions, fire water system protects all the system and buildings from fire hazard.

There is one electric motor-driven pump and three diesel engines driven pumps provided. A jockey pump is also provided in the system driven by electric motor. Each diesel engine drive has its own local control panel, which is equipped with automatic starting arrangement by means of pressure switches sensing low pressure on the pump header. The setting of the pressure switches is staggered so that motor-driven pump will start first and then the diesel-driven pumps depending upon the fire water requirement. All the pumps can be started manually also if required. The pumps can be switched off only manually. A hydropneumatic type pressure tank is provided in the system in order to keep the fire water system always pressurized. The seismic fault tree of the fire water system is shown in Figs. 4.67 and 4.68. The fragility of fire water system is shown in Fig. 4.69.

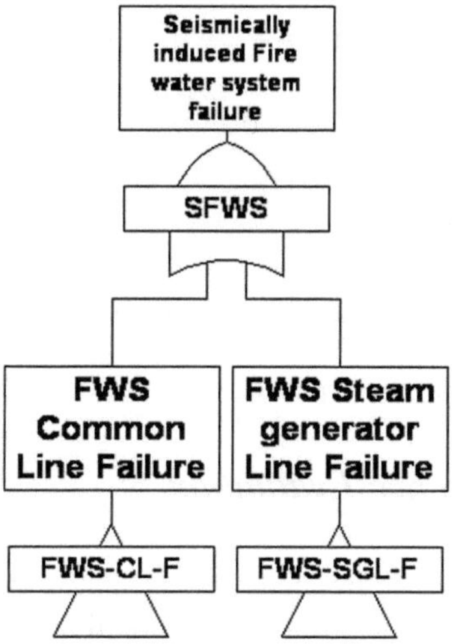

Fig. 4.67 Seismic fault tree of FWS

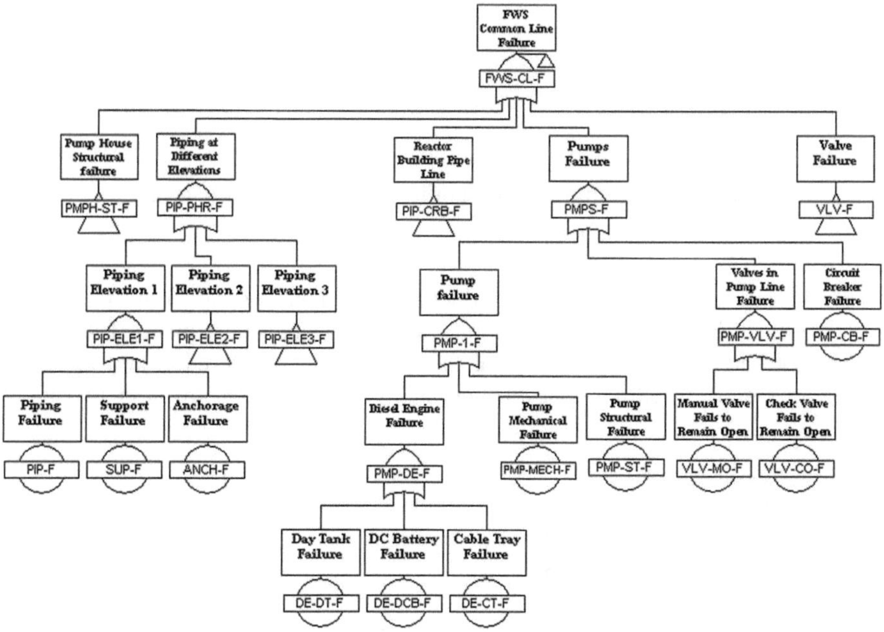

Fig. 4.68 Fault Tree for Common System

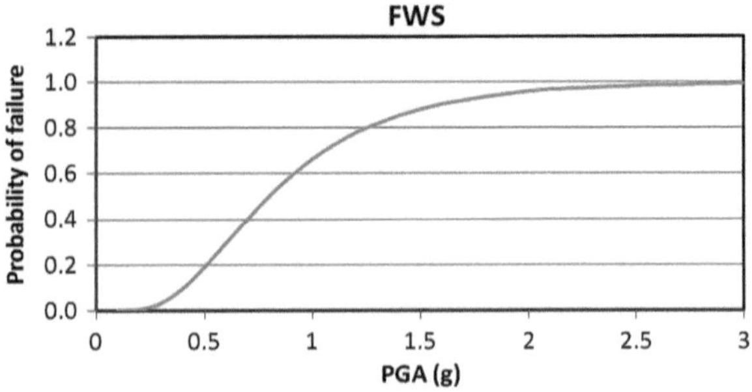

Fig. 4.69 Fragility of Fire water system

4.4.3 Accident Sequence Evaluation

In this step, accident sequence frequency is calculated by convoluting both seismic hazard curves and the fragilities of the corresponding systems as given in Eq. 4.43 and is also illustrated in Fig. 4.70 [1].

$$P_F = \int_0^\infty \left(\frac{\mathrm{d}H}{\mathrm{d}a}\right) p_f \mathrm{d}a$$

where

H = Hazard curve

a = PGA level

p_f = Conditional failure probability at a given PGA level

P_F = Total failure frequency (4.43)

The list of dominating accident sequences and their corresponding frequencies are given in Table 4.29. Each dominating accident sequence frequency is evaluated by convoluting the seismic hazard of the site under consideration and the corresponding system fragilities that are present in that sequence. In order to validate

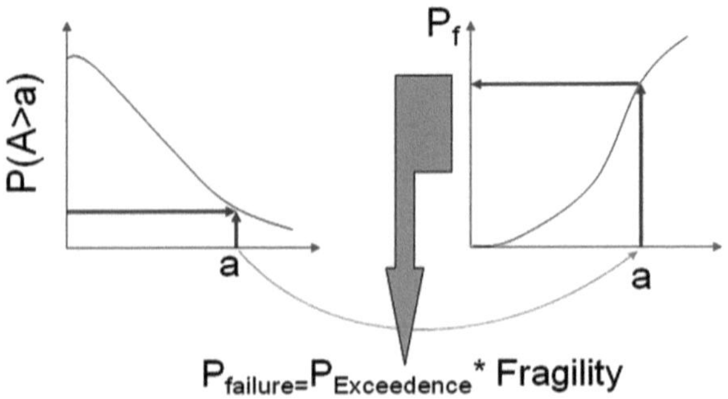

Fig. 4.70 Convolution of seismic hazard and fragility curves

Table 4.29 Accident sequences and their frequencies

S.no	Accident sequence	Frequency (/year)
1	Seismic-Class IV-RPS	9.31×10^{-8}
2	Seismic-Class IV-EPS-FWS	5.24×10^{-7}
3	Seismic-Class IV-ABFWS-SDC-FWS	5.06×10^{-12}

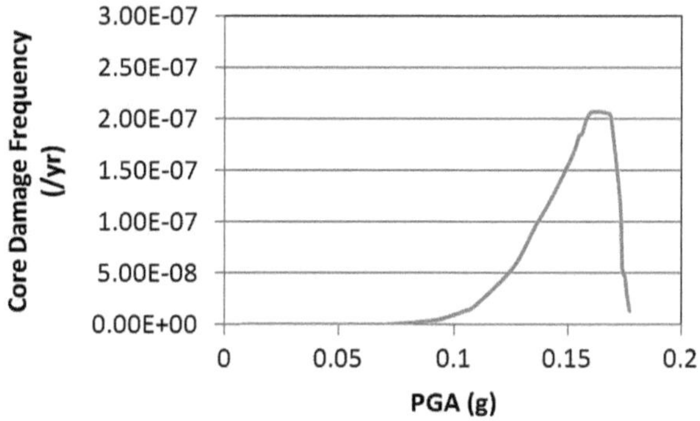

Fig. 4.71 Graph between PGA versus CDF

whether the PGA's of all the ranges have been considered in the analysis or not, a graph between PGA and CDF has been plotted as shown in Fig. 4.71. The CDF from the seismically induced Class-IV power supply failure is estimated as 6.40×10^{-7}/year.

References

1. IAEA-TECDOC-724 (1993) Probabilistic safety assessment for seismic events. International Atomic Energy Agency, Vienna
2. Cornell CA (1968) Engineering seismic risk analysis. Bull Seismol Soc Am 58(5):1583–1606
3. Kennedy RP, Cornell CA, Campbell RD, Kaplan S, Perla HF (1980) Probabilistic seismic safety study of an existing nuclear power plant. Nucl Eng Des 59:315–338
4. Kennedy RP, Campbell RD, Kassawara RP (1988) A seismic margin assessment procedure. Nucl Eng Des 107:61–75
5. AERB Safety Guide SG/S-11 (1990) Seismic studies and design basis ground motion for nuclear power plant site. AERB, Mumbai, India
6. Bommer JJ, Scott SG, Sarma SK (2000) Hazard-consistent earthquake scenarios. Soil Dyn Earthquake Eng 19:219–231
7. EPRI NP-4726 (1991) Seismic hazard methodology for the central and eastern United States. EPRI
8. Reiter L (1991) Earthquake hazard analysis: issues and insights. Columbia University Press, New York
9. IAEA-SSG-9 (2010) Seismic hazards in site evaluation for nuclear installations. International Atomic Energy Agency, Vienna

10. Iyenger RN, Raghukanth STG (2003) Unform hazard spectra for Tarapur and Mumbai. BRNS project: MCV/RNI/DAE/102, Interim progress report
11. Kramer SL (2003) Geotechnical earthquake engineering. University of Washington, Prentice-Hall International Series in Civil Engineering and Engineering Mechanics
12. McGuire RK (1995) Probabilistic seismic hazard analysis and design earthquakes: closing the loop. Bull Seismol Soc Am 85(5):1275–1284
13. NUREG/CR-6372 (1997) Recommendations for PSHA. USNRC
14. Kijko A, Dessokey MM (1987) Application of the extreme magnitude distributions to incomplete earthquake files. Bull Seismol Soc Am 77(4):1429–1436
15. Kijko A, Sellevoll MA (1989) Estimation of earthquake hazard parameters from incomplete data files Part I. Utilisation of extreme and complete catalogs with different threshold magnitudes. Bull Seismol Soc Am 79(3):645–654
16. Kijko A, Sellevoll MA (1992) Estimation of earthquake hazard parameters from incomplete data files Part II. Incorporation of magnitude heterogeneity. Bull Seismol Soc Am 82(1): 120–134
17. Shankar D, Sharma ML (1997) Statistical analysis of completeness of seismicity of the Himalayas and its effect on earthquake hazard determination. Bull Ind Soc Earth Technol 34 (3):159–170
18. Chandra U (1977) Earthquakes of Peninsular India—A seismotectonic study. Bull Seismol Soc Am 67(5):1387–1413
19. Bazzurro P, Allin Cornell C (1999) Disaggregation of seismic hazard. Bull Seismol Soc Am 89(2):501–520
20. Hwang H, Huo J-R (1997) Attenuation relations of ground motion for rock and soil sites in eastern United States. Soil Dyn Earthquake Eng 16:363–372
21. Kennedy RP, Ravindra MK (1984) Seismic fragilities for nuclear power plant risk studies. Nucl Eng Des 79:47–68
22. NUREG/CR-4659-V4 (1991) Seismic fragility of nuclear power plant components. USNRC
23. NUREG/CR-3558 (1985) Handbook of nuclear power plant seismic fragilities. USNRC
24. Harris SP (1989) Seismic Piping fragilities development based on data from the 1989 LOMA PRIETA earthquake. PVP, vol 220. Seismic engineering, ASME 1991, pp 297–303

Chapter 5
Reliability Analysis of Passive Systems

5.1 Introduction

In general, a power plant (nuclear, thermal, chemical, etc.) consists of operating and emergency safety systems. These systems vary from simple to very complex systems. A system normally consists of active components and passive components. The failure of any operating system will lead to a change in the state of the plant. The availability of the plant depends on the successful operation of the components of the operating systems. In order to ensure the availability of the plant, reliability of the systems/components should be ensured. In recent years most of the advanced nuclear reactors implemented passive systems, aiming at improved safety and availability, in order to reduce human error, in particular, and active component malfunctions. The passive systems play an important role in various innovative reactor safety systems and can result in substantial economic benefits and simplicity in operation over the active systems.

5.2 Active and Passive Systems

Active systems are those systems which need external power to operate and it involves human intervention. The unavailability of the system can be due to the failure of the subcomponents in the system or due to human error or due to failure of the external power. Even though the active systems are more reliable, in the absence of the external input power, they will not be available for the operation. Hence, in order to avoid these problems innovative reactors are employing passive systems along with the active systems.

A.K. Verma et al., *Risk Management of Non-Renewable Energy Systems*,
Springer Series in Reliability Engineering, DOI 10.1007/978-3-319-16062-7_5

A passive system can be defined as follows: As per IAEA definition [9, 11], a passive component is a component, which does not need any external input to operate, and a passive system is either a system, which is composed entirely of passive components, or a system, which uses active ones in a very limited way to initiate subsequent passive operation. Passive systems mainly depend on the natural laws (e.g., gravity, natural circulation) to accomplish their designated safety function. The deviations in the behavior of the system from the expected conditions can impair the performance of the system. Following natural thermophysical phenomena are adopted in the design of advanced nuclear reactors to fulfill the safety functions namely,

- Natural circulation due to thermosyphon phenomenon
- Stored internal energy (gravity or pressurized water storage, battery, spring force)
- Siphon phenomenon (establishment of vacuum)
- Natural draft (pressure gradient establishment)

Functionality of these passive phenomena depends on the usage of these systems in serving the safety functions. For example, passive heat removal by natural circulation may have to fulfill the function over an integral mission time.

5.3 Need for Passive Systems

Safety is the main concern in nuclear power plants. In evaluating the safety of a plant it is important to consider both the events internal and external to the plant. The main purpose of considering the external events is that one can identify the different systems which are needed to function during an extreme external event and these systems should be designed to withstand during that event. External events can disrupt external sources of power supply, service water required for cooling operations, other essential supplies and also prompt operator actions which are required to perform following the event. In such a situation, active components will not be functioning due to the failure of external power. Hence, there is always a need to develop systems which can function during the above described scenario. This can be achieved with the help of passive systems.

The passive systems can be provided within the inner containment or the protected reactor building which do not need any external power supply or external cooling water for the operation. Some innovative reactors are provided with large heat sinks within the containment to cool the reactor core without depending on externally powered pumps. In such situations the reactor can be maintained in a safe state without any operator intervention. Also the operator gets prolonged grace period during which emergency actions can be implemented [9].

Table 5.1 IAEA categorization of passive systems

Category	A	B	C	D
Input signal	No	No	No	Yes
Moving mechanical parts	No	No	Yes	Yes
Moving working fluid	No	Yes	N/A	N/A

5.4 Categorization of Passive Systems

Nowadays, passive systems are being implemented in advanced nuclear reactors in order to eliminate active components malfunctions and to reduce operating errors. The implementation of passive systems may result into economic benefits as well as simplicity in operation when compared to active components. According to IAEA-TECDOC-626 [11], passive systems are classified into four categories (A, B, C, D) as tabulated in Table 5.1.

Passive systems mainly depend on the natural laws (e.g., gravity, natural circulation) to accomplish their designated safety function. A passive system should be theoretically more reliable than the active one. Nevertheless the passive system may fail to fulfill its mission not only because of a consequence of classical mechanical failure of component (passive or active) of the passive system [19], but also due to the deviation of certain parameters, which affect the passive system performance, from its normal operating range. Hence, it is very important to ensure the reliability of the passive systems.

5.5 Various Passive Systems Used in Advanced Reactors

There are different passive safety features available in advanced reactors [10] for carrying out various safety functions such as decay heat removal from the core during shutdown condition or containment depressurization during high pressure scenarios, etc. These are explained briefly in the following subsections.

5.5.1 High Pressure Injection by Using Accumulators

These types of systems are used to remove the decay heat during the loss of coolant accident (LOCA) kind of situation when the reactor is shutdown. In the initial stage of the event the high pressure primary circulation fluid comes out from the break location and it depressurizes the primary system rapidly. Immediately after the event due to the high containment pressure and other related signals reactor automatically gets shutdown and the reactor power comes down to as low as 6 % of full power. However, decay heat always existed in the core and it needs to be removed. Since the primary fluid is continuously coming out from the pipe break location

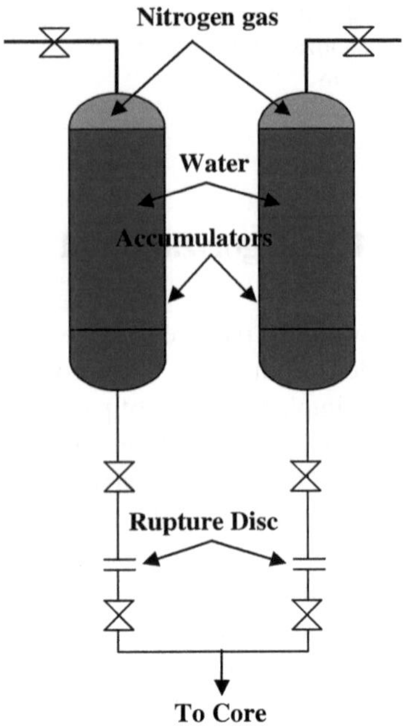

Fig. 5.1 High pressure accumulator injection system

there is loss of coolant in the core which is not sufficient to remove the decay heat from the fuel. In such cases one can utilize the passive safety features such as injection of coolant into the core by passive means (accumulators).

In this system accumulators will be prepressurized with the help of nitrogen or inert gases and this system is connected through check valves and rupture discs and directly or indirectly injects coolant into the core. This will be activated whenever there is a differential pressure observed across the rupture disc in the upstream and downstream side. If the differential pressure reaches some set point the rupture disc ruptures and coolant in the accumulators enters the core due to the high pressure in the accumulators. The schematic of the system is shown in Fig. 5.1.

5.5.2 Low Pressure Gravity Driven Water Tank

This type of system can be made available for LOCA kind of situations where the primary pressure comes down to a very low level. In this system a high-capacity water tank will be located at higher elevation so that during emergency situations the water can be injected into the core by means of gravity. This system is connected with check valves and rupture disc to a common header from where the water can be directly or indirectly injected into the core. In this case also when there is a differential pressure observed across the rupture disc in upstream and in downstream it gets ruptured and the water enters the core due to gravity. However,

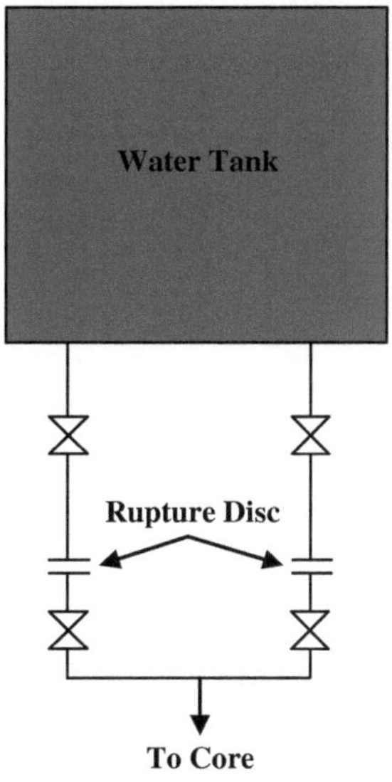

Fig. 5.2 High pressure accumulator injection system

in this scenario the rupture disc ruptures when the pressure in the primary loop comes down to a very low level unlike the case of high pressure injection. The schematic of the system is shown in Fig. 5.2.

5.5.3 Passive Isolation Condenser System

Passive isolation condenser system is used to remove the decay heat from the core during shutdown condition without operator intervention. The system consists of passive isolation condensers (PIC) submerged in a pool of water with appropriate valves, piping, and headers. The PICs consist of vertical tubes, joined at both ends to cylindrical headers and submerged in a water pool. The PICs are connected to the steam drum which collects the heat from the core in the form of steam. During normal reactor shutdown, decay heat in the form of steam enters the steam drum and into the ICs from the top, condenses in the tubes, and the condensate returns by gravity to the steam drum. This condensation of steam enables to bring down the core temperature in few hours. The schematic of the system is shown in Fig. 5.3.

5.5.4 Passive Containment Cooling System

During accident conditions whenever primary containment pressure increases passive containment cooling system will be able to cool the containment and keeps the pressure of the primary containment within design limits. It comprises of heat

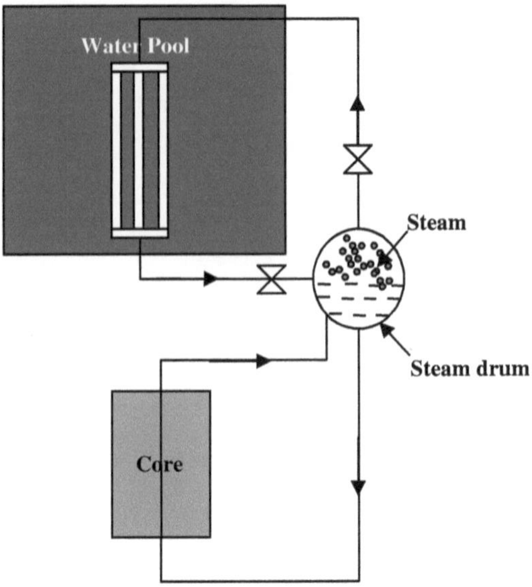

Fig. 5.3 Passive isolation condenser system

exchangers with inclined tubes connected between two headers below a large pool of water tank. The two headers, i.e., the inlet header and the outlet header are connected from the pool at an elevation difference. The heat exchanger receives cold water from the bottom of the pool through the inlet header, absorbs heat from the primary containment via the heat exchangers and rejects the heated water at a certain level in the pool through the outlet header. The reason for using the tubes at a certain inclination is to allow natural circulation to be established in the system. The schematic of the system is shown in Fig. 5.4.

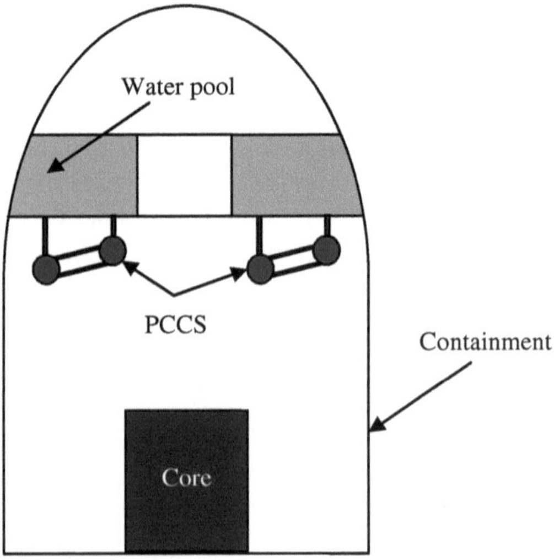

Fig. 5.4 Schematic of passive containment cooling system

5.5.5 *Passive Poison Injection System*

The system is used to inject poison into the moderator in order to cease the chain reaction which ultimately stops the core to generate heat. However, there is always decay heat existed in the core. In this system the poison is added to the moderator by passive means. It contains a helium storage tank which is kept in pressurized condition. A poison tank is connected to this helium storage tank through passive valves and piping header. Further, the poison tank is connected with valves and perforated tubes which are kept inside the calandria that contains moderator. On sensing the signals the passive valves between helium storage tank and poison tank open and helium gas pressurize the poison tank and insert the poison into the moderator through perforated tubes. The schematic of the system is shown in Fig. 5.5.

5.6 Issues Related to Passive Systems

Even though passive systems are simple to use there are some issues related to its functionality and its reliability estimation as listed below [1, 13]:

- Can passive system work under all the operating conditions?
- Given that geometric and material properties do not vary with time and considering the long-term reliable operation of the system can natural circulation stop? If yes, what is the probability that this occurs and consequently the reliability of the passive system?
- What is the definition of failure mode for a passive system?

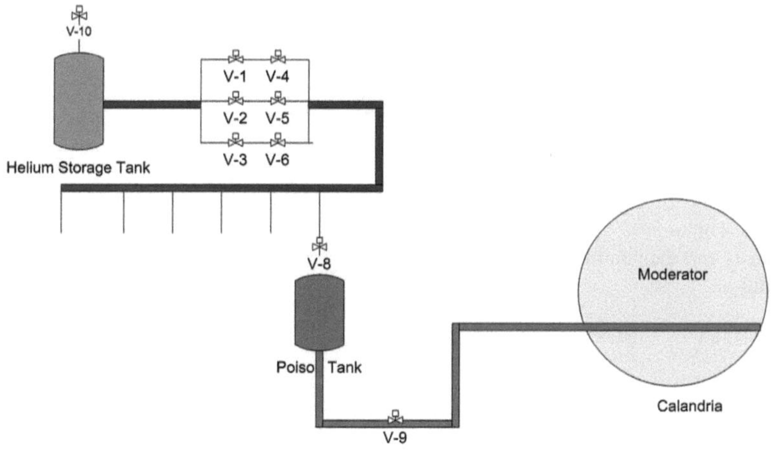

Fig. 5.5 Schematic of passive poison injection system

- How to treat the scarcity in data with respect to the experimental and the operational?
- How to implement the passive system reliability in PSA models?

In the present chapter the above-mentioned issues have been addressed and the solution strategies have been proposed.

5.7 Need for Estimating Reliability of Passive Systems

Consider a passive system as shown in Fig. 5.6. The purpose of the system is to reject the heat from the steam which is produced in a steam drum to the heat sink by condensing primary fluid in the heat exchanger tube bundle and the condensate will return to the steam drum. The process is entirely a natural circulation process.

In traditional reliability analysis the failure probability is estimated based on the actual components present in the system and their corresponding failure data information. The details of the components and their failure probabilities are shown in the following Table 5.2 for the present case. The corresponding fault tree is shown in Fig. 5.7.

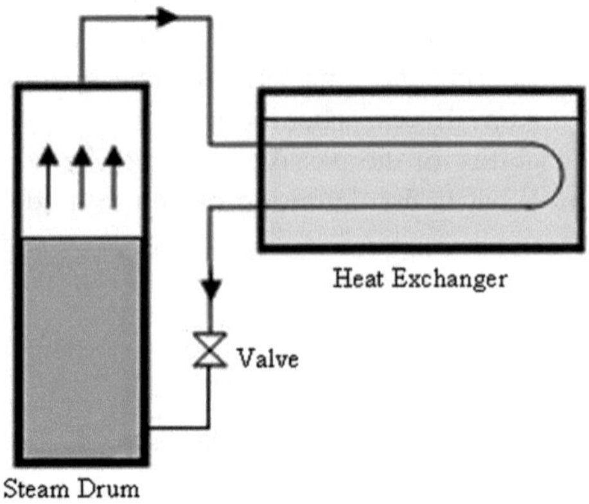

Fig. 5.6 Natural circulation loop

Table 5.2 Different components and their failure probabilities

S. no	Component	Failure probability	Value
1	Steam drum	P_{f1}	1×10^{-6}
2	Heat exchanger	P_{f2}	1×10^{-4}
3	Valve	P_{f3}	1×10^{-4}
4	Piping	P_{f4}	1×10^{-5}

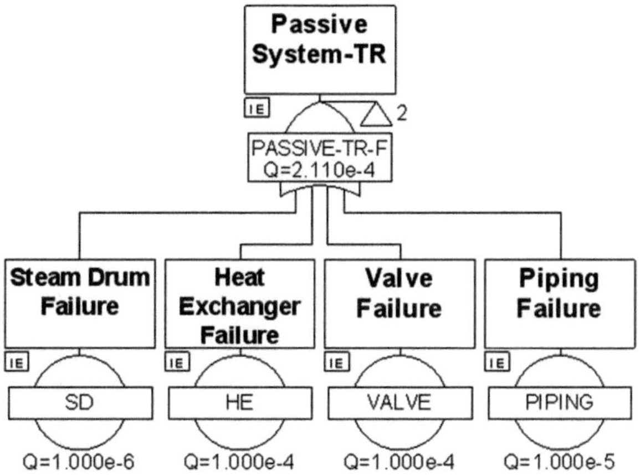

Fig. 5.7 Fault tree of passive system failure based on traditional reliability analysis

The failure probability of the system can be calculated based on the component configuration in the system. In the present case the failure of the system takes place if any of the components fails. Hence, the failure probability of the system can be written as

$$P_{fs} = P_{f1} + P_{f2} + P_{f3} + P_{f4} \tag{5.1}$$

In the present calculation only the actual components in failure contributions have been considered. A passive system may fail to fulfill its mission not only because of a consequence of classical mechanical failure of component (passive or active) of the passive system, but also due to the deviation from expected behavior due to physical phenomena mainly related to thermal hydraulic behavior of the system or due to different boundary or initial conditions. The failure as a result of the above phenomenon can be called as failure of virtual component [8], since it cannot be seen as a physical component in the system.

The failure of the virtual component depends on both the design and process parameters. Now consider the failure probability of the virtual component as P_{fvc} (assume 1×10^{-4}), then the system failure probability can be rewritten as

$$P_{fs} = P_{f1} + P_{f2} + P_{f3} + P_{f4} + P_{fvc} \tag{5.2}$$

The fault tree of the passive system failure including virtual component failure is shown in Fig. 5.8. If one compares the above two Eqs. 5.1 and 5.2 the failure probability value (3.11×10^{-4}) arising from Eq. 5.2 is higher as compared to Eq. 5.1 (2.11×10^{-4}). But if the failure probability contribution arising from virtual component is very low then the Eq. 5.2 can be approximated to Eq. 5.1. In the traditional reliability analysis, the failure probability contribution arising from virtual component is neglected and the entire failure probability is attributed to the actual component failure probabilities. To have a more realistic assessment of the failure probability one should include the failure probability contribution arising

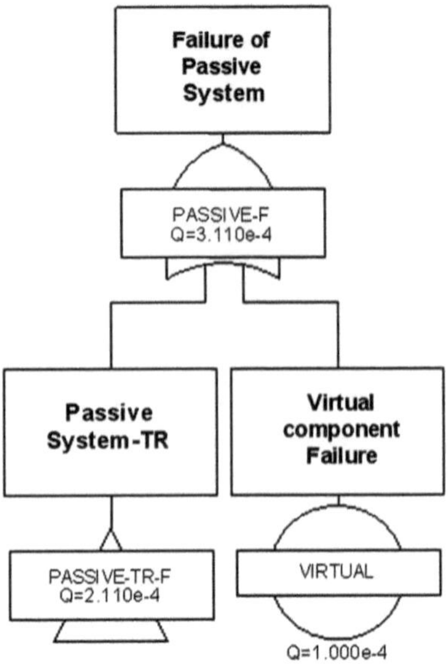

Fig. 5.8 Fault tree of passive system failure including virtual component failure

from virtual component as well. Similarly, the uncertainty band of the system failure probability with virtual component is more conservative compared to the case where virtual component is not involved (see Table 5.3). For the uncertainty analysis lognormal distribution with error factor 3 has been used for the component failure probabilities.

The above analysis can be referred as static reliability analysis since the time variance of reliability has not been considered. But, in real life situations the components undergo aging, fatigue, etc., and the failure probability of the components varies with time. Hence, one needs to consider the time-variant concept of reliability in the analysis. There are well-established techniques for estimating the reliability of active systems/components, whereas for passive systems very few techniques are available and the techniques are not yet frozen. Research on the passive system reliability has started just a decade ago. Many researchers, academicians, and engineers in various fields are working extensively to develop methods for carrying out reliability analysis of passive systems.

Table 5.3 Comparison of uncertainty limits of system failure probability

Percentile values	Without VC	With VC
5th	1.34×10^{-5}	1.97×10^{-4}
50th	2.87×10^{-4}	4.23×10^{-4}
95th	5.23×10^{-4}	7.71×10^{-4}

5.8 Passive System Reliability

Passive systems are characterized by their full reliance upon natural laws (e.g., gravity, natural circulation) to accomplish their designated safety function and dispense with active components. The reliability of passive systems refers to the ability of the system to carry out its safety function under the prevailing conditions for the given duration of time when required.

There are well-established techniques for estimating the reliability of active systems/components. Whereas, for passive systems very few techniques are available such as Reliability Evaluation of Passive Safety Systems (REPAS) [12], Reliability Methods for Passive Safety (RMPS) [13], Applying Passive System ReliAbility (APSRA) [8, 16] and the techniques are not yet frozen. These methods have their own advantages and disadvantages, these are explained briefly in the following subsections with case studies.

5.8.1 REPAS Methodology

In late 1990s several research organizations such as ENEA, the University of Rome, the University of Pisa, and the Polytechnic of Milan participated in the development of methodology for estimating reliability of passive systems known as Reliability Evaluation of Passive Safety Systems (REPAS) and later that was incorporated in the EU RMPS project. This methodology evaluates the probability of failure of a system based on the epistemic uncertainties of different parameters, such as geometric as well as physical, which can cause failure of the system. This methodology also considers the model uncertainties of the codes. The code prediction uncertainties are evaluated by code-to-code comparisons and by calculations of sensitivities to input parameters [12, 13].

The methodology has been applied to an experimental natural circulation test loop by Jafari et al. [12], Zio et al. [21] applied this methodology for reliability evaluation of an isolation condenser system. Marques et al. [13] proposed the integration of reliability of passive system obtained by REPAS in accident analysis. In evaluating the failure probability of the gas cooled fast reactor (GFR) natural circulation system simpler conservative codes have been used by Pagani et al. [18] but with similar approach. However, it was later identified that a large number of calculations with best estimate codes were required to assess the impact of uncertainties of the input parameters on the predicted performance of the passive system. A flowchart of the REPAS methodology is shown in Fig. 5.9 [12] and the different steps are explained briefly as follows:

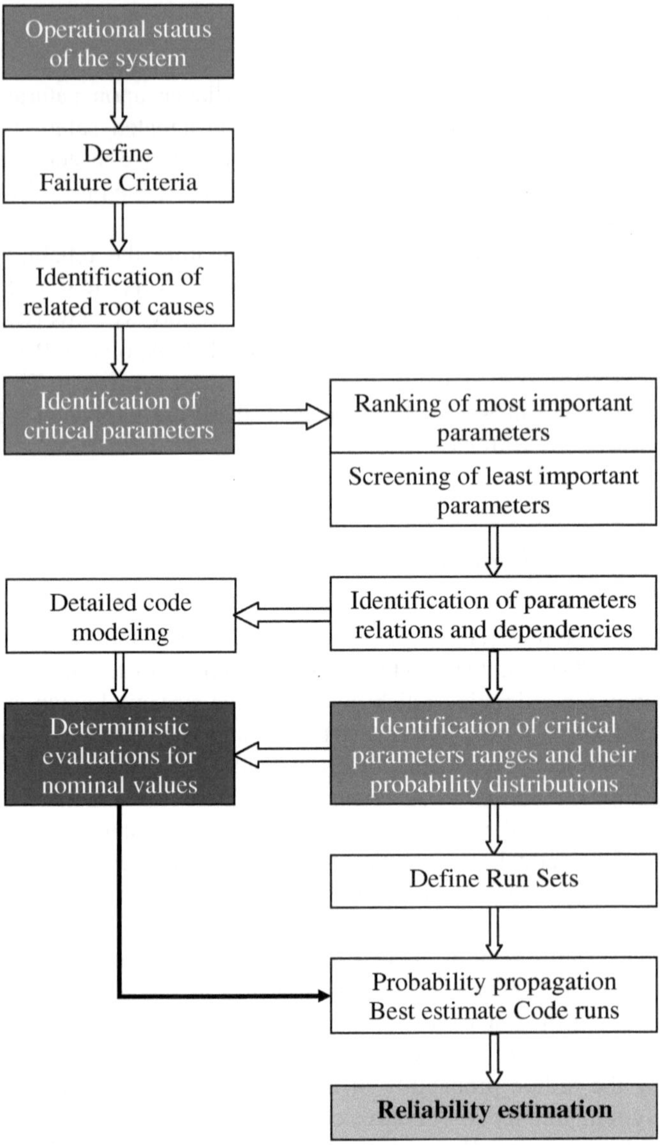

Fig. 5.9 Flowchart of REPAS methodology [12]

Step 1: Operational status of the system

In this step one should identify the passive system functioning, physical phenomenon involved (condensation, natural circulation, etc.), how the system is connected with the other systems, initial and boundary conditions and deign parameters (pressure, temperature, levels, diameter, length, etc.).

Step 2: Failure criteria
While defining the failure criteria one should consider the functioning of the system and mission time of the system. The failure criteria may be defined as

- Peak clad temperature should not exceed design limits
- Pressure should not exceed the critical values
- Flow rate of the fluid should not be less than the threshold limits

Step 3: Identification of root causes
In this step one should identify the root causes of failure of the system with respect to the failure criteria defined. This can be done through classical fault tree and event tree approaches.

Step 4: Identification of critical parameters
This step consists of identification of all the fundamental system parameters that will affect the system performance or process of the system. Systematic procedures should be applied to identify all the relevant parameters.

Step 5: Ranking and screening of important parameters
The purpose of this step is to identify those parameters which will have more effect on the system performance. In this way one can eliminate the parameters which are having insignificant effect on the system performance. Unless one performs this step it is not possible to curtail the number of parameters that enters the analysis and in some cases it will not be manageable for computer applications. Hence, the final goal in this step is to reduce the number of design and critical parameters.

Step 6: Identification of parameters relations and dependencies
Once the critical parameters are identified one should also clearly identify the dependency or independency between the parameters. For example, if pressure and temperature are critical parameters, then these two parameters should not be considered as independent variables, because as the pressure decreases the saturation temperature of the liquid also comes down. This will be very helpful in preparation of the sets for analysis, i.e., in one combination of the sets if pressure is low then in that combination temperature should not be considered higher value, it should also be considered a lower value. Otherwise that set will not be realistic in practice and it may give erroneous results.

Step 7: Detailed code modeling and deterministic evaluation
In this step one has to develop a model which will represent the system with all its initial and boundary conditions. Nodalization has to be developed which will communicate the geometric, material, and thermal hydraulic data with the computational tools. Once the proper model is developed then one can evaluate the performance of the system for the nominal values of the parameters by using the best estimate codes. In order to validate the codes one may have to perform the experiments but in some cases it will not be possible to conduct the experiments. Hence, in such situations only alternative will be the utilization of best estimate codes.

Step 8: Identification of critical parameters ranges and distributions
This step is very important for generation of different sets for carrying out the analysis. One needs to identify all the parameters ranges which will be seen in normal or abnormal operation of plant. Hence each parameter may vary in between some minimum and maximum values with its nominal values being observed in normal plant operation and extremes may be seen during some transient in the plant or abnormal situation.

Step 9: Defining run sets
In this step all the possible combinations of the critical parameters will be made and it will be used in carrying out the analysis. Care must be taken to incorporate the dependency between the variables.

Step 10: Best estimate code runs
Code runs have to be taken for all the sets generated. The outcome of each run should be related to the performance of the system and one should be able to tell whether the system performance is degraded or not based on the failure criteria. If peak clad temperature is the failure criteria then the output of each run will be temperature of the clad as function of time. If the temperature exceeds a critical value then one can infer that the system functioning is degraded or system fails to performs its intended function for a given period of time.

Step 11: Reliability Estimation
Finally, reliability of system can be estimated based on the results obtained in the previous step. The failure probability of the system can be estimated as follows:

$$\begin{aligned} P_F &= \frac{N_F}{N} \\ R &= 1 - \frac{N_F}{N} = 1 - P_F \end{aligned} \tag{5.3}$$

where N_F is the total number of times the peak clad temperature exceeded a critical value, N is the total number of code runs (run sets) carried out.

If out of 200 run sets three times the peak clad temperature exceeded a critical value then the probability of failure will be 3/200 = 0.015 and reliability of the system will be 0.985.

5.8.1.1 Limitations of REPAS Approach

- Engineering judgment is required in different phases of the methodology.
- It needs more sets of runs in finding out the system reliability.
- It does not utilize some of the advanced techniques (neural networks, response surface method, latin hypercube sampling, etc.) in minimization of sample runs.
- This approach does not mention about the implementation of passive system reliability into the accident sequence analysis.

Example 5.1 A Case Study on PDHRS

The above-described methodology has been illustrated with a case study on Passive Decay Heat Removal System (PDHRS) [5, 6] and the failure probability has been estimated for the same.

5.8.1.2 System Operating Mechanism

One of the passive systems used in typical NPP is passive decay heat removal system (PDHRS). The objective of providing PDHRS is to enhance reactor safety during the station blackout event by ensuring continuance of thermosyphon cooling of the reactor through steam generators in the absence of makeup to the Steam Generator (SG) drums. This can be accomplished by recirculating the steam through the PDHRS condensers provided along with each of the SGs. In this process heat is transferred to the inventory outside the tubes contained in the PDHRS tanks.

During normal shutdown, initially the decay heat is removed from Primary Heat Transport System (PHTS) in steam generators. The steam produced on the secondary side of the SGs is dumped to condenser or to atmosphere through atmospheric steam discharge valves (ASDVs). Feed water make up to steam generators is by main or auxiliary boiler feed water pumps. Further cool down of PHTS to room temperature is by shutdown cooling system. Station Blackout (SBO) in advanced reactors is characterized by simultaneous failure of class IV and class III power supplies. Under this condition, Primary Heat Transport and main boiler feed pumps which work on Class IV power will trip. Auxiliary boiler feed pumps will also be not available due to unavailability of Class III power supply. The normal heat removal through the steam generators (SGs) initially, and by shutdown cooling system later are affected during SBO condition. In case of station blackout, the available heat sink is the PDHRS for recirculating the steam generator side inventory through the U-tube condenser inside PDHRS tanks as shown in Fig. 5.10.

5.8.1.3 Setting up of Failure Criterion

In the event of station blackout the heat from the primary heat transport system is removed with the help of PDHRS by means of natural circulation both in primary as well as secondary side. If the natural circulation in secondary side does not function the heat from the PHT system will not be removed and it will increase the clad surface temperature beyond normal operating range. Hence, in this analysis the failure criteria has been considered as clad surface temperature exceeding 400 °C due to the failure of PDHRS.

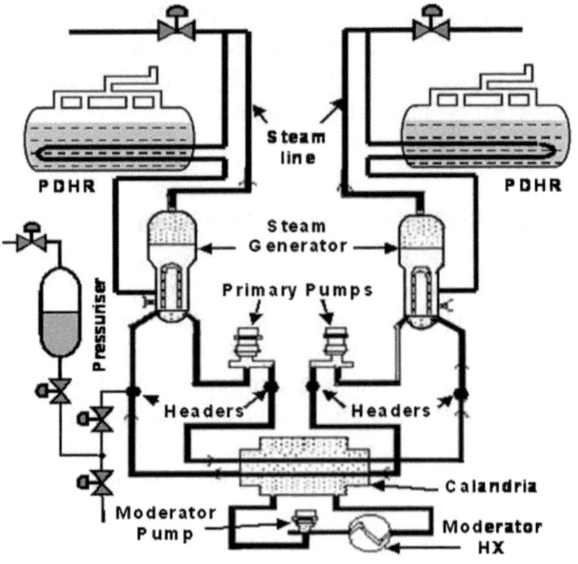

Fig. 5.10 Passive decay heat removal system

5.8.1.4 Critical Parameters that Affect the System Operation

More than 20 parameters (PHT system inventory, SG inventory, PDHR inventory, decay power, noncondensibles in PHT, PHT system pressure, SG pressure, primary coolant flow rate, thermal conductivities of various material involved, thermodynamic properties, reactor power, fouling on heat transport surfaces, SG feed water flow and temperature, noncondensible in SGs, etc.) that affect the system operation and in effect cause the failure of the system have been identified. Out of these parameters few key parameters have been identified based on the sensitivity studies which will have more impact on the failure of the system and are listed below.

- PHT inventory
- SG inventory
- PDHR Tank Inventory
- Decay heat
- Noncondensible gases in PHT
- Noncondensible gases in PDHR tube

5.8.1.5 Identification of Critical Parameters Ranges and Probability Distributions

The critical parameters ranges can be identified based on the expected deviations in the values of these parameters during normal operating conditions. This needs the collection of plant operating data of the critical parameters over a long period of time. In

Table 5.4 Critical parameters considered in the analysis

S. no	Key parameters	Variable
1	PHT inventory (tons)	x_1
2	SG level (m)	x_2
3	PDHR tank inventory (tons)	x_3
4	Decay heat (%)	x_4
5	Noncondensible gases in PHT (%)	x_5
6	Noncondensible gases in SG secondary side (%)	x_6

Table 5.5 Critical parameters range and their probability values

S. no	Key parameters	Range			Probability values		
		Min	Nominal	Max	Min	Nominal	Max
1	x_1	155	160	165	0.25	0.6	0.15
2	x_2	10.4	13.4	16.4	0.15	0.75	0.1
3	x_3	100	115	120	0.3	0.5	0.2
4	x_4	100	110	120	0.25	0.4	0.35
5	x_5	0	3.75	7.5	0.3	0.4	0.3
6	x_6	0	2.5	5	0.3	0.4	0.3

case the operating experience is not available one can depend on the experimental data or on analysis. Once the critical parameters ranges are identified in the next step proper probability distribution functions have to be assigned to each critical parameter. In REPAS methodology instead of defining the continuous probability density functions only discrete probability values are used. The different critical parameters ranges and their probability values are given in Tables 5.4 and 5.5.

5.8.1.6 Detailed Code Modeling

A thermal hydraulic model has been developed which represents the system with all its initial and boundary conditions. The system has been represented with proper nodalisation which will communicate the geometric, material, and thermal hydraulic data with the computational tools. Thermal hydraulic analysis has been carried out by using RELAP code for the nominal values of the parameters. The structure of RELAP model is shown in Fig. 5.11. Nodalisation of the system is shown in Figs. 5.12 and 5.13.

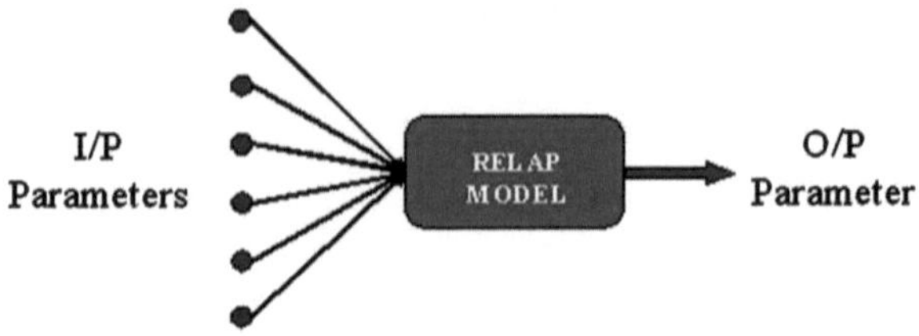

Fig. 5.11 Passive decay heat removal system

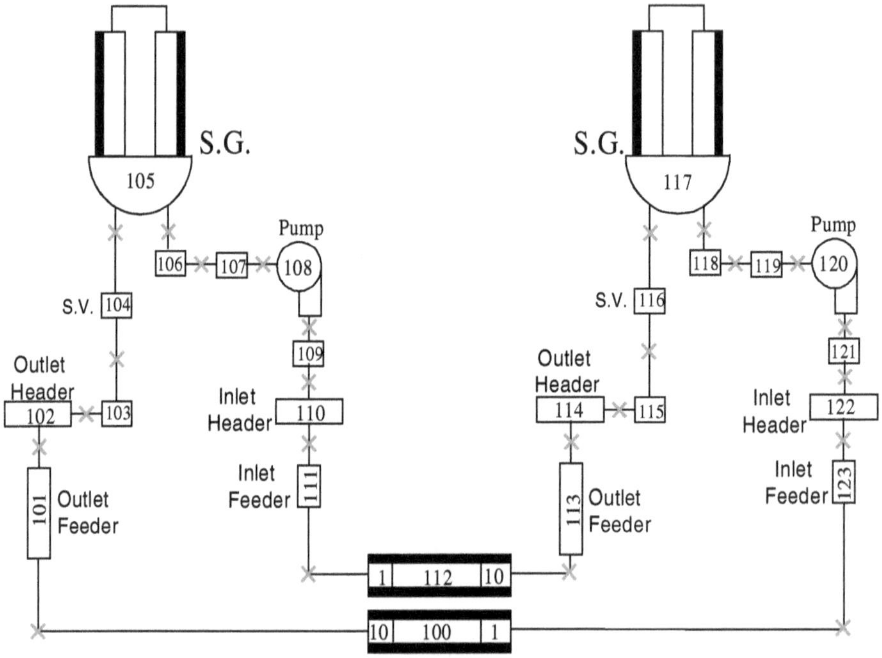

Fig. 5.12 RELAP5 nodalization for one of primary loops

5.8.1.7 Defining Run Sets and Best Estimate Code Run

In the present analysis we have six critical parameters that affect the clad surface temperature. Each parameter can take any of its three values as specified in Table 5.5. If one takes all the possible combinations there will be 3^6 (729) combinations. For all the combinations of the parameters, thermal hydraulics (process dynamics) RELAP5 simulation model runs have to be taken. Some of the sets used in the analysis are shown in Table 5.6.

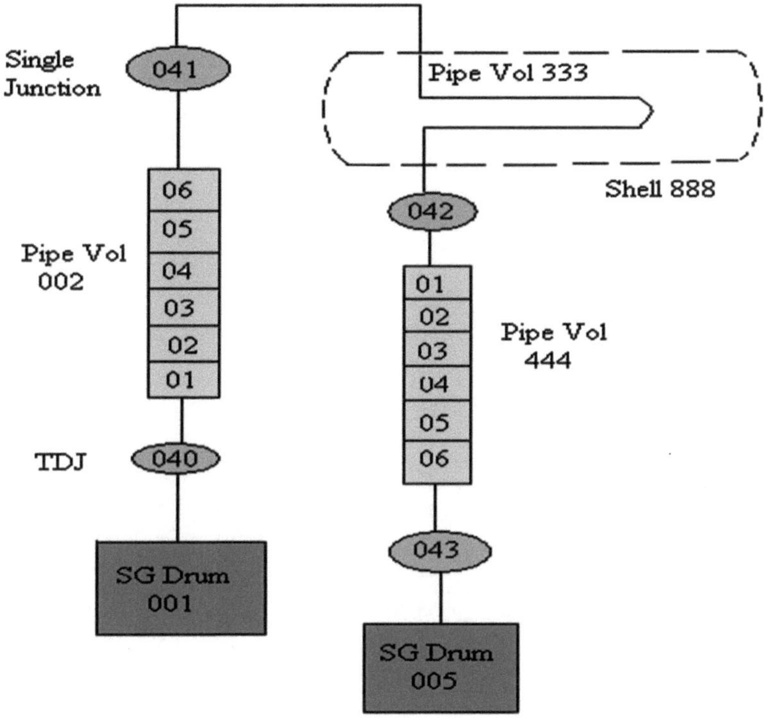

Fig. 5.13 Nodalisation for one of PDHRs

5.8.1.8 Reliability Estimation

Finally, reliability of system can be estimated based on the results obtained in the previous step. The failure probability of the system can be estimated as follows:

$$P_F = \frac{N_F}{N}$$

$$R = 1 - \frac{N_F}{N} = 1 - P_F$$

$$N_F = 15$$

$$N = 729$$

$$P_F = \frac{15}{729} = 0.020576$$

$$R = 1 - P_F = 0.979424$$

Hence, the reliability of the system has been estimated as 0.98. The reader should understand that this is only a case study explained for the purpose of

Table 5.6 Some of the sets used in the analysis

S.no	x_1	x_2	x_3	x_4	x_5	x_6	Probability
1	160	13.4	115	110	3.75	2.5	0.0144
2	155	10.4	100	100	0	0	0.000253
3	155	10.4	100	100	0	2.5	0.000338
4	155	10.4	100	100	0	5.0	0.000253
5	155	10.4	100	100	3.75	0	0.000338
6	155	10.4	100	100	3.75	2.5	0.00045
7	155	10.4	100	100	3.75	5.0	0.000338
8	155	10.4	100	100	7.5	0	0.000253
9	155	10.4	100	100	7.5	2.5	0.000338
10	155	10.4	100	100	7.5	5.0	0.000253
11	155	10.4	100	110	0	0	0.000405
12	155	10.4	100	110	0	2.5	0.00054
13	155	10.4	100	110	0	5.0	0.000405
14	155	10.4	100	110	3.75	0	0.00054
15	155	10.4	100	110	3.75	2.5	0.00072
16	155	10.4	100	110	3.75	5.0	0.00054
17	155	10.4	100	110	7.5	0	0.000405
18	155	10.4	100	110	7.5	2.5	0.00054
19	155	10.4	100	110	7.5	5.0	0.000405
20	155	10.4	115	100	0	0	0.000422

understanding the methodology and the value obtained here may not be true estimation due to some limitations.

5.8.2 RMPS Methodology

To overcome the limitations of REPAS approach, the Reliability Methods for Passive Safety functions (RMPS) [13] was developed. The RMPS approach identifies the different sources of uncertainties and determines the critical parameters that will affect the system performance. It considers probability density function (pdf) to treat variations of the critical parameters considered in the predictions of codes. Thermal hydraulic models will be utilized to propagate the uncertainties and estimate the reliability of passive systems. It also defines a methodology for incorporation of passive system reliability into accident sequence analysis. The RMPS methodology has been successfully applied to evaluate the reliability of the Passive Residual Heat Removal System of Light Water Reactors by Marquès et al. [13]. Figure 5.14 shows the different steps of RMPS methodology and are described briefly as follows:

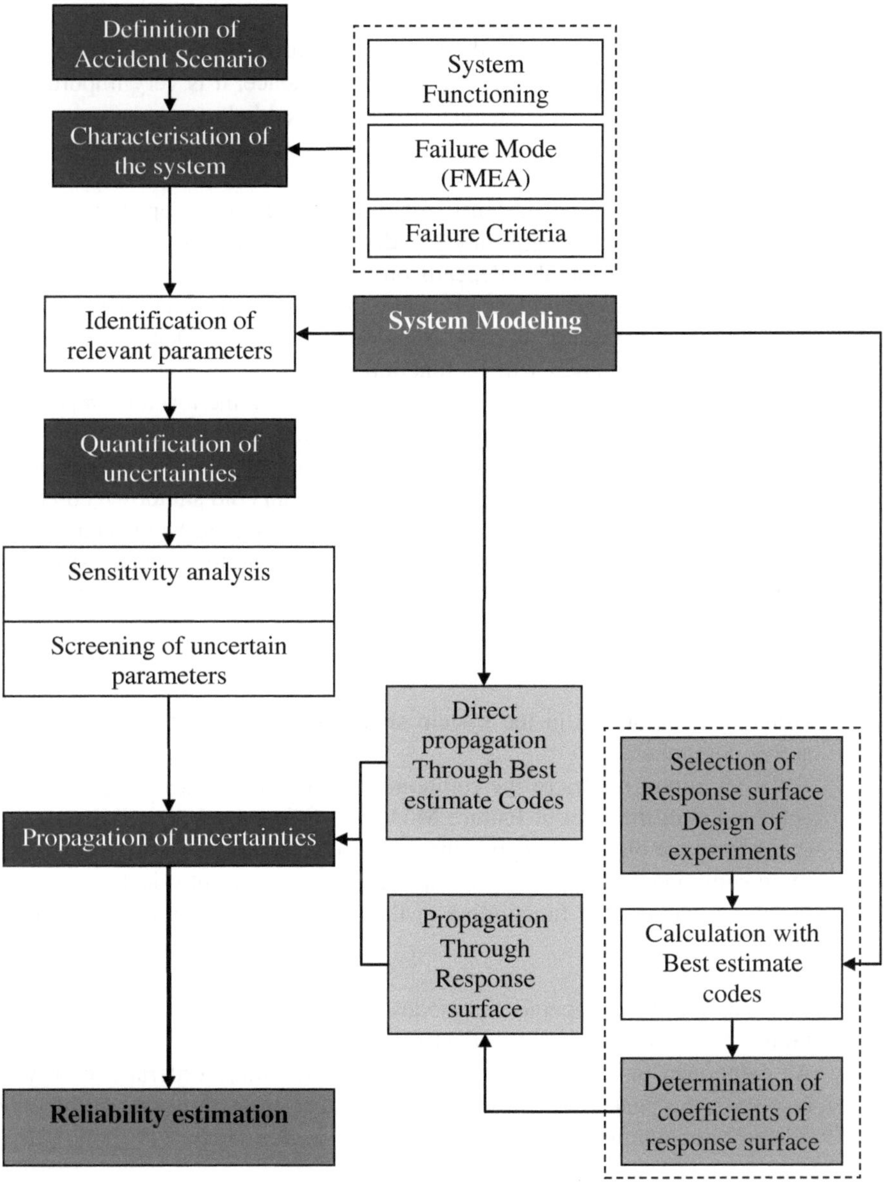

Fig. 5.14 Flowchart of RMPS methodology [13]

Step 1: Definition of Accident scenario
The very first step of the methodology is to identify the different accident sequences in which the passive system to be analyzed will be operated. This will be helpful in identification of particular failure criteria for a given scenario. Also depending on the scenario the mission time of the system will change.

Step 2: Characterization of the system
Depending on the accident scenario the passive system functioning, mission time, modes of operation, and failure criteria will change. Hence, it is very important to clearly define the passive system characteristics as listed below.

- System functioning
  - Passive system may be designed for different purposes depending on the scenario such as decay heat removal, depressurization, etc.
  - The mode of operation also depends on the scenario. For example, isolation condenser in an advanced reactor can be designed for maintaining either the hot shutdown condition or cold shutdown condition. Depending on the scenario different valves will be functioned.
  - During hot shutdown condition one should maintain the reactor temperature and pressure at full power conditions and this can be maintained with just a pressure relief valve whenever pressure raises the valve will be opened and it closes if the pressure reduces. Whereas to maintain cold shutdown condition lot of heat should be removed and different valves should be opened and the valve should be remained in open position to achieve the proper conditions.
  - Hence, based on the scenario the functioning of the system should be defined properly.
- Failure mode
  - Different components in the system should be identified both passive and active components.
  - Different failure modes of the components should be identified. This can be achieved with the help of Failure Mode and Effect Analysis (FMEA).
  - Apart from the physically appearing mechanical components there exists one more component called virtual component which is not physically appeared, but it can hamper the functioning of the system (e.g., natural circulation).
- Failure criteria
  - Based on the system operational modes and functioning one can establish the failure criteria for a given accident scenario.
  - As mentioned in the previous methodology the failure criteria can be with respect to the peak clad temperature of the fuel, mass flow rate in the core, related to containment pressure, etc.

Step 3: System modeling
In general reliability of a component or system is calculated based on the failure data obtained from the experiments. If the database is not available in such situations one can depend on the numerical modeling or analysis. In this case one should utilize best estimate codes to perform the analysis and the code should be able to simulate the actual system for different initial and boundary conditions. One can perform the analysis for the nominal values of the parameters and can be compared with the experimental results if exist.

Step 4: Identification of relevant parameters
Different parameters that will affect system performance have to be identified. The parameters can be related to initial and boundary conditions (e.g., temperature, pressure, water levels, material properties, etc.), geometry of the system.

Step 5: Quantification of uncertainties
In this step one should be able to identify the uncertainty existed in the parameters. Which will include the identification of the critical parameters ranges, within which the parameters may vary during the normal or abnormal operation of the plant. The uncertainty of a parameter can be well represented with the probability distributions and this will be very helpful in propagating the uncertainty from parameter level to the system performance level. Hence, It is very much important to identify the different probability distributions for the parameters. If the experimental or operational data is available one can easily develop the distributions for the parameters. However, if there is scarcity in data then one has to depend on the expert judgment. Also it is needed to identify the dependency between the parameters.

Step 6: Sensitivity analysis
Based on the sensitivity analysis one can clearly identify those parameters which have more impact on the performance of the system and can eliminate less important parameters. With this step one can reduce the total critical parameters for which can be treated as random variables and other can be treated as deterministic variables. In this way the total number of experiments one should be conducted can be reduced. This is because with more number of uncertain parameters more number of combinations will exist and so as the more number of code runs will have to be carried out.

Step 7: Propagation of uncertainty
In propagating the uncertainty of the input parameters there should be a well-defined function with which output parameter (with which the system performance is related) is related to the input parameters. This can be done in two different ways,

- One with the help of existing system model and best estimate codes one can run the analysis and can obtain the output which obviously depend on the several input parameters.
- The second method is to obtain a relation between output and input parameters with the help of response surface methodology.
  - In this method a proper model is selected in which output is a function of input parameters (either linear or nonlinear)
  - Depending on the ranges of input parameters different combinations of experiments will be designed.
  - These sets have to be run with the help of best estimate codes and output should be obtained.

- For each combination of inputs there exists an output and with the help of linear or nonlinear regression analysis one can obtain the coefficients of the response surface.
- Finally, one can obtain a clearly defined relation between input and output parameters.

Step 8: Reliability estimation
In this step with the help of response surface one can estimate the reliability of the system by applying proper failure criteria. The failure probability of the system can be defined as follows:

$$\begin{aligned} T_{\text{clad}} &= f(x_1, x_2, x_3, \ldots) \\ P_F &= P(T_{\text{clad}} > T_{\text{critical}}) \end{aligned} \tag{5.4}$$

One can apply different simulation or analytical techniques on the response surface to estimate the failure probability or the reliability of the system as listed below,

- Monte Carlo simulation
- Variance reduction techniques
- First Order/Second Order reliability techniques (FORM/SORM)

Step 9: Incorporation of passive system reliability in accident sequences
The objective of this step is to implement the passive system reliability into accident sequences that will be part of Probabilistic Safety Assessment (PSA). In PSA event trees are developed to identify the different accident sequences. Once the accident scenario is defined with the help of the event trees one can obtain the accident sequences where the passive system is involved. The accident sequence frequency can be obtained by incorporating initiating event frequency and different safety system failure probabilities.

5.8.2.1 Limitations of RMPS Methodology

- To apply the methodology, one needs to have the pdf values of the critical parameters.
- However, there are no guidelines on the pdf treatment of these parameters, which ultimately define the functional failure. To define the distribution for the parameter one should have the parameters of the distribution that one has to identify from the available data. In the absence of the data it is very difficult to define the distributions. In this methodology based on the expert judgment the distributions have been identified.
- Time-variant reliability is not considered in the analysis. It only estimates static reliability of the system.

Example 5.2 Consider the same Example 5.1 on PDHR system as explained in the REPAS methodology.

5.8.2.2 System Considered

Passive decay heat removal system (PDHR)

5.8.2.3 Definition of Accident Scenario

The accident scenario considered is station blackout event. During the station blackout PDHR will be used as decay heat removal system. The accident sequence can be given as

- Class IV failure followed by Class III failure followed by PDHR failure and further followed by firewater system failure.
- This sequence will lead to core damage situation.

5.8.2.4 System Characterization

- System consists of the following components:
  - Steam generator
  - Piping system
  - Upstream and downstream valves
  - Heat exchanger with U-tubes
  - Virtual component (natural circulation)
- The failure modes and failure criteria are similar to the one explained under REPAS methodology

5.8.2.5 System Modeling

- Similar to REPAS methodology

5.8.2.6 Identification of Relevant Parameters and Sensitivity Analysis

- Similar to REPAS methodology

Table 5.7 Critical parameters range and their probability density functions

S.no	Key parameters	Range	pdf
1	PHT Inventory (tons)	150–165	Normal
2	SG Level (m)	10.4–15.4	Uniform
3	PDHR tank Inventory (tons)	100–120	Normal
4	Decay heat (%)	0–20	Normal
5	Noncondensible gases in PHT (%)	0–7.5	Uniform
6	Noncondensible gases in SG secondary side (%)	0–5	Uniform

5.8.2.7 Quantification of Uncertainties

- Similar to REPAS methodology
- However, the uncertainties in the parameters are represented with a probability density function (pdf) as given in Table 5.7.

5.8.2.8 Propagation of Uncertainty

- Response surface methodology has been used
- The following response surface model has been used

$$\begin{aligned} T = a_0 &+ a_1x_1 + a_2x_2 + a_3x_3 + a_4x_4 + a_5x_5 + a_6x_6 \\ &+ b_1x_1^2 + b_2x_2^2 + b_3x_3^2 + b_4x_4^2 + b_5x_5^2 + b_6x_6^2 \\ &+ c_1x_1x_2 + c_2x_2x_3 + c_3x_3x_4 + c_4x_4x_5 + c_5x_5x_6 \end{aligned} \tag{5.5}$$

- In the present analysis we have six key parameters that affect the clad surface temperature. According to the central composite design (CCD) method one can get 77 ($2^6 + 2 \times 6 + 1$) combinations from these six key parameters. For all the 77 combinations of the key parameters, thermal hydraulics (process dynamics) RELAP5 simulation model runs were taken.
- After performing the regression analysis the regressor coefficients have been obtained and are given in Table 5.8.
- The response surface which is generated has been validated with the three different sets of input parameters and the corresponding experimental output. The results are shown in Table 5.9 along with the percentage of error from the calculated value.

Table 5.8 Regressor coefficients of the polynomial

Coefficient	Value	Coefficient	Value	Coefficient	Value
a_1	732.838	b_1	−1.8289	c_1	−10.191
a_2	2666.33	b_2	−45.296	c_2	2.5778
a_3	62.5492	b_3	0.6364	c_3	−2.3831
a_4	−156.79	b_4	1.9211	c_4	−1.3553
a_5	105.511	b_5	12.357	c_5	−20.623
a_6	42.4151	b_6	5.4610	a_0	−71639

Table 5.9 Sets used in the analysis for validation

S. no	x_1	x_2	x_3	x_4	x_5	x_6	T_C	T_E	% Error
1	160	16.4	115	1.0	0	0.0	177.4	173	2.55
2	160	13.4	115	1.2	0	2.5	395.2	390	1.34
3	155	16.4	100	1.0	0	2.5	358.4	362	0.98

5.8.2.9 Reliability Estimation

- In this step Monte Carlo simulation [see Appendix] technique has been applied on response surface to estimate the reliability of the system by applying failure criteria.
- The failure probability of the system is defined as follows:

$$P_F = P(T_{\text{clad}} > 400\,^{\circ}\text{C})$$

- The probability of failure is calculated as follows:

$$\begin{aligned} P_F &= \frac{N_F}{N} \\ N_F &= 169 \\ N &= 10^6 \\ P_F &= \frac{169}{10^6} = 1.69 \times 10^{-4} \end{aligned}$$

- where N is the total number of simulations carried out, i.e., 10^6.
- N_F is the number of simulations in which temperature exceeded 400 °C, i.e., 169.

5.8.3 APSRA Methodology

In the APSRA methodology [16, 17], the passive system reliability is evaluated from the evaluation of the failure probability of the system to carry out its desired function. In this approach, the deviation of critical parameters, which influence the system performance, during normal operation are considered to generate the failure surface. The causes of critical parameter deviations are identified through root cause analysis. These deviations are attributed to the failure of existing components in the system (e.g., valves and control systems). Finally, with the help of conventional system reliability techniques (such as fault tree analysis) the failure probability of the system is evaluated by utilizing component failure probabilities. The APSRA methodology has been applied to the Passive Containment Isolation System (PCIS) [16] and to the Isolation Condenser System (ICS) [17] of the advanced reactor. Different steps of APSRA methodology are explained as follows:

- **Step 1**: Passive system for which reliability assessment is to be done is considered.
- **Step 2**: Identification of parameters affecting the operation.
- **Step 3**: Definition of failure criteria.
- **Step 4**: Identification of key parameters which may cause failure.
- **Step 5**: Generation of failure surface and validation with test data.
- The steps from 1 to 5 are similar to the previous methodologies.
- **Step 6**: Root diagnosis to find deviation of key parameters for causing ultimate failure of system.
  - This step is different from the other methods.
  - In this step root causes of deviations of the critical parameters, which eventually result into system failure will be identified. This is done through a root diagnosis method.
  - Usually, a passive system fails to carry out its desired function not due to the failure of its driving mechanism, but due to deviations of some of the critical parameters. This happens either due to failure of some active components such as valves, pump, electric signal, etc., or due to failure of some passive components such as passive valves, relief valve, etc.
- **Step 7**: Evaluation of failure probability of the system
  - System failure probability is estimated with the help of fault tree method in which the root causes of the parameter deviations are logically represented with gates and propagate to the system failure.
  - The system failure probability will be estimated based on the failure data of the basic events. Basic events are nothing but the component failures.

5.8.3.1 Limitations of APSRA Methodology

- The probability values for different key parameter ranges do not change with the value. It assumes a constant value. This is due to the reason that root causes for a parameter deviation is same for different values of the parameter.
- For example, the root causes of water level of tank is lower than the nominal value is same for different levels. Assume that nominal value is 5 m. Now consider the level goes below 4 m and find out the root causes for this scenario. Then even if the level goes below 3 m the same root causes are applicable.
- It will not consider all the possible combinations of parameters.
- It does not consider failure probability variation with time.

Example 5.3 A case study on passive isolation condenser system

5.8.3.2 System Description

- Core decay heat removal system is used to remove the decay heat from the core, during shutdown condition for 3 days without operator intervention, by means of Isolation Condensers.
- The system is designed for cooling the Main Heat Transport System from 285 to 150 °C after the reactor is shut down.
- The system consists of Isolation Condensers (ICs) submerged in a pool of water called Gravity Driven Water Pool (GDWP) with appropriate valves, piping, and headers.
- The ICs consist of vertical tubes, joined at both ends to cylindrical headers and submerged in a GDWP.
- During normal reactor shutdown, decay heat in the form of steam enters the steam drum into the ICs from the top, condenses in the tubes, and the condensate returns by gravity to the steam drum.
- This condensation of steam enables to bring down the PHT temperature from 285 to 150 °C in few hours. The schematic of the system is shown in Fig. 5.15.

5.8.3.3 Failure Criteria

- In the present study, PIC fails to remove decay heat from the core during hot shutdown condition has been considered as the failure criteria.

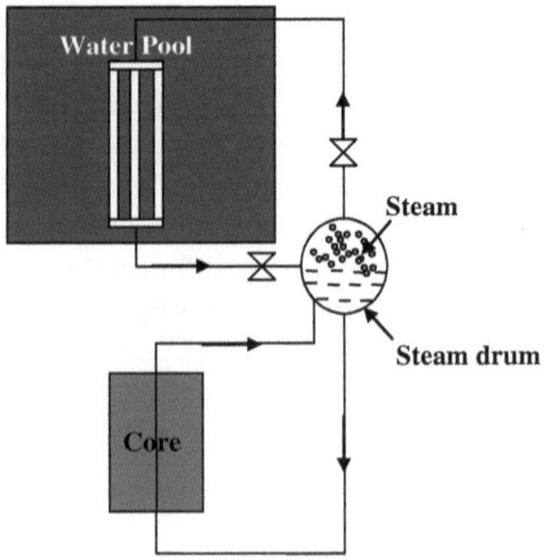

Fig. 5.15 Passive isolation condenser system

5.8.3.4 Key Parameters

- The above situation can arise when there is any deviation in system parameters.
- The key parameters to cause the failure of the system from the passive system reliability analysis point of view are
  - Presence of noncondensable gases in the IC tubes
  - Low water level in the GDWP
  - GDWP high temperature
- Each of these key parameter deviation was analyzed for their cause using Failure Mode Effect Analysis (FMEA) (Table 5.10).
- This will act as the input to further thermal hydraulic analysis to evaluate the system performance for various ranges of identified parameter values.

5.8.3.5 Thermal Hydraulic Analysis

Thermal hydraulic analysis has been carried out to generate the failure surface, keeping the criterion as peak clad temperature exceeding 400 °C. The failure surface is shown in Fig. 5.16.

5.8.3.6 Failure Probability Estimation

- The low water level in GDWP can arise from different sources like, failure of makeup circuit, failure of header to pool valves, etc. Fault trees are drawn for

Table 5.10 FMEA for parameter deviations in IC system

S.no	Deviation	Causes
1	Presence of noncondensable gases	• Failure of Vent valves
2	Low water level in GDWP	• Failure of makeup circuit • Failure of header to pool valves
2.1	Make up circuit failure	• Failure of Maintenance valves • Failure of pump • Check valve failure
3	GDWP high temperature	• Recirculation Loop failure • Pool to header valves failure • Header to pool valves failure
3.1	Recirculation loop failure	• Demineralization loop failure • MOV failure • Pump failure • Check valve fails to remain open • Heat exchanger failure • Valves failure • Human Error
3.2	Demineralization loop failure	• Filter failure • Valves fail to remain open

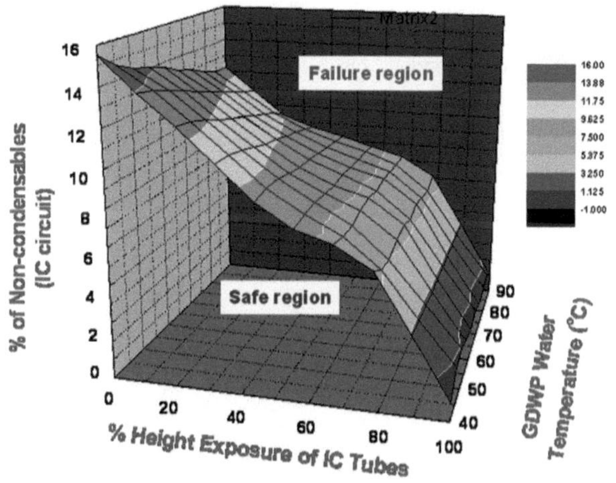

Fig. 5.16 Failure surface generated from thermal hydraulic studies

each of these parameter deviations. Some of the fault trees are shown in Figs. 5.17, 5.18, 5.19, 5.20, 5.21 and 5.22.

- Probability values are attributed for each of these causes to reach a final estimate of probability of occurrence of these parameter deviations.

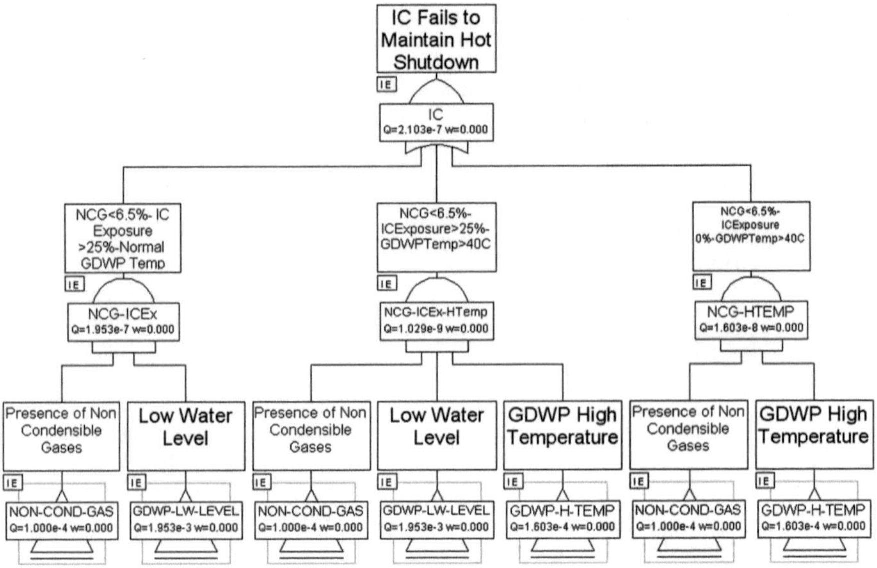

Fig. 5.17 Fault tree for IC failure

Fig. 5.18 Fault tree for presence of noncondensable gases

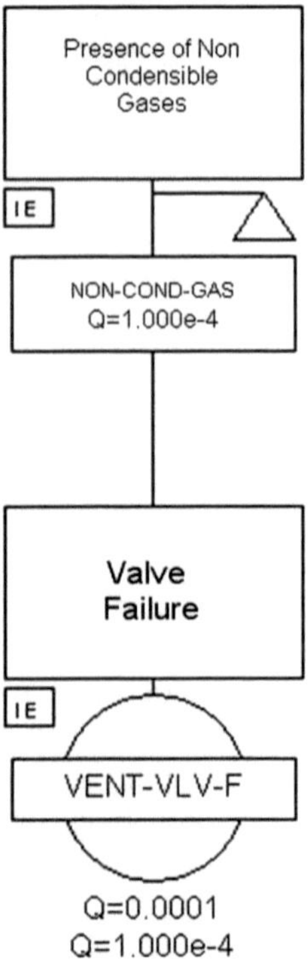

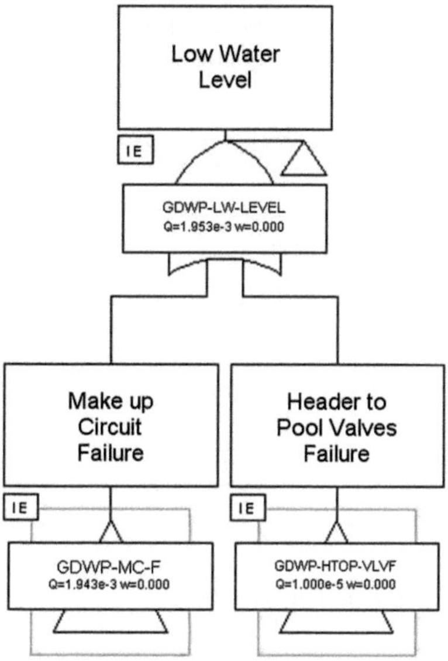

Fig. 5.19 Fault tree for low water level in GDWP

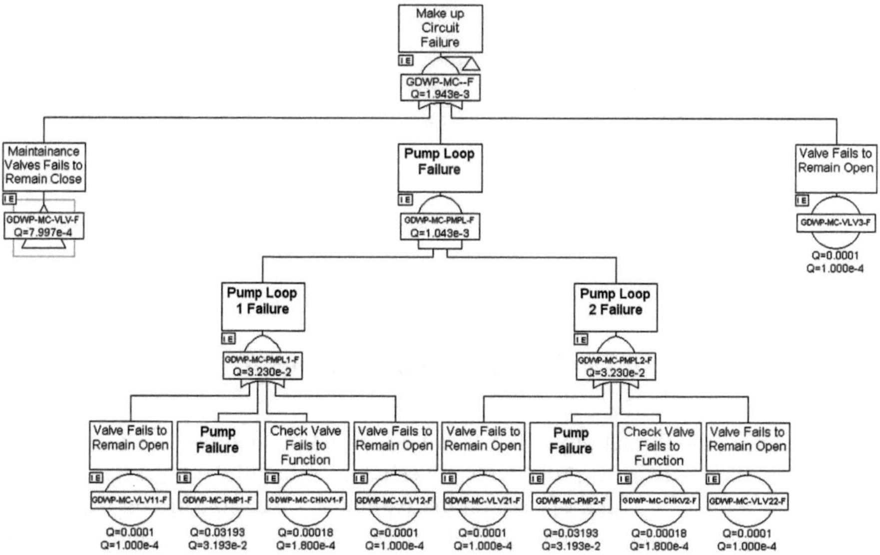

Fig. 5.20 Fault tree for makeup circuit failure

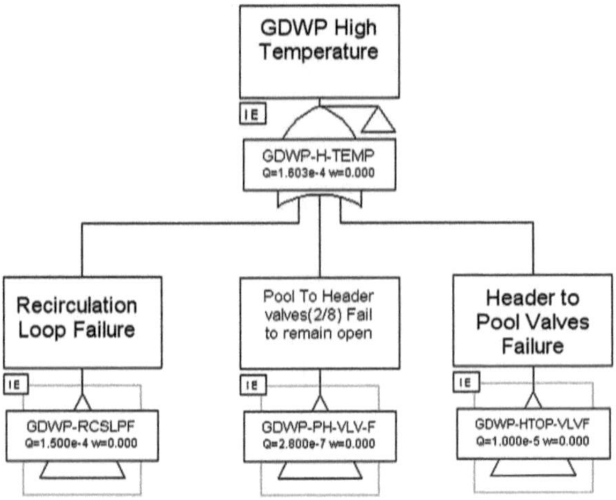

Fig. 5.21 Fault tree for GDWP high temperature

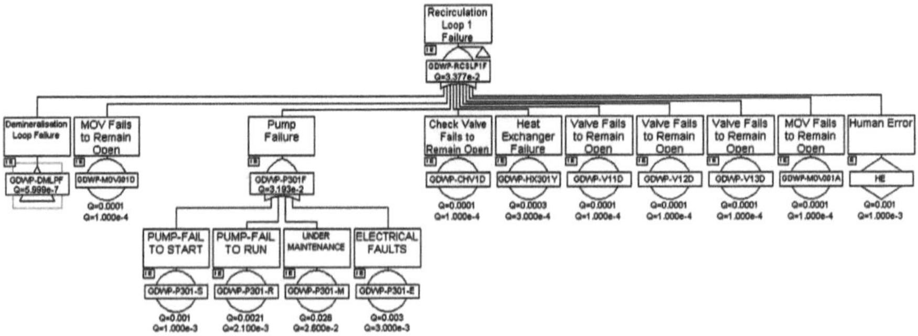

Fig. 5.22 Fault tree for Recirculation loop failure-Contd

5.8.4 Fuzzy Monte Carlo Simulation Approach

By considering the limitations in the above discussed methodologies there is a need to develop much realistic procedures to estimate reliability of passive systems, mainly when there is a lack of data on the parameters that consider in the analysis and also implementing the dynamic reliability concepts. This dynamic reliability analysis will be helpful in estimating the risk of a plant as a function of time. There are several components which contribute to the failure of the passive systems. These are explained with a fault tree as shown in Fig. 5.23 [6, 8]. The failure contribution can come from both static and time-variant analysis. In the static analysis the following component failures cause the failure of the system.

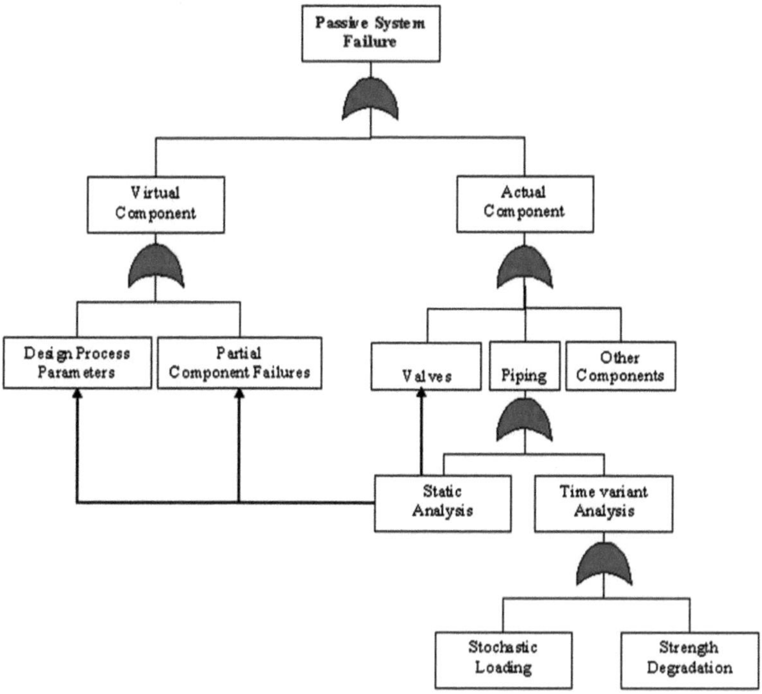

Fig. 5.23 Fault tree for passive system reliability

- Virtual component
  - Design and process parameters
  - Partial component failures
- Active component
  - Valves
  - Piping
  - Other Components

In the time-variant analysis the failure probability contribution come from the following factors:

- Stochastic loading
- Strength degradation

In the present chapter different aspects of static reliability analysis have been explained with an appropriate methodologies and a case study. The methodology to treat the partial component failures is explained under the section mechanistic models. Finally, the time-variant reliability analysis has been explained in Chap. 6.

5.8.4.1 Static Reliability Analysis

The failure of the passive systems can be attributed to two aspects; one failure contribution from the component failures in the system and the other contribution from the failure of the virtual component. The virtual component indirectly refers to the failure of the process itself. Since the functioning of the passive system depends mainly on the physical laws such as natural circulation, gravity, temperature difference and density difference, the process failure refers to the failure of the natural circulation in the system which is caused by the deviations of the process parameters, design parameters and sometimes on the partial failures of the active components. Hence, to evaluate the passive system reliability one has to address all the above three mentioned aspects. The accuracy of the reliability analysis depends on the failure data and models used in the analysis. Hence, failure data is also important from reliability point of view. Figure 5.24 shows the structured methodology for the evaluation of reliability of passive systems. The methodology is explained in the following steps:

Step 1: Passive System to be analyzed
In the first step, it is required to identify the passive system for which the reliability analysis has to be carried out.

Step 2: Identification of Operational Mechanism
In this step one should identify the purpose of the system, the accident scenario in which the system will function and clearly bring out the operational mechanism of the passive system. The mission time and modes of operation also depend on the accident scenario. Passive system may be designed for different purposes depending on the scenario such as decay heat removal, depressurization, etc. Hence, based on the scenario the functioning of the system should be defined properly and the designer should have a clear understanding of the operational mechanism of the passive system, its modes of operation, and its failure criteria.

Step 3: Identification of Failure Criteria
In this step one has to identify the failure criteria with which the system performance will be affected. As discussed earlier, as long as the heat source and sink are available, natural circulation always continues or prevails. However, the flow rate may not be sufficient to fulfill the desired objective of the system. The system designer may consider the system to fail if any of the following criteria is not met either during the normal operation or during any operational transients.

- Adequate removal of heat without exceeding certain limiting clad surface temperature; or
- No flow oscillations, etc.

Step 4: Key parameters that affect the system operation
As discussed before, the performance characteristics of the passive system is greatly influenced by some critical parameters. Some of the critical parameters which influence the natural circulation flow in a boiling two-phase natural circulation system are

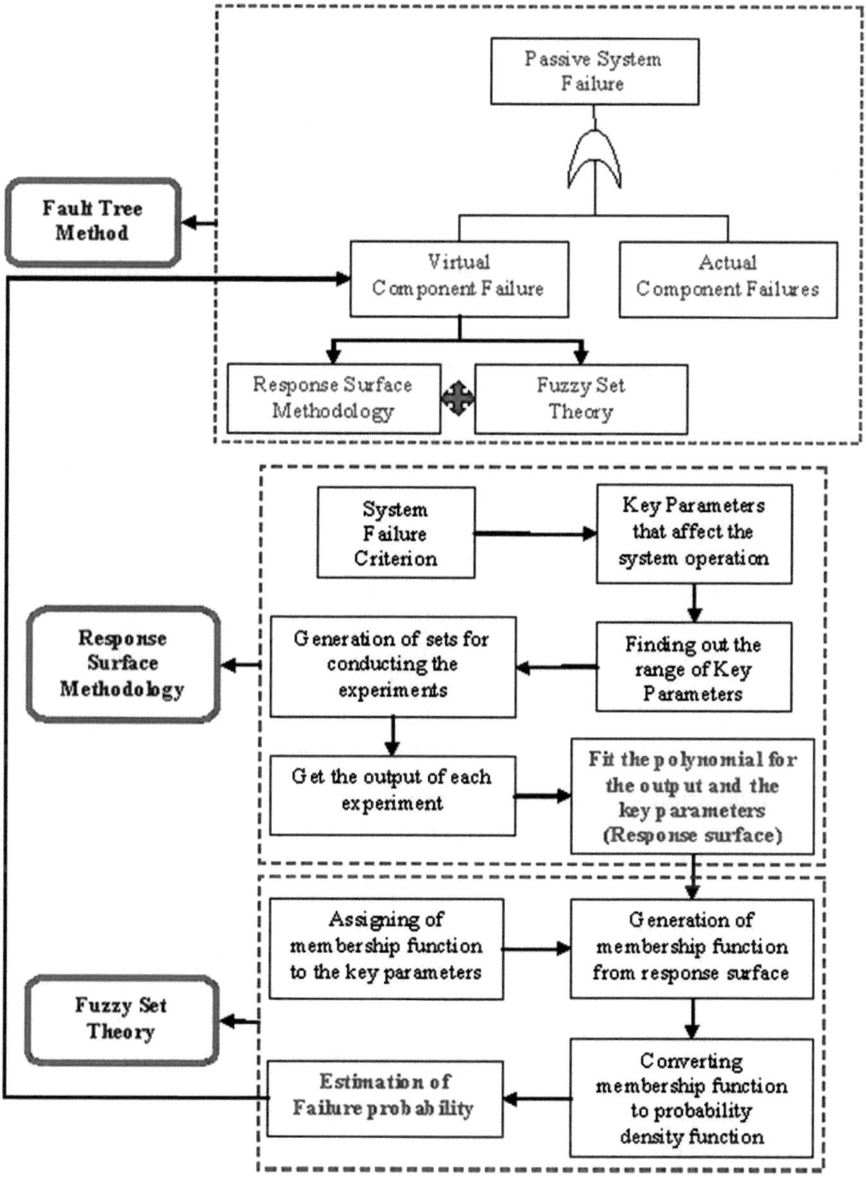

Fig. 5.24 Procedure for estimating the passive system reliability [8]

- System pressure;
- Heat addition rate to the coolant;
- Water level in the steam drum or downcomer;
- Feed water temperature or core inlet subcooling;

- Presence of noncondensible gases;
- Flow resistances in the system.

This step will help in understanding the performance characteristics of the system due to variation of the critical parameters. But, all the critical parameters may not cause much influence on the system performance, this can be identified by performing sensitivity analysis. From the sensitivity analysis one can identify the key parameters that affect the system performance.

Step 5: Identification of key parameter ranges and their membership functions
After identification of the key parameters that affect the system performance, then next step is identification of key parameter ranges and their corresponding membership functions. These ranges can be identified based on the operating experience. Actually, the key parameters can be treated as random variables and the problem can be solved using probabilistic analysis. But, the application of the probabilistic analysis will become difficult in certain situations especially when one has limited information about the parameters. This is because the probabilistic analysis needs information about the probability density function and its basic parameters (mean, standard deviation, etc.) of the parameter under consideration. In most of the cases only the maximum and minimum values for uncertain parameter can be obtained and therefore, uncertain parameter can be expressed only with interval between minimum and maximum. Under this condition, one can utilize the concept of fuzzy set theory in which uncertain parameter can be expressed as a fuzzy number. In the fuzzy analysis it is required to assign the membership function to the uncertain parameter. This can be done with the help of available key parameter ranges.

Step 6: Generation of Response Surface
In this analysis response surface is generated based on the Response Surface Methodology (RSM) [15]. The most extensive applications of RSM are in the particular situations where several input variables potentially influence some performance measure or quality characteristic of the process. Thus, performance measure or quality characteristic is called the response. The input variables are sometimes called independent variables. The final output of this analysis is an empirical formula which describes the relation between response of the system and its input parameters or key parameters. To develop these relations it is very much needed to design the experiments from which the output can be extracted for different combinations of the input parameters. In the absence of the experiments one can use the best estimate codes to simulate the process. In the present case one has to perform the thermal hydraulic analysis for different cases. Based on the results obtained from thermal hydraulic analysis for all the cases response surface can be generated by using nonlinear regression analysis.

Step 7: Generation of Membership Function
Since in the present analysis the variables are assumed as fuzzy variables one has to generate the membership function for the output variable so as to utilize in the

failure probability calculations. This can be done with the help of the response surface. Since the response surface is a function of different input parameters (key parameters) and further their membership functions are known, by using the fuzzy arithmetic one can generate the membership function for the output variable.

Step 8: Estimation of Failure Probability
In this step the failure probability can be estimated by using the simulation methods or by directly from probability density function of the output variable. Both approaches need probability distributions of the output variable. Hence, it is required to transfer the fuzzy membership function into an equivalent probability distribution function. Once the transformation is done by using the simulation methods/direct method failure probability of the system can be evaluated.

Example 5.4
Consider the same Example 5.1 on PDHR as discussed under REPAS methodology

- The initial steps are similar to the RMPS methodology except the definition of probability distributions. In this case membership functions have been used and are listed in Table 5.11.
- Response surface has been developed similar to that of RMPS methodology and is given as follows:

$$\begin{aligned} T = a_0 &+ a_1x_1 + a_2x_2 + a_3x_3 + a_4x_4 + a_5x_5 + a_6x_6 \\ &+ b_1x_1^2 + b_2x_2^2 + b_3x_3^2 + b_4x_4^2 + b_5x_5^2 + b_6x_6^2 \\ &+ c_1x_1x_2 + c_2x_2x_3 + c_3x_3x_4 + c_4x_4x_5 + c_5x_5x_6 \end{aligned} \quad (5.6)$$

- The regression coefficients obtained from the analysis is shown in Table 5.12. Once the response surface is available, next step is the estimation of failure probability of the system. In this case Fuzzy Monte Carlo simulation approach has been used. This is explained in detail in the next section.

Table 5.11 Key parameters range and their membership functions

S. no	Key parameters	Range	Membership
1	PHT Inventory (tons)	150–165	Normal
2	SG Level (m)	10.4–15.4	Triangular
3	PDHR tank Inventory (tons)	100–120	Normal
4	Decay heat (%)	0–20	Normal
5	Noncondensible gases in PHT (%)	0–7.5	Triangular
6	Noncondensible gases in SG secondary side (%)	0–5	Triangular

Table 5.12 Regressor coefficients of the polynomial

Coefficient	Value	Coefficient	Value	Coefficient	Value
a_1	732.838	b_1	−1.8289	c_1	−10.191
a_2	2666.33	b_2	−45.296	c_2	2.5778
a_3	62.5492	b_3	0.6364	c_3	−2.3831
a_4	−156.79	b_4	1.9211	c_4	−1.3553
a_5	105.511	b_5	12.357	c_5	−20.623
a_6	42.4151	b_6	5.4610	a_0	−71639

5.8.4.2 Failure Probability Estimation

- As discussed earlier failure probability has been estimated by using fuzzy set theory.
- In the fuzzy set theory the objective is to obtain the membership function for the output parameter that is obtained from the response surface.
- To obtain this function one should have the information about the input key parameters and their respective membership functions.
- Once the membership functions for all the key parameters are available one can apply fuzzy arithmetic [2, 3, 20] concepts on response surface to obtain the output parameter membership function.
- Finally, the output parameter membership function can be transferred to probability density function and the failure probability can be estimated based on the failure criteria. This is pictorially shown in Fig. 5.25.
- In this study the above-discussed approach has been utilized for evaluating the failure probability of passive system which uses fuzzy set theory and Monte Carlo simulation/direct method [14].
- In this approach uncertain parameters are considered as fuzzy numbers and the corresponding membership functions have been derived from available information (which is given in Table 5.11) and are shown in Fig. 5.26.
- By using this information one can obtain the membership function for the clad surface temperature which is shown in Fig. 5.27.
- Later, this membership function has been converted to probability density function [8] which is shown in Fig. 5.28, with which by using Monte Carlo simulation and direct approaches failure probability has been estimated. If one uses the direct method (finding out the probability of exceedence directly from the cumulative distribution function) one can eliminate the simulation time.
- For the present analysis the estimated failure probability (contribution from virtual component failure) obtained is 5.203×10^{-4}. The failure probability contribution from different components (Process parameters, partial failure of components, active component failures) is tabulated in Table 5.13 and is found to be 7.513×10^{-4}.

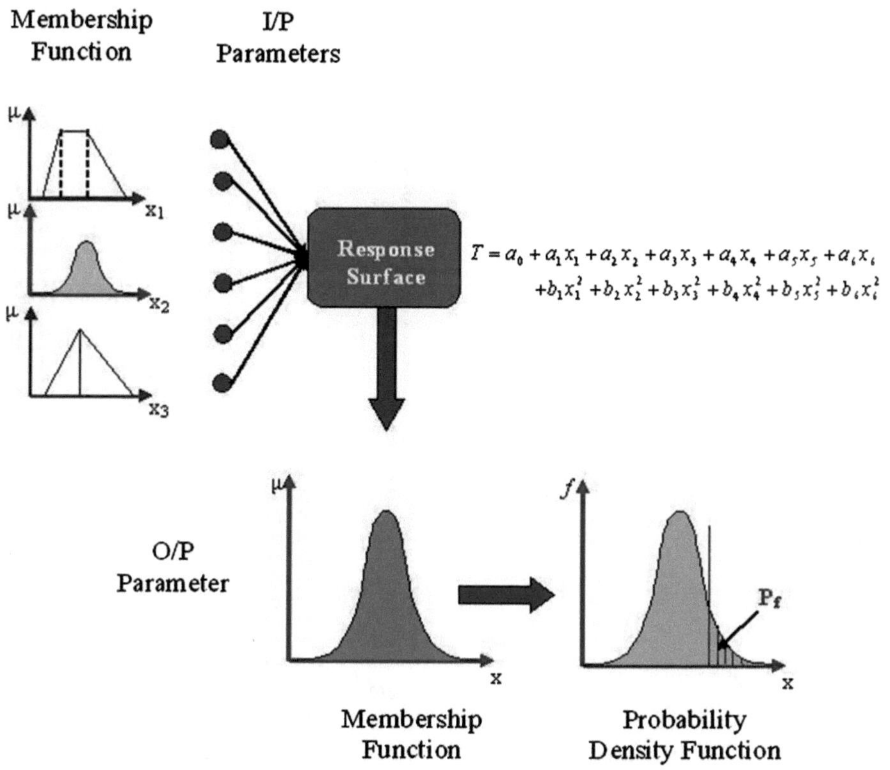

Fig. 5.25 Concept of Fuzzy set theory

5.9 Mechanistic Modeling Approach

The traditional reliability analysis, considers components to be in binary state, either functional or faulty, and does not consider the concept of multistate or intermediate states between these two binary states [6]. However, there are several components, which need to be operated in different states and their failure criterion also depend on these states. Hence, when dealing with these types of components one should use multistate concept. This can be achieved by modeling the components with mechanistic models, which can give a new dimension for reliability analysis for multiple states. The mechanistic model approach is based on the first principles of science and engineering which provides details about the various failure mechanisms and thereby improved understanding of the associated root causes of the failure and driving forces responsible for component failures.

Reliability estimation of components has two components. The first one is deterministic, which deals with finding various failure modes and causes of failure, while the second one is probabilistic where probability of failure and uncertainty in

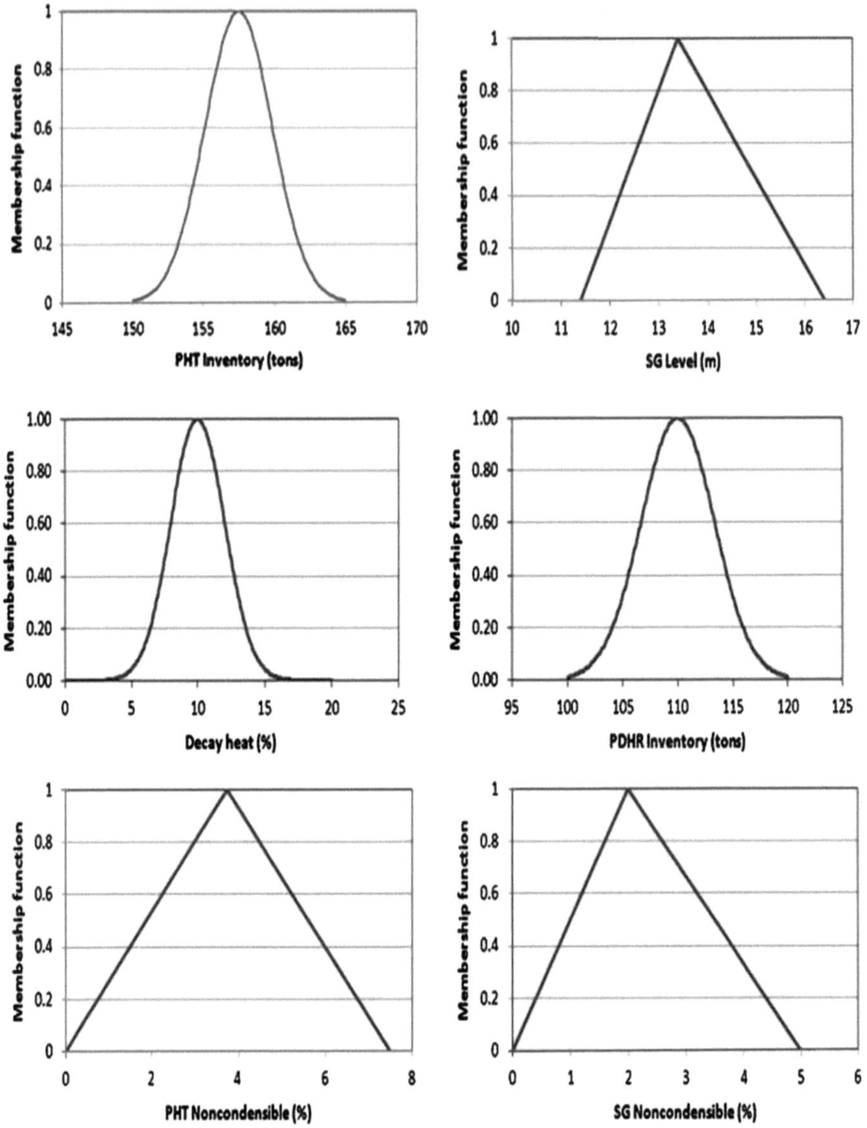

Fig. 5.26 Different membership functions of key parameters

parameters is estimated. These two aspects together make reliability prediction more accurate and complete. In conventional reliability analysis only two states of component are considered namely functional or faulty, and do not consider any intermediate states between these two states. However, there are several components, which need to be operated in different states. The one approach which deals with multistate problems is mechanistic modeling approach. The mechanistic model

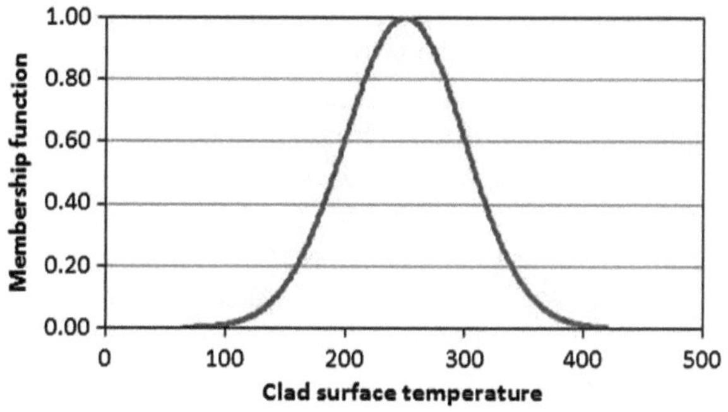

Fig. 5.27 Membership function of clad surface temperature

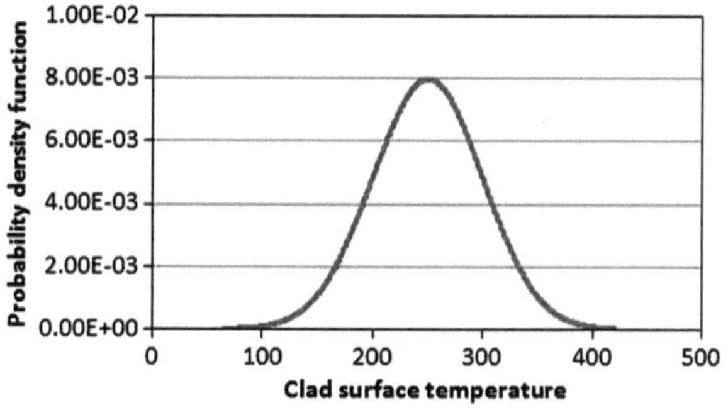

Fig. 5.28 Probability density function of clad surface temperature

Table 5.13 Failure probability contribution from different components

S.no	Component		Value
1	Virtual	Design and process parameters	5.203×10^{-4}
	Component	Partial failure of Valve 2	0.900×10^{-4}
2	Actual	Valve 1	1.000×10^{-4}
	Component	Piping	4.100×10^{-5}
Total failure probability			7.513×10^{-4}

approach provides details about the various failure modes and failure mechanisms of a component. Failure mode can be defined as a manner in which the component or equipment failure takes place. Components can fail in different ways based on geometry, load direction, environmental conditions, and other variables. In mechanical components failures can occur due to complex set of interactions between stresses that act on (within the system) and the materials that the system is comprised of.

A material can fail when stresses acting on the material exceeds the strength of the material. The stresses can accumulate irreversibly to cause damage to the material. There are different failure mechanisms with which a material can fail.

Some examples of failure mechanisms of mechanical components are buckling (failure of a long, slender column that has been subjected to a compressive, axial load), corrosion, creep (slow deformation of a solid material over time due to applied loads and often increased temperatures and can result in changes in material properties and part geometries that can cause failures), fatigue, fracture, impact, rupture, thermal shock, wear, yielding, etc. Each failure mode and failure mechanism will call for different performance function since the way the component fails differ from one failure mode to other and also one failure mechanism to other.

In this section a general methodology for modeling the components with mechanistic models has been explained and is further illustrated with an example component. A case study on feed water system (consisting of control valves and other mechanical components) of a typical nuclear reactor has been presented.

5.9.1 The Approach

The mechanistic model approach is based on the first principles of science and engineering and will give realistic estimation to the reliability of a component. The mechanistic model approach uses an understanding of the failure mechanisms involved such as crack growth, corrosion, creep, fatigue, impact, etc. Mechanistic model approach encourages innovative, cost-effective design through the use of realistic reliability assessment. This approach tends to overcome the limitations that are inherent in traditional approaches as it facilitates estimation of life and root causes of failure, employing models that are based on physics and engineering, which address deterministic as well as probabilistic applications, incorporation of operational load profiles of the component, evaluation of associated failure mechanisms, and detailed modeling for identified dominant failure mechanisms and modeling for wear out phenomenon as part of life cycle loads. The basic steps involved in the approach are explained as follows [4, 7]:

- **Step 1**: The first step is identification of component to be analyzed.
- **Step 2**: Once the component is identified next step is collection of input data which includes material properties, design features of the component, operational and environmental load profiles (number of cycles of operations), test and maintenance provisions, etc.
- **Step 3**: In the next step one has to identify the different failure modes in which the component fails and also one has to establish the failure criteria of the component, based on the operational and engineering requirements of the system.
- **Step 4**: Once the failure modes are identified one has to evaluate associated component failure mechanisms for each failure mode considering basic phenomenon involved in degradation/failure of components.
- **Step 5**: In this step one has to establish limit state function (failure function) for each failure mode.

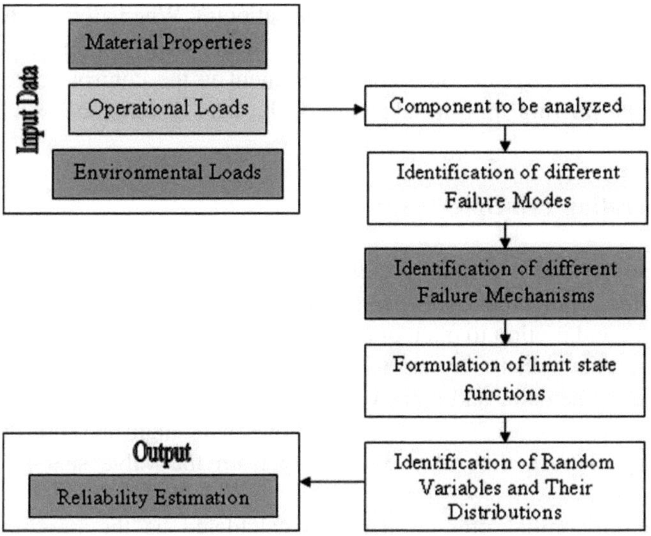

Fig. 5.29 Flowchart of the analysis [6]

- **Step 6**: In this step one has to identify the different random variables in the limit state function and their corresponding distribution based on the available information.
- **Step 7**: The final step is estimation of reliability of the component based on simulation or analytical approaches.

A flowchart explaining the above steps are shown in Fig. 5.29. An example on control valves explaining the above steps are given in the following section.

5.9.2 Control Valves

In nuclear power plants, to ensure the quality of the process it is important to keep some process variables (pressure, flow, level, temperature, etc.) within the operating range. These process variables are in general affected by the load disturbances in the plant. However, the effect of load disturbances on the process variables can be controlled with the help of control loops. A typical nuclear power plant consists of hundreds of control loops. There will be several sensors and transmitters which collect process variables information and passes it to a controller. The controller process the collected information and checks with the set points of each process parameter. The set points of process parameters normally varies with the load disturbances. Hence, whenever there is a load disturbance the process parameter should be brought back to the corresponding set point. When all the calculations are done there should be some control element which will implement the strategy selected by controller. One such control element is the control valve which will manipulate the flowing fluid to take care of the load disturbances.

5.9.2.1 Operating Principle

A typical control valve assembly consists of valve body, valve seat, valve plug, bonnet, sealing arrangement between the bonnet and valve stem, an actuator which supplies force and motion to open or close a valve and the force can be generated by hydraulic, by pneumatic or by electrical means. The valve also consists of spindle connected to the actuator and other valve accessories. A single seat two-port globe valve is shown in Fig. 5.30. As the fluid flows with high pressure difference the fluid will try to push the valve plug away from the valve seat in the upward direction.

Here the differential pressure (ΔP) is calculated as the pressure difference between upstream (P_1) and downstream (P_2) of the valve. In order to keep the valve plug fully sitting on the valve seat some force needs to be applied from actuator and this force can be calculated as a function of differential pressure across the valve and friction allowance as given in Eq. 5.7.

$$F = A\ \Delta P + \text{Friction Allowance} \tag{5.7}$$

where 'A' is the valve seating area. The movement of the valve plug is controlled by actuator with the necessary force applied to it depending on the flow requirements.

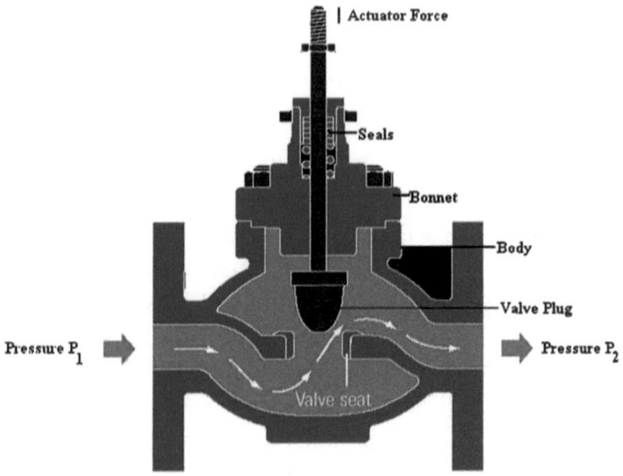

Fig. 5.30 Flow through a single seat two-port globe valve

5.9.2.2 Failure Modes

A component can fail in different ways, the manner in which the component fails is defined as failure mode. A control valve can be treated as failed if the valve fails to perform its function completely or partially. There are different failure modes with which a control valve can fail but the dominating one is related to complete valve failure, speed of operation, or leakage. The different failure modes of control valve can be listed as: Leakage, fails to open partially, fails to open fully, and fails to close fully. The leakage can be treated as breach of the pressure boundary. The leakage can happen in two different ways either leakage in the sealing arrangement between the bonnet and valve stem or leakage in the valve body itself. Valve body leakage can happen if there exists some type of damage mechanism which will affect the pressure boundary (corrosion, etc.). Different failure modes of control valve and their causes are listed in Table 5.14.

5.9.2.3 Reliability Estimation

In general one can obtain the failure probability of the control valve through generic failure data sources. However, for plant specific conditions one can obtain the failure probability based on the mechanistic models. In order to find the reliability or failure probability of the control valve first one has to select the failure mode. In this section the calculations are explained for fails to open of a control valve to a specified lift of the valve plug. This can happen if the required pressure to lift the valve plug is greater than the supplied pressure, in other words if the supplied pressure is less than the required pressure then the valve will not open for the required lift. This can be mathematically represented with a failure function as given in Eq. 5.8.

Probability of failure = P (Required Pressure > Supplied Pressure)

$$P_f = P(\sigma_R > \sigma_S) \tag{5.8}$$

The required pressure for a diaphragm type valve can be calculated as follows, the spring force required to lift the valve to a height of H is

$$F = KH \tag{5.9}$$

Table 5.14 Different failure modes of Control valve

S.no	Failure mode	Causes
1	Leakage	Mechanical damage, corrosion
2	Fails to close	Line pressure too high
3	Fails to open	Corrosion, dirt/debris trapped in valve seat

where, K is the stiffness of the spring. The lift of the valve can be calculated based on the flow requirements from the Eq. 5.10.

$$H = \frac{\ln\left[\frac{\dot{V}}{\dot{V}_{\max}}\tau\right]}{\ln\tau} \tag{5.10}$$

where

τ Valve rangeability (ratio of maximum to minimum controllable flow rate).
$\dot{V}$ Volumetric flow through the valve at lift H
$\dot{V}_{\max}$ Maximum Volumetric flow through the valve

From the above discussions the required pressure that should act on the diaphragm to open the valve to a certain height can be calculated from the Eq. 5.11 as follows:

$$\begin{aligned} \sigma_R &= \frac{KH}{A_D} \\ \text{where } A_D &= \frac{\pi}{4}D^2 \\ D &= \text{Diameter of diaphragm} \end{aligned} \tag{5.11}$$

Supplied pressure depends on the compressor and control signals. Hence, the probability of failure [6] can be written as follows:

$$\begin{aligned} P_f &= P\left(\frac{4KH}{\pi D^2} > \sigma_S\right) \\ P_f &= P\left(\sigma_S < \frac{4KH}{\pi D^2}\right) \end{aligned} \tag{5.12}$$

From the Eq. 5.12 one can find out the probability that the control valve will fail to open for a given height (H) as it can be considered as constant and other parameters such as K, D, σ_S can be treated as random variables. By applying the structural reliability methods (SORM) one can find out the failure probability. One should be careful in selecting the failure probability as these values differ for different failure modes. For the instance the leakage probability will be different from that of fails to open or fails to close as the failure functions of these failure modes are formulated based on different principles. However, in order to estimate the total failure probability of the control valve one has to consider all the failure modes. This methodology can be applied in all the process control loops where the functioning of control valve is sought, not only fully open or closed but also partial opening (% of opening).

5.9.3 A Case Study on Feed Water System

Feed water system of a typical nuclear reactor [6] has been considered for the illustration purpose. The feed water system handles the liquid phase of the main heat transport inventory. After the steam is cooled in condensers, the inventory is received by the feed water system, which enhances its properties before delivering it back to the steam drum. The feed water system consists of condensate system and feed system. The function of condensate system is to draw condensate from the condenser hot well and deliver it to the feed system through purification system and moderator heat exchanger system. The feed system purifies and heats the condensate and raises its pressure sufficiently to permit injection into the steam drums. The steam drum level is controlled on the basis of steam flow, drum water level, and feed flow. The schematic of feed water system is shown in Fig. 5.31. The boundary for the analysis of feed system has been considered as main condenser at the inlet side and the steam drum at the outlet side. A fault tree has been developed for the feed water system as a part of reliability analysis and is shown in Fig. 5.32 Failure probability has been estimated as 9.243×10^{-2}.

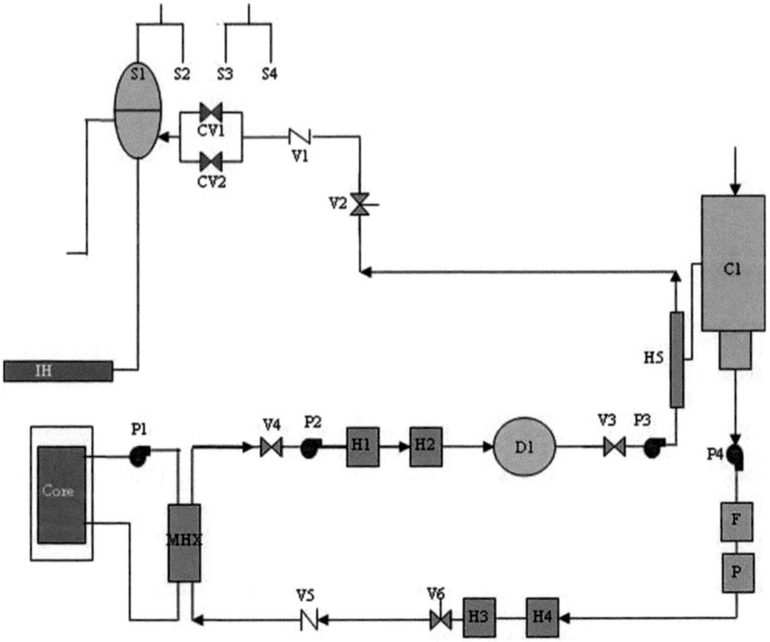

Fig. 5.31 Schematic of feed water system

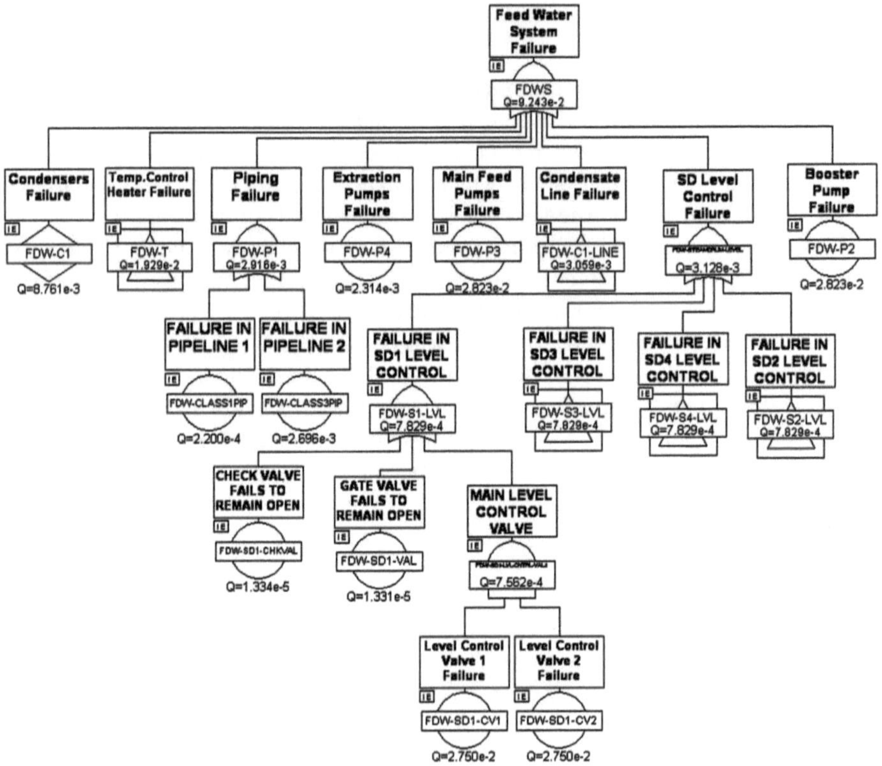

Fig. 5.32 Fault tree of feed water system

References

1. DAuria F, Galassi GM (2000) Methodology for the evaluation of the reliability of passive systems. DIMNP NT 420 (00), University of Pisa, Pisa, Italy
2. Ferson S, Kuhn R (1992) Propagating uncertainty in ecological risk analysis using interval and fuzzy arithmetic. Computer Techniques in Environmental Studies IV. Elsevier Applied Science, London
3. George JK, Yuan B (1995) Fuzzy sets and fuzzy logic. Prentice Hall of India Pvt. Ltd., New Delhi
4. Hari Prasad M, Rami Reddy G, Srividya A, Verma AK (2010) Reliability estimation of control valves through mechanistic models. J Mech Eng Res 2(3):52–57
5. Hari Prasad M, Gaikwad AJ, Rao VVSS, Lele HG, Vaze KK (2010) Reliability studies of decay heat removal system. In: International conference on reliability safety and hazards—2010, 14–16 Dec 2014, Mumbai, India
6. Hari Prasad M, Gaikwad AJ, Srividya A, Verma AK (2011) Failure probability evaluation of passive system using fuzzy Monte Carlo simulation. Nucl Eng Des 241:1858–1866
7. Hari Prasad M, Rami Reddy G, Srividya A, Verma AK (2011) Applying mechanistic models to reliability evaluation of mechanicl components-An illustration. Ann Nucl Energy 38:1447–1451

8. Hari Prasad M, Rami Reddy G, Srividya A, Verma AK (2012) Reliability estimation passive systems using fuzzy fault tree approach. Int J Syst Assur Eng Manag 3(3):237–245
9. http://www-pub.iaea.org/MTCD/Publications/PDF/te_1487_web.pdf
10. IAEA-TECDOC-1624 (2009) Passive safety systems and natural circulation in water cooled nuclear power plants. International Atomic Energy Agency, Vienna
11. IAEA-TECDOC-626 (1991) Safety related terms for advanced nuclear plants. International Atomic Energy Agency, Vienna
12. Jafari J, Auria FD (2003) Reliability evaluation of a natural circulation system. Nucl Eng Des 224:79104
13. Marques M, Pignatel JF (2005) Methodology for the reliability evaluation of a passive system and its integration into a probabilistic safety assessment. Nucl Eng Des 235:2612–2631
14. Marquez AC, Heguedas AS, Iung B (2005) Monte Carlo-based assessment of system availability. Reliab Eng Syst Saf 88:273–289
15. Myers RH, Montgomery DC (2002) Response surface methodology: process and product optimization using designed experiments, 2nd edn. John Wiley, New York
16. Nayak VJ, Gartia MR, Hari Prasad M et al (2008) Reliability assessment of passive containment isolation system using APSRA methodology. Ann Nucl Energy 35:2270–2279
17. Nayak VJ, Gartia MR, Hari Prasad M et al (2009) Reliability assessment of passive isolation condenser system of AHWR using APSRA methodology. Reliab Eng Syst Saf 94:1064–1075
18. Pagani LP, Apostolakis GE, Hejzlar P (2005) The impact of uncertainties on the performance of passive systems. Nucl Technol 149:129–140
19. http://www.iaea.org/INPRO/CPs/PGAP/Final_PGAP_report.pdf
20. Zadeh LA (1965) Fuzzy sets. Inf Control 8(3):338353
21. Zio E, Cantarella M, Cammi A (2003) The analytic hierarchy process as a systematic approach to the identification of important parameters for the reliability assessment of passive systems. Nucl Eng Des 226:311336

Chapter 6
Time-Variant Reliability Analysis

6.1 Introduction

Structural reliability analysis aims at computing the probability of failure of a structural system with respect to a prescribed failure criterion by accounting the uncertainties arising in the model description (geometry, material properties) or the environment (loading). When the behavior of the system under consideration evolves in time, the reliability problem is referred to as time-variant. In general just mentioning a value for the reliability does not give any meaning without specifying the period of time for which it was derived.

In general, the loading or stresses on structures and also the material properties are not constant, but changes with time. In the standard structural reliability analysis only the random variables are used in the failure probability calculations. If the loading process is stochastic in nature one should use random processes instead of random variables. As a result the failure probability is no longer just a single number, but also a function of the time. There are two classes of random processes, one is stationary and another one is non-stationary random processes. In stationary processes the statistical properties (mean, standard deviation, etc.) do not change with time. Whereas in the case of non-stationary processes these properties change with time. There are different ways in which the loading can take place on a structure and it can be either discrete in nature or continuous in nature. And also the loading can be time invariant or time variant. Hence, there are four categories with respect to sample space and time space. Appropriate reliability models have to be used to find the failure probability of structures/components subjected to these types of loadings.

Several researchers [1–4] have investigated different time-variant reliability models. Torres and Ruiz [5] proposed an approach to evaluate the reliability of structures that considers degradation of capacity over the lifetime of the structure. This reliability analysis considers events of different intensities by means of environmental hazard curves. Czarnecki and Nowak [6] developed a time-variant

A.K. Verma et al., *Risk Management of Non-Renewable Energy Systems*,
Springer Series in Reliability Engineering, DOI 10.1007/978-3-319-16062-7_6

reliability-based model for evaluating reliability of steel highway bridges subjected to corrosion, in which the limit state functions were formulated by considering both load and resistance parameters as time-variant random variables. Becker et al. [7] proposed a theory of dynamic reliability, which was extended to incorporate random changes of the state variables at the time points of transition between the discrete states of the Markovian component of the model. Streicher and Rackwitz [8] proposed a method for reliability-oriented time-variant structural optimization of separable (independent) series systems using first-order reliability methods in standard space. Kopustinskas et al. [9] developed an innovative dynamic reliability model based on stochastic differential equations. Wen and Chen [10], Melchers [11], developed the stress–strength interference (SSI) model to analyze the reliability of mechanical structures under time-varying loads. Huang [12], has developed a generalized reliability model based on SSI considering stochastic loading and strength aging degradation. Xie et al. [13], have used the concept of extended stress–strength interference model to calculate fatigue reliability under cyclic stress with uncertainty in stress amplitude. Wang and Xie [14], have developed dynamic reliability model of components using order statistics and probability differential equations. Newman [15], provided crack opening stress equations for fatigue crack growth.

In these problems, the time dependency may be of two kinds, one is loading may be randomly varying in time and second one is material properties may be decaying in time. The elementary reliability problem in "stochastic" (i.e., time variant) terms with a resistance $R(t)$ and a load effect $S(t)$, at time t is given as follows:

$$P_F(t) = P(R(t) \le S(t)) \tag{6.1}$$

In this chapter the solution of the above-mentioned Eq. 6.1 is presented depending on the nature of the load and resistance with example problems.

6.2 Different Types of Load Actions

Generally, the load application on a component or structure is random in nature. There are different ways in which the loading can take place and it can be either discrete in nature or continuous in nature. And also the loading can be time invariant or time variant. Hence, there are four categories with respect to sample space and time space. The following classification is usually made with respect to loading actions:

- Permanent loading (Dead weight, etc.)
- Variable loading (live loads, wind, snow, etc.)
- Exceptional loading (impact, fire, explosion, avalanches, etc.)

In the case of permanent loads there is of course no need for time-dependent modeling. It can well be modeled by a set of deterministic and random variables.

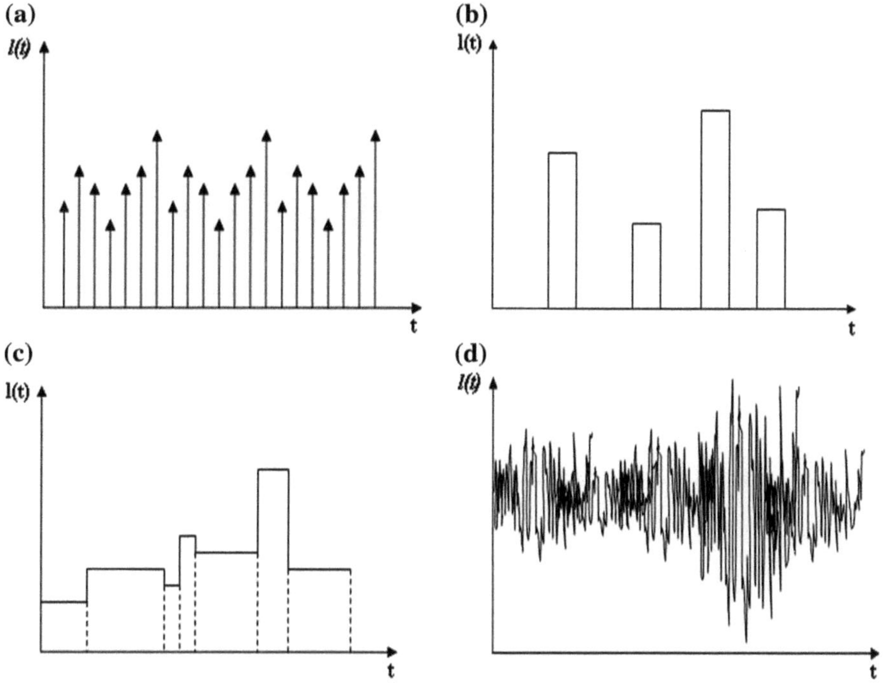

Fig. 6.1 Different types of load actions

For variable and exceptional loadings one has to develop time-dependent models. The appropriate time-variant models for action parameters may vary very much depending on the nature of the load. Some typical and useful process models are [16] (see Fig. 6.1):

- Random sequence
- Point pulse process with random intervals
- Rectangular wave process with random intervals
- Continuous and differentiable process.

6.3 Failure Probability Formulations

The general case in structural reliability theory is the case where one has only simple random variables (not time dependent) in the limit state function. These types of problems can be represented by two horizontal lines in a load resistance time diagram as shown in Fig. 6.2a. The probability of failure for a given period of time "*t*" can be given as follows:

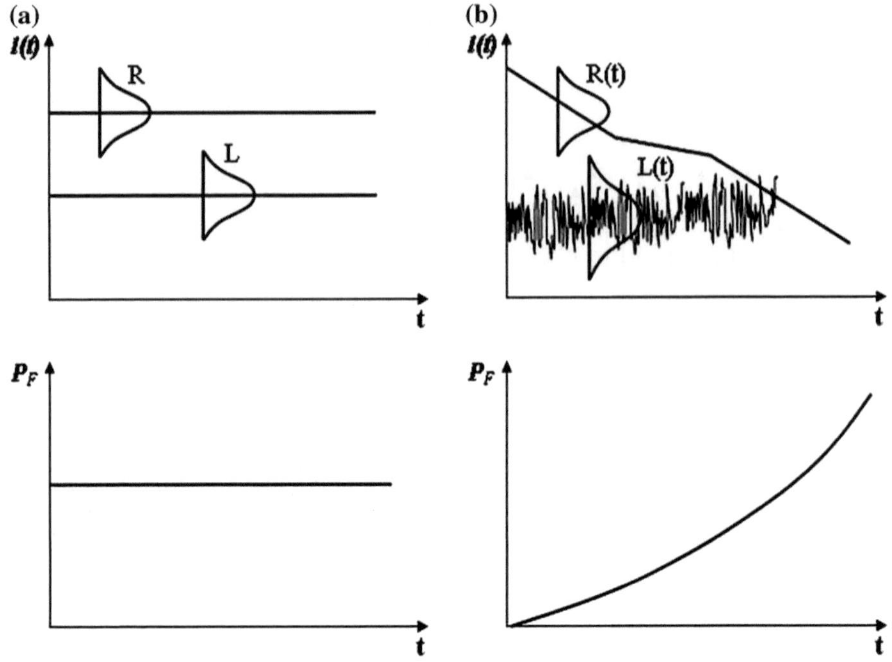

Fig. 6.2 Time invariant and variant reliability problems

$$P_F(t) = P(Z<0) \tag{6.2}$$

where Z is the limit state margin which can be written as follows:

$$Z = g(L, R) = R - L \tag{6.3}$$

where
g limit state function
R random variable representing the resistance
L random variable representing the load

Since in the above formulation neither load nor resistance is time variant hence the probability of failure is also time invariant and it will be constant for a given period of time as shown in Fig. 6.2a. One can solve these types of problems with available methodologies like first-order and second-order reliability methodologies.

Now consider a different problem in which the loading varies with time. In this case loading can be discrete in nature (cases a and b in Fig. 6.1) or it can be continuous in nature (cases c and d in Fig. 6.1). For such cases the probability of failure for a given period of time "t" can be given as follows:

$$\begin{aligned} P_F(t) &= P(Z(t) < 0) \\ Z(t) &= R - L(t) \end{aligned} \tag{6.4}$$

In this case the failure will occur if the limit state function is negative at one or more points of time in the time interval $0 < t < T$. If the loading is simple sequence of point loadings or spikes as shown in (a) and (b) of Fig. 6.1 then one can solve the problem with the help of extreme value theory. Suppose that the loading is taking place n times over a period of time t, then the component or system under these loadings will survive if it does not fail under the maximum among these loadings. One can find out the maximum load distribution based on the extreme value theory [17–19]. According to this the cumulative distribution function (CDF) of the equivalent load [20, 14] can be expressed as follows:

$$\begin{aligned} P(L_{\max} < l) &= P(L_1 < l \cap L_2 < l \cap L_3 < l \ldots \cap L_n < l) \\ F_X(l) &= P(L_1 < l)P(L_2 < l)P(L_3 < l)\ldots P(L_n < l) \\ &= F_L(l)F_L(l)F_L(l)\ldots F_L(l) \\ F_X(l) &= [F_L(l)]^n \\ F_X(x) &= [F_L(x)]^n \end{aligned} \tag{6.5}$$

The probability density function of the equivalent load can be expressed as follows:

$$\begin{aligned} f_X(x) &= \frac{\mathrm{d}F(x)}{\mathrm{d}x} \\ &= \frac{\mathrm{d}}{\mathrm{d}x}[F_L(x)]^n \\ &= n[F_L(x)]^{n-1}\frac{\mathrm{d}F_L(x)}{\mathrm{d}x} \\ f_X(x) &= n[F_L(x)]^{n-1}f_L(x) \end{aligned} \tag{6.6}$$

where $f_L(x)$ is the PDF of the loading.

Example 6.1 Assume that load on a structure follows normal distribution with mean = 40 MPa and standard deviation = 10 MPa. Find the probability density function of the maximum load assuming that structure is subjected to such loading 1, 10, 100, 500 times in its lifetime at different points of time.

Solution:

- Load is assumed to follow normal distribution with following distribution parameters
  - Mean = 40 MPa
  - Standard deviation = 10 MPa

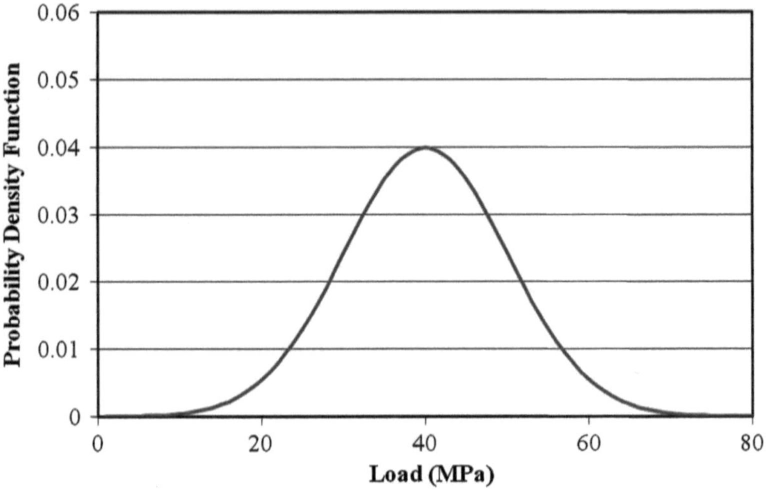

Fig. 6.3 Probability density function of load

- The probability density function of the load is shown in Fig. 6.3.
- Now consider the load is acting only once in the lifetime of the structure. In that case pdf of equivalent load will be same as the pdf of load which is same as the one shown in Fig. 6.3.

$$f_X(x) = n[F_L(x)]^{n-1} f_L(x)$$
$$n = 1$$
$$f_X(x) = 1[F_L(x)]^{1-1} f_L(x)$$
$$f_X(x) = f_L(x)$$

- Now consider the structure is subjected to load for 10 times then pdf of equivalent load is given as follows and is shown in Fig. 6.4.

$$n = 10$$
$$f_X(x) = 10[F_L(x)]^9 f_L(x)$$

- Similarly, if the structure is subjected to load for 100 or 500 times then pdf of equivalent load is shown in Fig. 6.5.

$$n = 100$$
$$f_X(x) = 100[F_L(x)]^{99} f_L(x)$$
$$n = 500$$
$$f_X(x) = 500[F_L(x)]^{499} f_L(x)$$

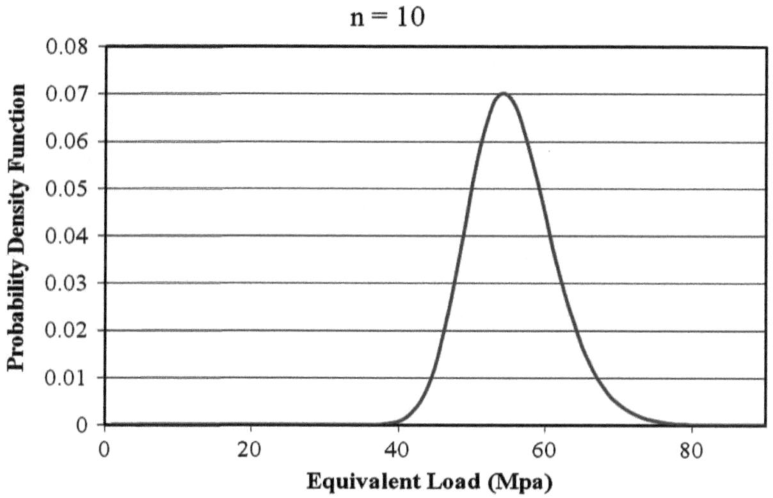

Fig. 6.4 Probability density function of equivalent load for $n = 10$

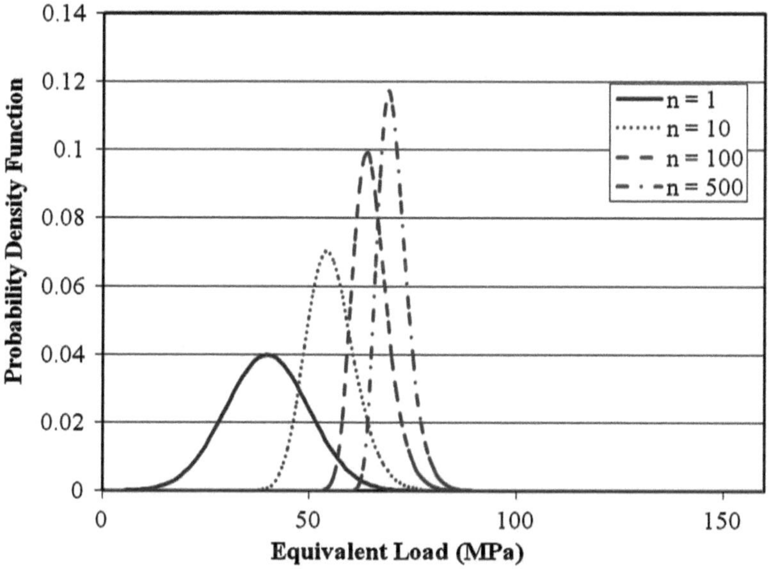

Fig. 6.5 Probability density function of equivalent load for different n values

- One can observe from the above results that as the number of times load action increases the mean value of the equivalent load increases as well as standard deviation decreases.

6.4 Stress–Strength Interference Model

Stress–Strength Interference (SSI) model [20] has been widely used for reliability design of structures. Figure 6.6 shows a simple case considering two variables one relating to the demand on the structure (load or stress on the structure) S, and the other relating to the capacity (strength or resistance) of the structure R. Both S and R are random in nature; their randomness is characterized by their means, μ_S and μ_R; standard deviations, σ_S and σ_R; and corresponding probability density functions, $f_S(s)$ and $f_R(r)$ as shown in Fig. 6.6. Structure failure takes place when load exceeds the resistance of the structure. Hence, the probability of failure of the structure can be represented as

$$\begin{aligned} P_f &= P(\text{failure}) = P(R < S) \\ &= \int_0^{\infty} \left[\int_0^{s} f_R(r)\,\mathrm{d}r \right] f_S(s)\,\mathrm{d}s \\ P_f &= \int_0^{\infty} F_R(s) f_S(s)\,\mathrm{d}s \end{aligned} \tag{6.7}$$

where $F_R(s)$ is the cumulative distribution function (CDF) of R evaluated at s. The Eq. 6.7 states that when the load is $S = s$ the probability of failure is $F_R(s)$, and since the load is a random variable, the integration needs to be carried out for all the possible values of S, with their respective likelihoods represented by the PDF of S. The CDF of

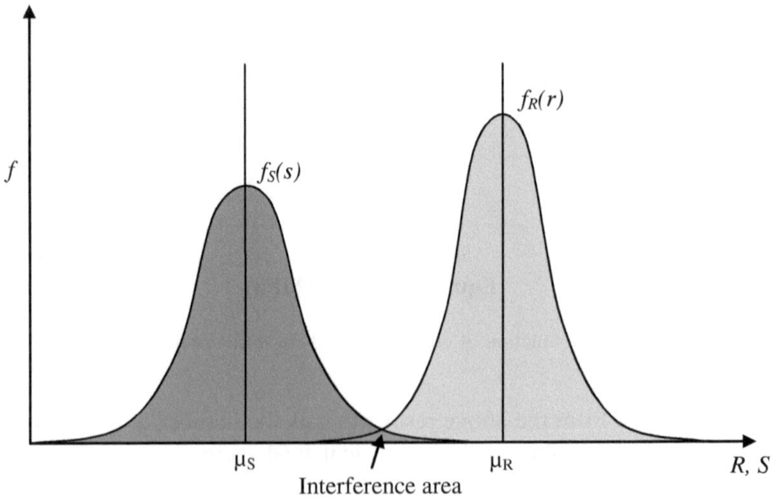

Fig. 6.6 Representation of stress–strength interference model

R or the PDF of S may not be always available inexplicit form and thus the integration of Eq. 6.7 may not be practical.

From the stress–strength interference model one can find out the failure probability of the structure subjected to loading for n number of times. Substituting the equivalent load in the stress–strength interference model, probability of failure can be estimated as

$$\begin{aligned} f_X(x) &= n[F_L(x)]^{n-1} f_L(x) \\ f_S(s) &= n[F_L(s)]^{n-1} f_L(s) \\ P_f &= \int_0^{\infty} F_R(s)\, n[F_L(s)]^{n-1} f_L(s)\,\mathrm{d}s \\ P_f &= n \int_0^{\infty} F_R(s)\, [F_L(s)]^{n-1} f_L(s)\,\mathrm{d}s \end{aligned} \tag{6.8}$$

Example 6.2 Assume that the strength (resistance) of the structure also follows a normal distribution with mean as 100 MPa and standard deviation as 15 MPa. Then find the failure probability of the structure if the structure is subjected to loading for 10, 100, and 500 times during its lifetime. Consider the loading distribution as mentioned in Example 6.1.

Solution:

- Load follows normal distribution
  - Mean = 40 MPa
  - Standard deviation = 10 MPa
- Resistance follows normal distribution
  - Mean = 100 MPa
  - Standard deviation = 15 MPa
- Pdf of load and resistance are shown in Fig. 6.7.
- Now consider the equivalent load distribution as obtained in Example 6.1 for n = 10, 100, and 500 times
- The pdf can be shown in Fig. 6.8 along with resistance.
- Probability of failure can be obtained as follows:

$$P_f = n \int_0^{\infty} F_R(s)\, [F_L(s)]^{n-1} f_L(s)\,\mathrm{d}s$$

- It is difficult to integrate the above equation. However, one can apply numerical integration and get the results as follows:

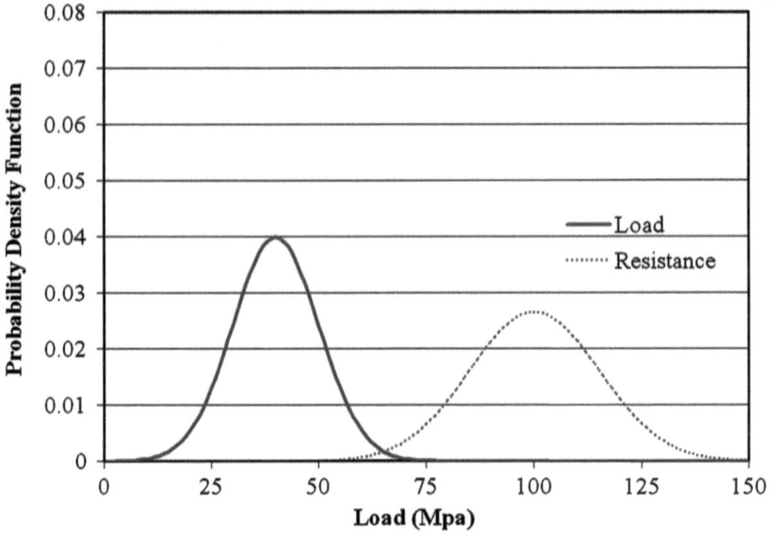

Fig. 6.7 Pdf of load and resistance

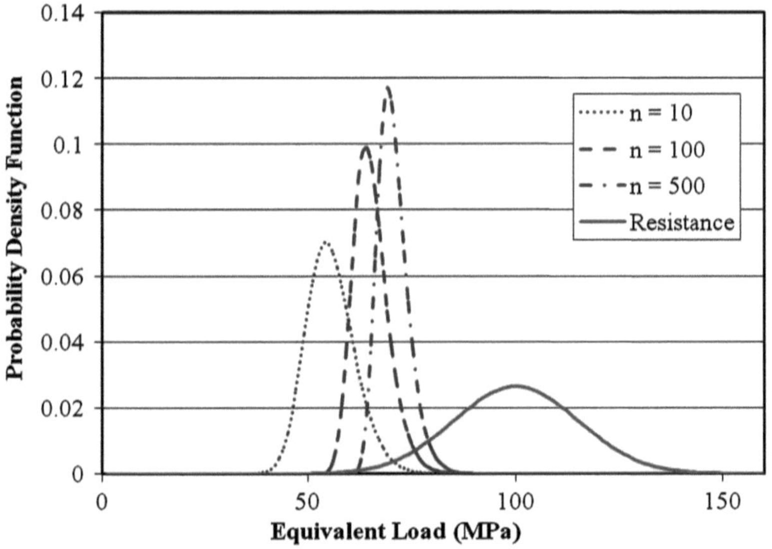

Fig. 6.8 Pdf of equivalent load and resistance

$$P_f = n\sum_s F_R(s)\,[F_L(s)]^{n-1} f_L(s)\,\Delta s$$

- For $n = 10$

$$P_f = 10\sum_s F_R(s)\,[F_L(s)]^{9} f_L(s)\,\Delta s$$

- Now, one can divide the load range into very small intervals assuming that $\Delta s = 1$ MPa.
- For the purpose of illustration consider $s = 70$ MPa and $\Delta s = 1$ MPa. Then

$$P_f = 10\left[F_R(70)\,[F_L(70)]^9 f_L(70)\,\times 1\right]$$

- From the standard normal tables one can find

$$\begin{aligned}
F_X(x) &= \Phi\left(\frac{x-\mu}{\sigma}\right)\\
F_R(70) &= \Phi\left(\frac{70-\mu_R}{\sigma_R}\right) = \Phi\left(\frac{70-100}{15}\right)\\
&= \Phi\left(\frac{-30}{15}\right) = \Phi(-2) = 0.0228\\
F_L(70) &= \Phi\left(\frac{70-\mu_L}{\sigma_L}\right) = \Phi\left(\frac{70-40}{10}\right)\\
&= \Phi\left(\frac{-30}{10}\right) = \Phi(-3) = 0.9986\\
f_L(s) &= \frac{1}{\sigma_L\sqrt{2\pi}}\mathrm{e}^{-\frac{1}{2}\left(\frac{s-\mu_L}{\sigma_L}\right)^2}\\
f_L(20) &= \frac{1}{10\sqrt{2\pi}}\mathrm{e}^{-\frac{1}{2}\left(\frac{20-40}{10}\right)^2}\\
&= \frac{1}{10\sqrt{2\pi}}\mathrm{e}^{-2^2} = 4.43\times 10^{-4}\\
P_f &= 10\left[4.82\times 10^{-8}\times\ [0.0228]^9\times 4.43\times 10^{-4}\right]\\
P_f &= 9.961\times 10^{-5}
\end{aligned}$$

- The reader should remember that the above calculated value is only a part contribution from entire range of loading. These calculations should be repeated for all the values of loading in its range.
- The probability of failure for different n values is obtained as follows:

$$P_f = \begin{cases} 3.023 \times 10^{-3} & n = 10 \\ 1.292 \times 10^{-2} & n = 100 \\ 2.786 \times 10^{-2} & n = 500 \end{cases}$$

- It can be seen from the above results that as the number of loadings increases the probability of failure also increases. This can also be seen from Fig. 6.8 as n increases the mean value of equivalent load increases and the interference area between load and resistance also increases. Hence, the probability of failure also increases.
- Assume that if the rate of occurrence of number of loadings is constant and it is 10 per year. Then in 1 year the structure can see the loading for 10 times. Similarly, it will take 10 years to see 100 times and in 50 years it will see 500 times. Hence, the above probability values can be related to time as well.
- That means the probability of failure of the structure with respect to time can be given as follows:

$$P_f = \begin{cases} 3.023 \times 10^{-3} & t = 1\,\text{year} \\ 1.292 \times 10^{-2} & t = 10\,\text{years} \\ 2.786 \times 10^{-2} & t = 50\,\text{years} \end{cases}$$

- The above calculations can be repeated for different values of n (considering rate of occurrence as constant and is 10 per year) and one can obtain probability of failure of the structure as a function of time and it is shown in Fig. 6.9.

6.5 Poisson Process

In the previous section it is assumed that the structure can be subjected to a random loading for any number of time in its lifetime. Also calculations have been carried out for fixed number of events arriving in a year. This can be further elaborated by

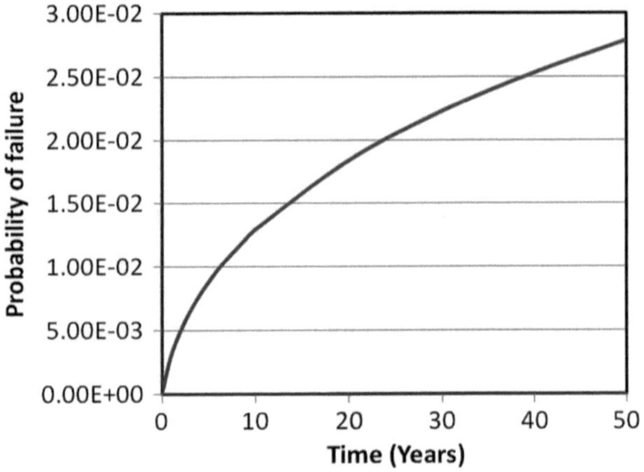

Fig. 6.9 Probability of failure as a function of time

assuming that events are occurring with Poisson arrival process. According to Poisson process the rate of occurrence of events is constant and no two events will occur in a very small interval of time. The occurrence of one event is independent of other event. The probability of occurrence of n events in a time period of "t" can be given as follows:

$$P(N = n) = \frac{e^{-\lambda t}(\lambda t)^n}{n!} \tag{6.9}$$

where λ is the rate of occurrence of events per year. This equation is valid when the rate of occurrence is considered as a constant (homogeneous) and it does not change with time. From this one can say that in a given period of time any number of events can occur. However, as the time increases probability of occurrence will also increase. Now, one can find the probability of failure of a structure subjected to Poisson loading process as follows [14]:

$$\begin{aligned} P_f &= \sum_{n=0}^{\infty} P(N = n) P_f(n) \\ P_f &= \sum_{n=0}^{\infty} \frac{e^{-\lambda t}(\lambda t)^n}{n!} n \int_0^{\infty} F_R(s)\,[F_L(s)]^{n-1} f_L(s)\,\mathrm{d}s \end{aligned} \tag{6.10}$$

6.6 Stochastic Fatigue Loading

A different way of dealing with the discrete random loading process is presented in this section on the basis of stochastic fatigue loading concept. When a metal is subjected to repeated cycles of stress or strain, it causes its structure to break down, ultimately leading to fracture, this behavior is called fatigue. When a body (specimen or structural component) is subjected to cyclic loads, the process of rupture starts with damage nucleation (voids, slip lines, micro cracks, etc.) at stress concentrators of the body [21]. After some time, a crack appears that propagates until fracture of the body. Generally, the maximum stress values are less than the ultimate tensile strength limit and may be below the yield strength limit of the material.

The nature of the failure apparently results from the fact that there are microscopic regions, usually on the surface of the member, here the localized stress becomes much greater than the average stress acting over the cross section. As this higher stress is cycled, it leads to the formation of minute cracks. Occurrence of these cracks causes a further increase of stress at their tips or boundaries, which in turn causes a further extension of the cracks into the material as the stress continues to be cycled. Eventually, cross-sectional area of the member is reduced to the point where the load can no longer be sustained, and as a result sudden fracture occurs. The material, even though known to be ductile, behaves as if it were brittle.

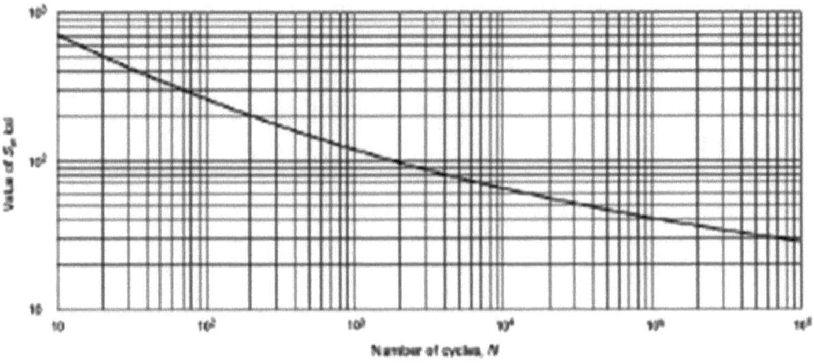

Fig. 6.10 S-N diagram for a typical metal

In the traditional fatigue analysis the fatigue data is represented as a S-N diagram. In this approach a series of specimens are each subjected to a specified stress and find the number of cycles to failure. The results are plotted as a graph representing the stress S as the ordinate and the number of cycles to failure N as the abscissa. Most often the values of N are plotted on a logarithmic scale since they are generally quite large. A typical S-N diagram for austenitic steel [21] is shown in Fig. 6.10. However, if a large number of specimens are tested for a given stress amplitude there will be large scatter in the number of cycles to failure for a particular value of the stress amplitude. Hence, one gets a distribution for the number of cycles to failure for a given stress amplitude. The uncertainty in the number of cycles to failure can be attributed to variation in material properties. This type of analysis leads to the introduction of probability of failure P with other quantities S and N and the S-N diagrams will be represented with P-S-N diagrams.

In this chapter a model developed by Hari Prasad et.al. [16] for reliability calculation of the system under stochastic cyclic loading is discussed. In this model stress amplitude, number of cycles to failure for a given stress amplitude and the number of cycles over a given period of time have been considered as uncertain parameters.

6.6.1 Representing Random Sequence Loading as Random Cyclic Loading

Consider the random loading as shown in Fig. 6.11a. In this case the loads are occurring at different times with different amplitudes. Now consider a NPP site in which the earthquakes occur at Poisson arrival process. The measure of the earthquake can be taken as peak ground acceleration (PGA). The systems in the plant can be subjected to different levels of earthquakes. Consider a particular level of earthquake with which the PGA at the site is calculated as 0.01 g and this level of

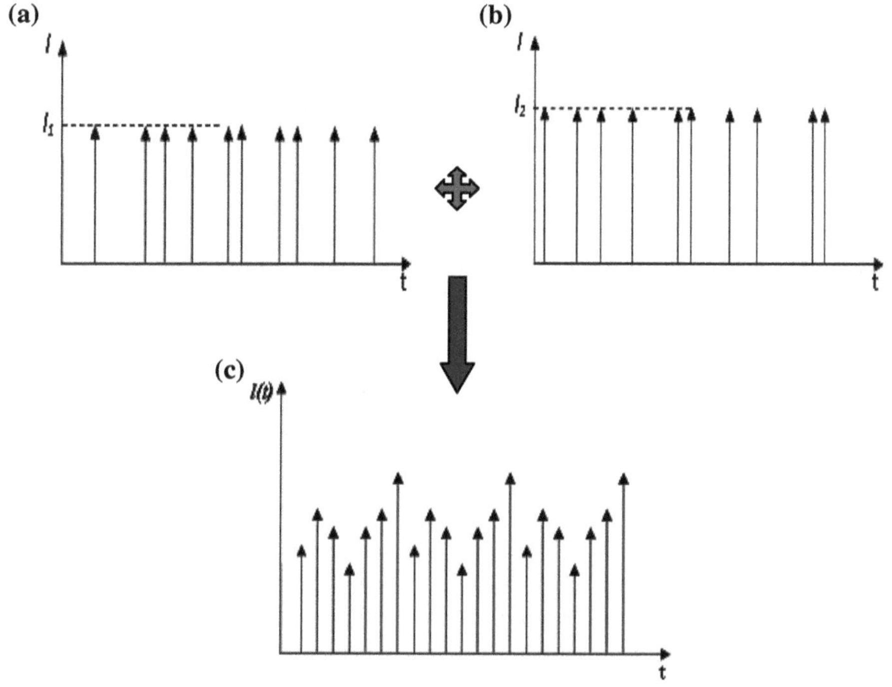

Fig. 6.11 Representing random loading as random cyclic loading

earthquake occurs at different points of time over the plant lifetime as shown in Fig. 6.11a.

In this diagram the first loading represents the peak ground acceleration of one earthquake similarly the second loading represents the PGA of second earthquake. If earthquakes of same PGA are plotted one can obtain Fig. 6.11a, but they will occur at different points of time. Now consider another level of earthquake 0.02 g which also occurs with a particular arrival rate over the plant lifetime as shown in Fig. 6.11b. In this way the NPP site can see different levels of earthquakes over the given plant lifetime. If one combines all these loadings one can obtain random sequence type of loading as shown in Fig. 6.11c and is same as the one shown in Fig. 6.11a. So far these types of loadings have been solved by using the concept of extreme value theory. But with this method one is neglecting the damage accumulation that will take place from one loading to another loading. Hence, in this chapter the concept of stochastic fatigue loading proposed in [16] is discussed to deal with the discrete stationary random processes.

In fatigue loading one has to find the stress amplitude. Now consider a piping system which is subjected to the normal loads such as deadweight, pressure, and thermal loads during the normal operation of a plant. Let s_{OP} be the stresses due to all these loadings. Whenever the system is subjected to a particular level of earthquake the stresses in the system change. Let s_1 be the minimum stress developed and s_2 be the maximum stress developed in the structure. The average on the difference of these stresses will give the stress amplitude due to a particular level of earthquake. Similarly for different levels of earthquakes, the stress amplitudes can be calculated. Now the question here is how to calculate the probability distribution function for different levels of earthquakes. This can be obtained from hazard curve analysis for a particular site; this is further discussed in the following section.

In the traditional fatigue analysis, the fatigue data is represented as S-N diagram and to take care of the uncertainty in the number of cycles to failure, which can be attributed to variation in material properties, the approach of P-S-N diagrams are used [21]. Let $f_N(n, s_i)$ denote the probability density function of number of cycles to failure for a given stress amplitude s_i. Then the probability of failure for a given number of cycles for a given stress amplitude can be given as follows:

$$P_F(N, s_i) = \int_0^N f_N(n, s_i)\mathrm{d}n \tag{6.11}$$

Now consider the variation in the stress amplitude. One can represent the variation in the parameter by a probability density function $f_S(s)$, it means, the stress amplitude can take the values of s_i ($i = 1, 2, \ldots, m$) with the respective probabilities of p_i ($i = 1, 2, \ldots, m$) depending on the probability density function. Now the failure probability can be given as (see Fig. 6.12):

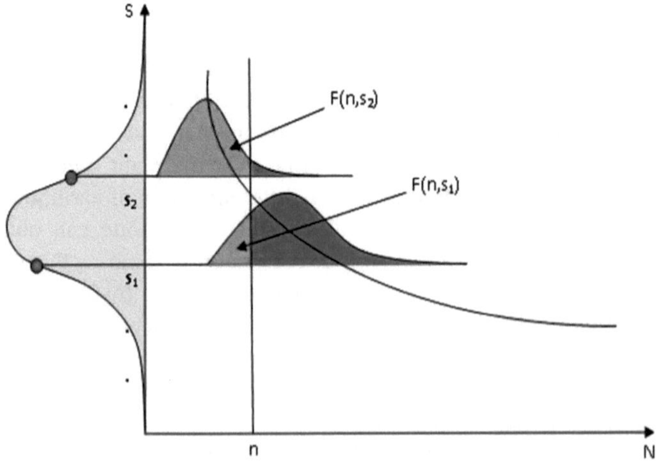

Fig. 6.12 Random stress amplitude and corresponding failure probabilities

$$P_F = \sum_{i=1}^{m} p_i F_N(n, s_i)$$
$$P_F = \sum_{i=1}^{m} p_i \int_0^N f_N(n, s_i) \mathrm{d}n \tag{6.12}$$

When the stress amplitude is a random variable with pdf $f_S(s)$, the weights p_i (i = 1, 2, …, m) in Eq. 6.12 can be approximately replaced by $f_S(s_i)\Delta s_i$, i.e.,

$$p_i = f_S(s_i)\Delta s_i \; (i = 1, 2, \ldots, m) \tag{6.13}$$

where s_i are the arbitrarily assigned discrete values of the random stress s, Δs_i are the associated stress intervals. Accordingly, Eq. 6.12 can be rewritten as:

$$P_F = \sum_{i=1}^{m} f_S(s_i)\Delta s_i \int_0^N f_N(n, s_i) \mathrm{d}n \tag{6.14}$$

Let $\Delta s_i \rightarrow 0$, $m \rightarrow \infty$ one can obtain the equation of fatigue failure probability in the situation of randomly distributed stress amplitude with the pdf $f_S(s)$ as:

$$P_F = \int_0^\infty \int_0^N f_S(s) f_N(n, s) \mathrm{d}s\mathrm{d}n \tag{6.15}$$

This expression can be correlated to stress–strength interference model. Hence, one can utilize the concepts of structural reliability methods or simply by discretising the range of the random variables solve the problem. This is the case when stresses are acting on the structure randomly, but if the stresses act in a Poisson arrival process, then one has to solve the problem by considering the arrival rate of the stresses and also probability of occurrence of the events as mentioned in the previous section.

6.6.2 A Case Study on Nuclear Piping

The procedure that has been explained in the previous sections is shown in a flow chart in Fig. 6.13 and is applied to a fire water piping system of a nuclear power plant. Fire water system is very important for a NPP from safety point of view. The main objective of fire water system is to ensure availability of water for firefighting under normal operating conditions and anticipated operational occurrences. The system has been designed to be available under OBE (operation-based earthquake)

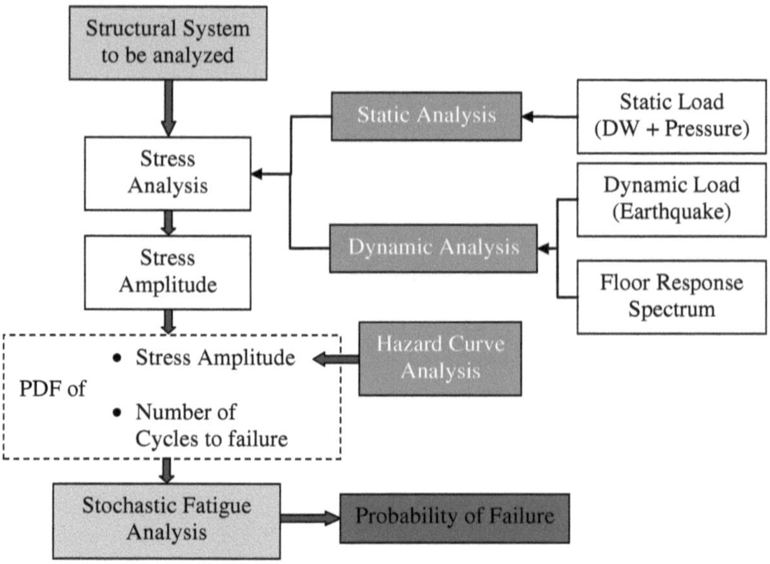

Fig. 6.13 Flow chart of the analysis [16]

and SSE (Safe shutdown earthquake) for firefighting of related equipment or system.

It has been proved by safety analysis that fire water system is the ultimate heat sink (supply fire water to the steam generator) in case of station blackout condition when no other decay heat removal system is available. In the present study the piping system that is connected to steam generator line has been analyzed. The piping system comprises of 250 NB schedule-40 carbon steel pipes and pipefitting. For the fatigue analysis it is important to find the stresses that act on the piping system. The piping system has been analyzed for pressure, dead weight, thermal, and earthquake loads. Under dead weight analysis, self-weight of pipe filled with water has been considered. Under the pressure loading, uniform maximum pressure has been considered at respective design/operating temperature.

- The material data that has been used in the present analysis is shown in Table 6.1.
- The dynamic analysis has been performed to evaluate the seismic response of the system under OBE condition. In the response spectrum analysis the percentage of critical damping 1 % for OBE has been taken.
- Finite element model of the piping systems is shown in Fig. 6.14 and the corresponding floor response spectra used in the analysis are shown in Fig. 6.15.

Table 6.1 Material data used in the analysis

Parameter	Value
Material	250 nb Sc40 carbon steel
Outer diameter	273.05 mm
Wall thickness	9.271 mm
Coefficient of thermal expansion (α)	10.8×10^{-6}/°C
Modulus of elasticity	202 KN/mm^2
Operating pressure	50 kg/cm^2
Temperature	40 °C

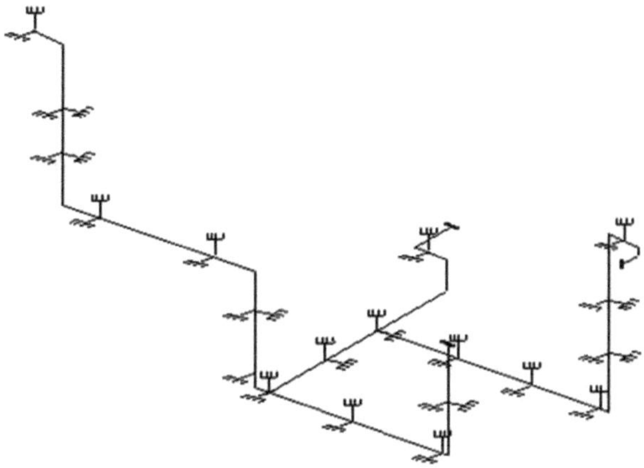

Fig. 6.14 Finite element model of steam generator piping system

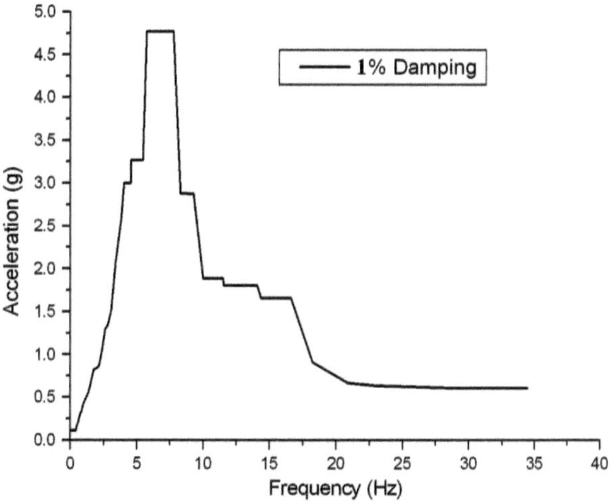

Fig. 6.15 Floor response spectra corresponding to steam generator line

Table 6.2 Maximum stress for operating and OBE loading

Case	Load	Max. stress (N/m^2)
1	Dead weight + Pressure	89.340×10^6
2	OBE	269.53×10^6
Stress amplitude		90.00×10^6

- Maximum stress obtained for sustained and OBE loading is shown in Table 6.2. The stress amplitude corresponding to the OBE level (0.075 g) is obtained as 90 MPa.
- Since the PGA is indirectly representing the stress amplitude, the probability density function of stress amplitude has been considered as lognormal distribution with the median as 90 MPa and an error factor as 1.875 and the same is shown in Fig. 6.16.
- The upper and lower limits of the distribution have been taken as 95th and 5th percentile values calculated from the error factor as follows.
- Let X denotes a random variable (PGA or Stress amplitude) and it follows lognormal distribution whose probability density function is given as follows:

$$f_X(x) = \frac{1}{\sigma\sqrt{2\pi}x} e^{\left[-\frac{1}{2}\left(\frac{\ln x - \mu_{\ln x}}{\sigma_{\ln x}}\right)^2\right]} \quad 0<x<+\infty$$

- Cumulative distribution function of X can be given as:

$$F_X(x) = \int_0^x f_X(x)\,dx = \Phi\left(\frac{\ln x - \mu_{\ln x}}{\sigma_{\ln x}}\right)$$

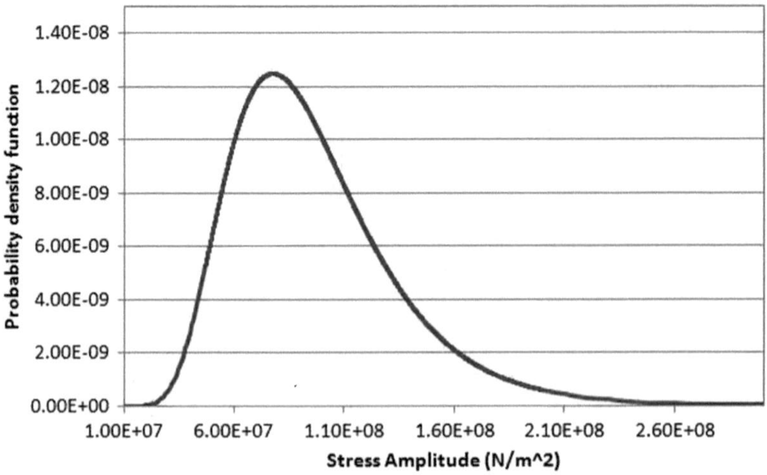

Fig. 6.16 Probability density function of stress amplitude

- Let $x = x_{50}$ (median) then the cumulative distribution function value corresponding to the median value will be 0.5. Now substitute this value in the cumulative distribution function of X. Then the equation becomes

$$\begin{aligned} F_X(x_{50}) &= \Phi\left(\frac{\ln x_{50} - \mu_{\ln x}}{\sigma_{\ln x}}\right) = 0.5 \\ \frac{\ln x_{50} - \mu_{\ln x}}{\sigma_{\ln x}} &= \Phi^{-1}(0.5) = 0 \\ \ln x_{50} - \mu_{\ln x} &= 0 \\ \mu_{\ln x} &= \ln(x_{50}) \end{aligned}$$

- Let $x = x_{95}$ (95th percentile value) then the cumulative distribution function value corresponding to this value will be 0.95. Now substitute this value and $\mu_{\ln x}$ in the cumulative distribution function of X. Then the equation becomes

$$\begin{aligned} F_X(x_{95}) &= \Phi\left(\frac{\ln x_{95} - \ln x_{50}}{\sigma_{\ln x}}\right) = 0.95 \\ \frac{\ln x_{95} - \ln x_{50}}{\sigma_{\ln x}} &= \Phi^{-1}(0.95) \\ \frac{\ln\left(\frac{x_{95}}{x_{50}}\right)}{\sigma_{\ln x}} &= 1.646 \\ \sigma_{\ln x} &= \frac{\ln\left(\frac{x_{95}}{x_{50}}\right)}{1.646} \end{aligned}$$

- Now define a term called Error Factor (EF) as the ratio of the 95th percentile value to the 50th percentile value or ratio of the 50th percentile value to the 5th percentile value and can be given as follows:

$$EF = \frac{x_{95}}{x_{50}} = \frac{x_{50}}{x_5}$$

- Then $\sigma_{\ln x}$ can be written as follows:

$$\sigma_{\ln x} = \frac{\ln(EF)}{1.646}$$

- Once the median value and error factor of the distribution are known one can estimate the lower and upper limits of the distribution, i.e., 95th and 5th percentile values as follows:

$$x_{95} = EF \times x_{50}$$
$$x_5 = \frac{x_{50}}{EF}$$
$$x_{50} = \sqrt{x_{95} \times x_5}$$

- In the present case

$$x_{50} = 90\,\text{MPa}$$
$$EF = 1.875$$
$$x_{95} = EF \times x_{50} = 1.875 \times 90 = 168.75\,\text{MPa}$$
$$x_5 = \frac{x_{50}}{EF} = \frac{90}{1.875} = 48\,\text{MPa}$$

- This means stress amplitude can take any value in its range with corresponding probability values.
- Now the range has been divided into 1000 equal intervals with interval size as 1.2075×10^5 N/m^2 (Δs). By using the S-N diagram for the given material one can obtain the number of cycles to failure.
- But this is also considered as a random variable and it follows lognormal distribution with median value as obtained number of cycles to failure and error factor as 10.
- From the above distribution, one can obtain $P(N < n)$, which is nothing but the cumulative distribution value and it represents the probability that the component will fail for the given number of cycles.
- Similarly one can obtain the probability of failure value for the different number of cycles (e.g., 10, 50, 100, etc.). Now consider annual rate of occurrence of earthquakes of magnitude greater than 3 and above as 6.
- This information can be calculated from Gutenberg and Richter recurrence relationship for a given NPP site. The recurrence relationship is generally represented by Gutenberg and Richter's Law as given below.

$$\log_{10} N_m = a - bm$$
$$N_m = 10^{a-bm}$$

- Where N_m is the cumulative number of earthquakes of a given magnitude or larger that are expected to occur during a specified period of time, "*a*" is the log of the number of earthquakes of magnitude zero or greater expected to occur during the same time and "*b*" is the slope of the curve which characterizes the proportion of large earthquakes to small earthquakes.
- The values of "*a*" and "*b*" for a given region are determined from the earthquake occurrence records of that region.
- For the NPP site which is discussed in the case study the values of "*a*" and "*b*" are estimated as 4.067 and 1.096, respectively. Substituting these values in Gutenberg–Richter's relation as discussed in Chap. 4, the equation will give

$$\log_{10} N_m = 4.067 - 1.096 \times m$$
$$N_m = 10^{4.067 - 1.096 \times m}$$
$$\text{let } m = 3$$
$$N_m = 10^{4.067 - 1.096 \times 3} = 6.012$$

- From the above the frequency of earthquakes of magnitude 3 (below which significant ground motion will not be produced at the site) and above can be estimated as 6.012/year (approximately 6 per year). This value has been considered in the present analysis.
- According to the design philosophy each earthquake is equivalent to 10 fatigue cycles of a given stress amplitude.
- Now if one considers a plant lifetime of 50 years on an average it will see more than 300 earthquakes. In this way one can calculate the probability of failure of the piping system as a function of time such as probability of failure for 1 year (60 cycles), 2 years (120 cycles), 10 years (600 cycles), and so on.
- For the present case on piping system the failure probability at 10 and 50 years has been estimated as 3.84×10^{-8} and 7.22×10^{-6}, respectively, and the same is shown in Fig. 6.17.

6.7 Out Crossing Approach

Now consider a problem where both resistance and load are varying with time. A somewhat different way of looking at the problem is to consider the safety margin associated with Eq. 6.1.

$$Z(t) = R(t) - S(t) \tag{6.16}$$

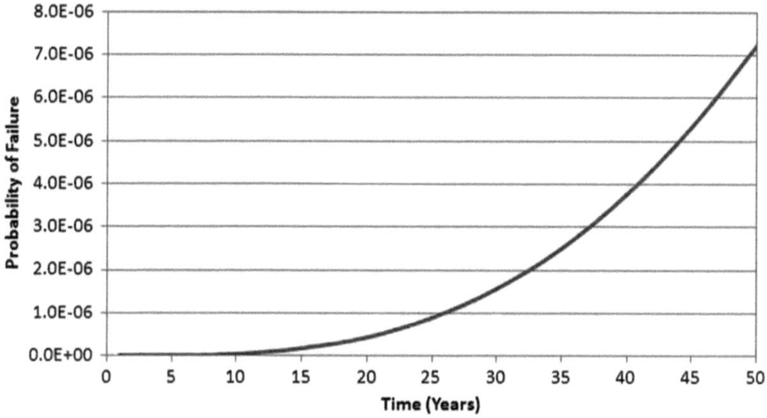

Fig. 6.17 Failure Probability of piping system as a function of time

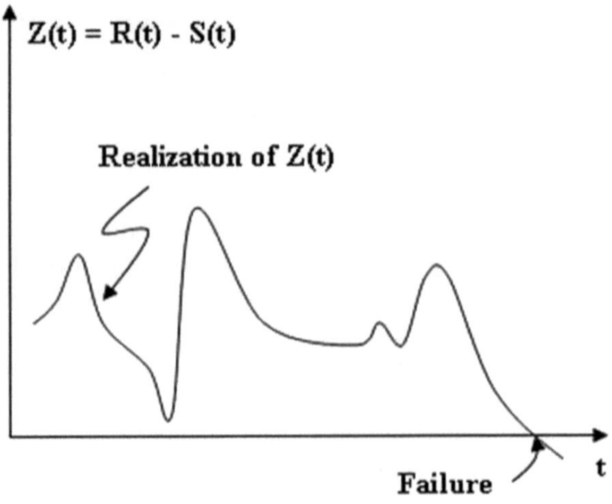

Fig. 6.18 Realization of safety margin process $Z(t)$ and time to failure

Here failure probability is defined as the probability that $Z(t)$ becomes zero or less in the lifetime t_L of the structure. This constitutes a so-called "crossing" problem [1, 22–25]. The time at which $Z(t)$ becomes less than zero for the first time is called the "time to failure" (Fig. 6.18) and is a random variable. The probability that $Z(t) \leq 0$ occurs during t_L is called the "first passage" probability. The corresponding situation in two-variable space is shown in Fig. 6.19. The probability that the process $X(t)$ will leave the "safe" region $G(X) > 0$ (i.e., the probability that an "outcrossing" will occur) during the structural lifetime t_L, is again the so-called "first-passage" probability. The first passage concept is more general than the classical approaches. In time-variant reliability one is only rarely interested in the instantaneous failure probability which one best interprets as some kind of

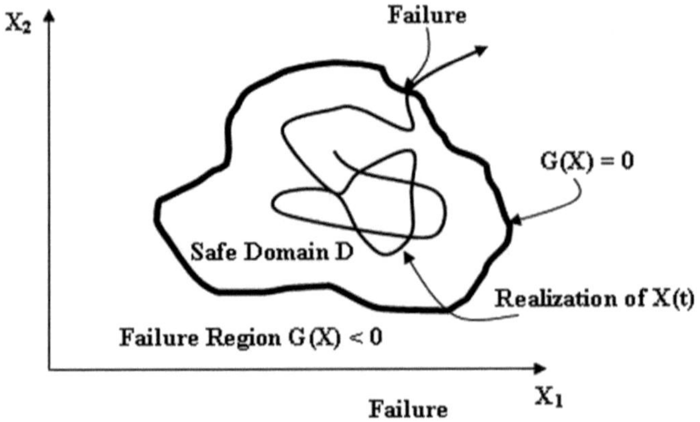

Fig. 6.19 Out crossing of vector process $X(t)$ [20]

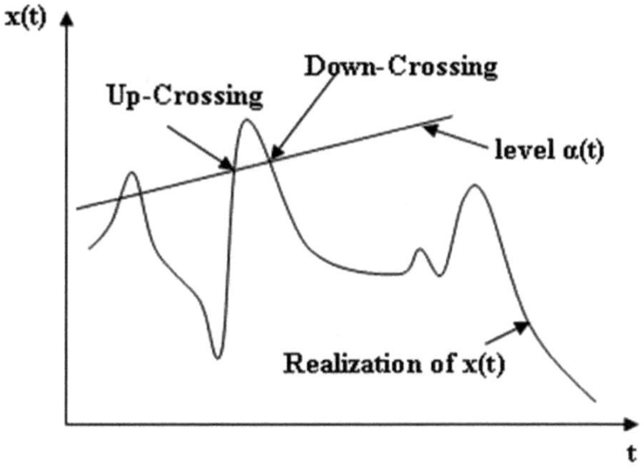

Fig. 6.20 Typical realization and level crossing of process $X(t)$

nonavailability. It is rather the distribution of the time of first passage of the trajectory of the time-dependent state function into the failure domain. But such distributions are known only for some few very special problems. In such cases the concept of out crossing/level crossing approach can be used [20].

Consider a continuous random process $X(t)$, $t \in T$, that is differentiable at least once. A sample function of the random process is shown in Fig. 6.20 [20]. The crossings of the level $X(t) = \alpha(t)$ with a positive slope (up-crossings) and negative slope (down-crossings) are shown in Fig. 6.20. Clearly, the number of crossings of the level $X(t) = \alpha(t)$ in the time interval T is a discrete random variable. Consider the segment of sample function $x(t)$ shown in Fig. 6.21 between the times t_1 and $(t_1 + \mathrm{d}t)$, where $\mathrm{d}t \rightarrow 0$. The time $(t_1 + \mathrm{d}t)$ may be considered to be the time at which the level crossing occurs. Also for $\mathrm{d}t$ sufficiently small the curves between t_1 and $t_1 + \mathrm{d}t$ can be taken as straight lines.

The random process which cross the level $X(t) = \alpha(t)$ during this interval must start below $\alpha(t)$ at t_1 and have sufficient slope $x(t)$ at t_1 to pass through $\alpha(t)$ during $\mathrm{d}t$. Then,

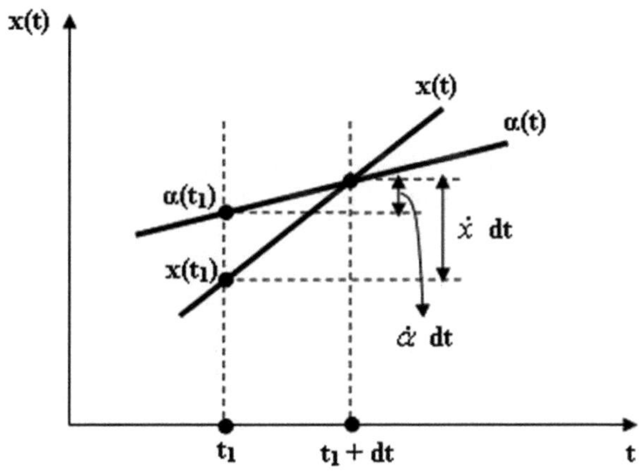

Fig. 6.21 Segment of sample function $x(t)$ crossing level $\alpha(t)$ in $\mathrm{d}t$ [20]

$$\begin{aligned}
& x(t+\mathrm{d}t) \geq \alpha(t+\mathrm{d}t) \\
& x(t+\mathrm{d}t) \geq x(t)+\dot{x}(t)\,\mathrm{d}t, \quad \dot{x}(t) > \alpha(t) \\
& \alpha(t+\mathrm{d}t) \geq \alpha(t)+\dot{\alpha}(t)\,\mathrm{d}t \\
& x(t)+\dot{x}(t)\,\mathrm{d}t \geq \alpha(t)+\dot{\alpha}(t)\,\mathrm{d}t \\
& x(t) \geq \alpha(t)-(\dot{x}(t)-\dot{\alpha}(t))\,\mathrm{d}t
\end{aligned} \tag{6.17}$$

Hence, the limits can be given as follows:

$$\begin{aligned}
& x(t) \geq \alpha(t)-(\dot{x}(t)-\dot{\alpha}(t))\,\mathrm{d}t \\
& x(t_1) \leq \alpha(t_1) \\
& \dot{x}(t) > \dot{\alpha}(t)
\end{aligned} \tag{6.18}$$

Let the probability of crossing the level $X(t) = \alpha(t)$ with a positive slope in the interval t_1 and $(t_1 + \mathrm{d}t)$ be denoted by $N_a^+(\mathrm{d}t)$ then

$$\begin{aligned}
& N_\alpha^+\mathrm{d}t = P[(\alpha(t)-(\dot{x}(t)-\dot{\alpha}(t))\,\mathrm{d}t \leq x(t) \leq \alpha(t)) \cap (\dot{x}(t) > \dot{\alpha}(t))] \\
& N_\alpha^+\mathrm{d}t = \int\limits_{\dot{\alpha}}^{\infty} \int\limits_{\alpha-(\dot{x}-\dot{\alpha})\mathrm{d}t}^{\alpha} f_{X\dot{X}}(x,\dot{x})\,\mathrm{d}x\,\mathrm{d}\dot{x}
\end{aligned} \tag{6.19}$$

Since the limits of integration of the second integral differ by an infinitesimal quantity, one can use the mean value theorem to set

$$\int\limits_{\alpha-(\dot{x}-\dot{\alpha})\mathrm{d}t}^{\alpha} f_{X\dot{X}}(x,\dot{x})\,\mathrm{dx}\,\mathrm{d}\dot{x} = f_{X\dot{X}}(x,\dot{x})\,(\dot{x}-\dot{\alpha})\mathrm{d}t \tag{6.20}$$

Substitute (2.19) into (2.18) and cancel dt from both sides:

$$N_\alpha^+ = \int\limits_{\dot{\alpha}}^{\infty} (\dot{x}-\dot{\alpha}) f_{X\dot{X}}(x,\dot{x})\,\mathrm{d}\dot{x} \tag{6.21}$$

which is due to Rice [26]. As a special case if α is not time dependent $\dot{\alpha} = 0$. Substituting this into Eq. 6.21, then the equation becomes

$$N_\alpha^+ = \int\limits_{0}^{\infty} \dot{x} f_{X\dot{X}}(x,\dot{x})\,\mathrm{d}\dot{x} \tag{6.22}$$

Example 6.3 In the special case when *X*(*t*) is a stationary normal process, $f_{X\dot{X}}(x,\dot{x})$ is given by

$$f_{X\dot{X}}(x,\dot{x}) = \frac{1}{2\pi\sigma_X\sigma_{\dot{X}}}\mathrm{e}^{-\frac{1}{2}\left[\left(\frac{x-\mu_X}{\sigma_X}\right)^2+\left(\frac{\dot{x}}{\sigma_{\dot{X}}}\right)^2\right]}$$

in which *X*(*t*) is normally distributed $N\left(\mu_X,\sigma_X^2\right)$ and $\dot{X}(t)$ is $N\left(0,\sigma_{\dot{X}}^2\right)$. The mean of $\dot{X}(t)$ is zero for a stationary process. Hence,

$$f_{X\dot{X}}(\alpha,\dot{x}) = \frac{1}{2\pi\sigma_X\sigma_{\dot{X}}}\mathrm{e}^{-\frac{1}{2}\left[\left(\frac{\alpha-\mu_X}{\sigma_X}\right)^2+\left(\frac{\dot{x}}{\sigma_{\dot{X}}}\right)^2\right]}$$

$$N_\alpha^+ = \int_0^\infty \dot{x} f_{X\dot{X}}(x,\dot{x})\,\mathrm{d}\dot{x}$$

$$N_\alpha^+ = \int_0^\infty \dot{x}\frac{1}{2\pi\sigma_X\sigma_{\dot{X}}}\mathrm{e}^{-\frac{1}{2}\left[\left(\frac{\alpha-\mu_X}{\sigma_X}\right)^2+\left(\frac{\dot{x}}{\sigma_{\dot{X}}}\right)^2\right]}\,\mathrm{d}\dot{x}$$

$$N_\alpha^+ = \frac{1}{2\pi\sigma_X\sigma_{\dot{X}}}\mathrm{e}^{-\frac{1}{2}\left(\frac{\alpha-\mu_X}{\sigma_X}\right)^2}\int_0^\infty \dot{x}\mathrm{e}^{-\frac{1}{2}\left[\left(\frac{\dot{x}}{\sigma_{\dot{X}}}\right)^2\right]}\,\mathrm{d}\dot{x}$$

Let

$$z = \frac{1}{2}\left(\frac{\dot{x}}{\sigma_{\dot{X}}}\right)^2$$

$$\frac{\mathrm{d}z}{\mathrm{d}\dot{x}} = \frac{1}{2}\frac{2\dot{x}}{\sigma_{\dot{X}}^2}$$

$$\mathrm{d}\dot{x} = \frac{\sigma_{\dot{X}}^2}{\dot{x}}\mathrm{d}z$$

$$N_\alpha^+ = \frac{1}{2\pi\sigma_X\sigma_{\dot{X}}}\mathrm{e}^{-\frac{1}{2}\left(\frac{\alpha-\mu_X}{\sigma_X}\right)^2}\int_0^\infty \dot{x}\mathrm{e}^{-z}\frac{\sigma_{\dot{X}}^2}{\dot{x}}\mathrm{d}z$$

$$N_\alpha^+ = \frac{1}{2\pi\sigma_X\sigma_{\dot{X}}}\mathrm{e}^{-\frac{1}{2}\left(\frac{\alpha-\mu_X}{\sigma_X}\right)^2}\sigma_{\dot{X}}^2\int_0^\infty \mathrm{e}^{-z}\mathrm{d}z$$

$$\int_{0}^{\infty} e^{-z} dz = 1$$

$$N_{\alpha}^{+} = \frac{\sigma_{\dot{X}}}{2\pi\sigma_X} e^{-\frac{1}{2}\left(\frac{\alpha-\mu_X}{\sigma_X}\right)^2}$$

But

$$f_X(\alpha) = \frac{1}{\sigma_X\sqrt{2\pi}} e^{-\frac{1}{2}\left(\frac{\alpha-\mu_X}{\sigma_X}\right)^2}$$

$$N_{\alpha}^{+} = \frac{\sigma_{\dot{X}}}{\sqrt{2\pi}} f_X(\alpha) \tag{6.23}$$

6.8 Strength Degradation

So far in this chapter in the failure probability calculations of structure it is considered that only the load changes with time and variation of strength (resistance) of the material with respect to time is not considered. But in actual practice strength of the material changes due to some degradation mechanisms that the material undergoes during its service lifetime. Hence, while calculating the probability of failure of a component one should consider both loading and strength variation with respect to time. This is further explained in the following subsections. Some of the degradation mechanisms are listed as follows.

Piping failures typically result from degradation mechanisms and loading conditions (e.g., IGSCC, flow accelerated corrosion, thermal stratification, etc.) not anticipated in the original design. Since the likelihood of a piping failure is strongly dependent upon the presence of an active degradation mechanism in combination with service conditions and transient loading conditions, it has been established that the relative rupture frequency of a pipe segment can be determined by evaluating the type of degradation mechanism present in a pipe segment during any mode of operation and by considering the associated loading and service conditions. The different degradation mechanisms that are present are stress corrosion cracking, erosion corrosion, flow accelerated corrosion, crevice corrosion & pitting, erosion–cavitation, thermal fatigue, vibration fatigue, water hammer, etc. These are explained in detail below [30].

6.8.1 Stress Corrosion Cracking

Stress Corrosion Cracking (SCC) refers to cracking caused by the simultaneous presence of tensile stress and a corrosive medium. The important variables affecting SCC are temperature, solution composition, metal composition, stress, and metal structure. Both intergranular (crack proceeds along the grain boundary) and transgranular (crack growth is not affected by grain boundaries) cracking have been observed. The criteria for stresses are that they may be simple tensile and of sufficient magnitude. These stresses may be due to any source, such as applied, residual, thermal, or welding. The minimum threshold tensile stress depends on the alloy-environment conditions.

6.8.2 Intergranular Stress Corrosion Cracking

Intergranular Stress Corrosion Cracking (IGSCC) results from a combination of sensitized materials (caused by a depletion of chromium in the regions adjacent to the grain boundaries in weld heat affected zones), high stress (residual welding stresses), and a corrosive environment (high level of oxygen or other contaminants). IGSCC is encountered most frequently in austenitic stainless steels that have become sensitized through the welding process and are subjected to operating environments.

6.8.3 Primary Water Stress Corrosion Cracking

Primary water Stress Corrosion Cracking (PWSCC) occurs when high temperature primary water is the corrosive medium and is present in combination with susceptible materials and high stress. Piping attachments are considered susceptible that is cold worked and welded without subsequent stress relief, are exposed to primary water.

6.8.4 Erosion Corrosion or Flow Accelerated Corrosion

Flow Accelerated Corrosion (FAC) is a complex phenomenon that exhibits attributes of erosion and corrosion in combination. Factors that influence, whether FAC is an issue, are velocity, dissolved oxygen, pH, moisture content of steam, and material chromium content. Carbon steel piping with chromium content greater than 1 % and austenitic steel piping are not susceptible to degradation from FAC.

6.8.5 Crevice Corrosion and Pitting

The electrochemical reaction caused by a corrosive or oxygenated media within a piping system can lead to cracking. Regions containing crevices (narrow gaps) that can result in oxygen depletion and relatively high concentration of chloride ions or other impurities are considered susceptible to crevice corrosion cracking. Pitting is an extremely localized attack that results in the formation of small holes/pits in pipe component wall. Pitting can occur in carbon as well as in stainless steel. In nuclear plants, pitting is observed most often in ferritic service water systems.

6.8.6 Erosion–Cavitation

This degradation mechanism represents degradation caused by turbulent flow conditions, which erode the pipe wall by cavitation. Cavitation damage is the result of the formation and instantaneous collapse of small voids within fluid subjected to rapid pressure and velocity changes as it passes through a region where the flow is restricted (e.g., a valve, pump, or orifice)

6.8.7 Thermal Fatigue

In this damage mechanism, alternating stresses caused by thermal cycling of a component result in accumulated fatigue usage and can lead to crack initiation and growth. Austenitic and carbon steel piping segments with operating temperatures less than 270° F and 220° F, respectively, are not susceptible to significant degradation by thermal fatigue. In nuclear power plants, piping components having operating temperatures greater than these values, two forms of thermal cycling have been observed—thermal transient cycles and thermal stratification cycling/striping. Significant thermal transient conditions occur when piping components are subjected to reoccurring rapid temperature changes greater than 150–220° F. Pipe nozzles and branch connections subject to cold shock operating conditions tend to be most susceptible.

In some pipe segments, large radial temperature differences are caused by the coexistence of hot and cold fluids in the pipe. This condition, referred to as thermal stratification, tends to bow pipe section. When flow conditions leading to thermal stratification are unsteady rapid movement of the hot–cold interface inside the pipe, the result can be thermal cycling at the inside pipe surface. This form of cycling is capable of producing thermal fatigue cracking and crack growth leading to leakage. A special form of thermal cycling that occurs under thermal stratification conditions is referred to as striping. The effects of thermal striping are typically limited to the pipe material in the immediate vicinity of inside surface.

6.8.8 *Vibration Fatigue*

Vibration fatigue failures are normally a result of poor component design and fabrication practice. Once a crack initiates, failure quickly follows. Consequently, this mechanism does not lend itself to typical periodic in-service examinations as a means of managing this degradation mechanism.

6.8.9 *Water Hammer*

The majority of water hammer events are caused by acceleration or deceleration of fluid flow caused by a pump start, pump trip, rapid control valve or isolation valve operation, check valve closure, safety, and relief valve operation. These events typically occur mostly in a single-phase flow environment and result in pressure waves that propagate along the pipe at sonic velocity. These water hammer events are well understood and typically anticipated and accounted for in the pipe design. Consequently, damage from these type water hammer loads is usually insignificant.

It is recognized that individual failure events may involve combinations of loading conditions and damage mechanisms. In the present chapter stress corrosion cracking has been explained briefly.

6.9 Stress Corrosion Cracking

Generally, piping components in nuclear power plants are subjected to different types of degradation mechanisms depending on the material composition, operating conditions, environment, and location. One of the degradation mechanisms that is important from piping point of view is stress corrosion cracking (SCC). This can be observed mostly in alloyed steels, especially in austenitic stainless steels. Due to the combined action of tensile stresses and corrosive environment SCC mechanism initiates cracks in the material.

In general, piping components failure due to SCC are rare compared with failures due to other degradation mechanisms such as vibration fatigue, thermal fatigue, erosion corrosion, etc. But, when it occurs it can be costlier and destructive. Normally, it is difficult to detect SCC in a piping component. Several case studies have reported damage involving cracking, wall thinning, and leakage in nuclear pipelines that are attributed to SCC [27]. SCC is a corrosion mechanism that forms cracks in susceptible material in the presence of an aggressive environment and tensile stresses. SCC can be intergranular or transgranular in nature depending on the material, level of stress, and environment. SCC occurs when the following three conditions occur simultaneously [28]:

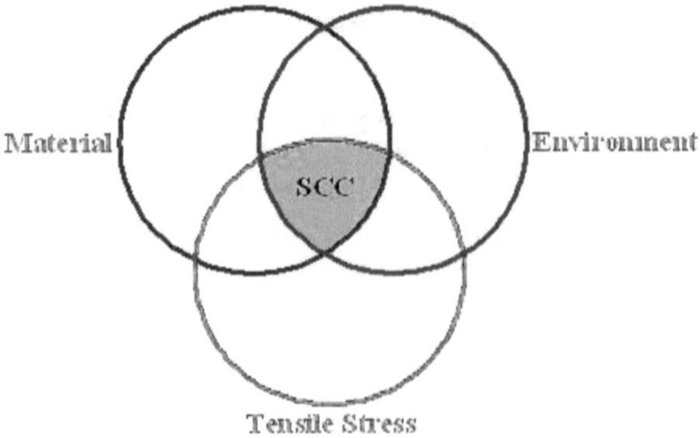

Fig. 6.22 Three-parameter representation of SCC

- Susceptible material;
- Tensile stress (applied and residual);
- An environment that can provide the chemical driving force for corrosion reaction as shown in Fig. 6.22.

In the nuclear power plant context, piping components are designed to withstand different types of loadings such as static loads (dead weight), operating pressure, thermal loads, seismic loads, and other transient loads. Even though the piping systems are designed for the above-mentioned loads it can be damaged due to SCC which occurs in a susceptible material, in a corrosive environment in the presence of high temperature and high applied and residual stresses. Since, the environmental conditions and operating conditions vary during the lifetime of the power plant, the safety of the piping component against SCC can be assessed in a probabilistic sense.

In this chapter, a piping component made of SS304 stainless steel is considered for the analysis, because this type of steel is highly susceptible to SCC, especially the material near the welds is more susceptible to SCC. This is because welding in pipelines results in considerable residual stress, and also increases the degree of sensitization due to the microstructural changes brought about in the material by spending more time at critical temperature. Austenitic stainless steels are commonly used in power generating industries because of high ductility and fracture toughness. This type of steel is used in applications where corrosion resistance is an important characteristic. However, under specific conditions localized attack in the form of pitting or SCC may occur.

General methodology recommended in PRAISE [28] has been used to estimate the failure probability of piping system subjected to SCC. Empirical relations are used to model the initiation and early growth rate of cracks. Fracture mechanics concepts are used deterministically in this study. The operating conditions like pressure and temperature and microstructural properties of the material show variations during the lifetime of the pipelines. To account for these variations, applied stress, degree of sensitization, initiation time of cracks, crack growth velocity after initiation, and initial crack length have been considered as random variables. Failure probabilities of the component at different times against failure criteria, namely, based on net section stress and detectable leak rate (as recommended in PRAISE) are computed using Montecarlo simulation technique and the results are discussed.

6.9.1 Time to Initiation

SCC is time-dependent phenomena. SCC may not occur as soon as the piping system put into the operation. It will take some time to initiate in the system. The time to initiation of SCC depends on the tensile stress, environment (characterized by the temperature, oxygen concentration, and/or the concentration of ionic species), alloy chemistry, and micro structure. Since, the microstructural properties of the material, environmental, and operating conditions vary during the lifetime of the piping system, the time to initiation needs to be treated as a random variable.

In PRAISE [28], the time to initiation of SCC is considered as a function of the damage parameter, *D*, which represents the effects of the loading, environment, and material variables on SCC. The statistical properties of initiation time were determined mainly based on data from laboratory experiments on specimens of 2 in length. Based on the experimental data it is assumed that the time to initiation (t_I) follow lognormal distribution with mean and standard deviation of log t_I as follows:

$$\begin{aligned} \mu_{\log t_I} &= B_0 + B_1 \log(D) \\ \sigma_{\log t_I} &= B_2 + B_3 \log(D) \end{aligned} \tag{6.24}$$

where, B_0, B_1, B_2, and B_3 are constants which can be calculated from experimental data and generally depend on the loading conditions (i.e., constant or varying load) and material type. Damage parameter (*D*) includes three multiplicative terms representing the three conditions necessary for initiation of SCC. The damage parameter is given by [28, 29]

$$D = f_1(\text{material}) \cdot f_2(\text{environment}) \cdot f_3(\text{loading}) \tag{6.25}$$

where f_1, f_2, and f_3 are given by

$$f_1 = \mathrm{C}_1 (P_a)^{\mathrm{C}_2} \tag{6.26}$$

where P_a is a measure of the degree of sensitization known as electrochemical potentiokinetic reactivation (C/cm^2).

$$f_2 = \mathrm{O}_2^{\mathrm{C}_3} \cdot e^{\frac{\mathrm{C}_4}{T+273}} \cdot \log\left(\mathrm{C}_5 \gamma^{\mathrm{C}_6}\right) \tag{6.27}$$

where O_2 is the oxygen concentration (ppm), T the temperature (°C), and "γ" is the water conductivity (μs/cm). The loading term f_3 is considered to be a function of stress if the applied load is constant and can be given by

$$f_3 = \left(C_8 \sigma^{C_9}\right)^{C_7} \tag{6.28}$$

where σ is stress in ksi. The constants C_1–C_9 can be calculated from experimental data and generally depend on material type.

6.9.2 Crack Size at Initiation

In the present study the shape of the surface crack initiated due to SCC is considered as semi-elliptical as assumed in PRAISE code. The geometry of the part-through initial crack considered is shown in Fig. 6.23. The size of the crack at initiation is assigned according to the recommendations in PRAISE. Surface length of initiated cracks, ($l = 2b$), is assumed to be lognormally distributed with a median value of 1/8 in and standard deviation of ln(b) as 0.89. The depth of initiated crack is taken to be 0.001, since identification of cracks with depths less than this value involves large uncertainty and is highly subjective because of the difficulty in distinguishing very shallow cracks from general surface roughening.

6.9.3 Crack Propagation Due to SCC

In the present study, it is assumed that the initiated cracks grow at a constant velocity (initiation velocity) until conditions are appropriate for treating the crack growth by fracture mechanics. The statistical properties of initiation velocity are determined using the expressions given in PRAISE, which are obtained by

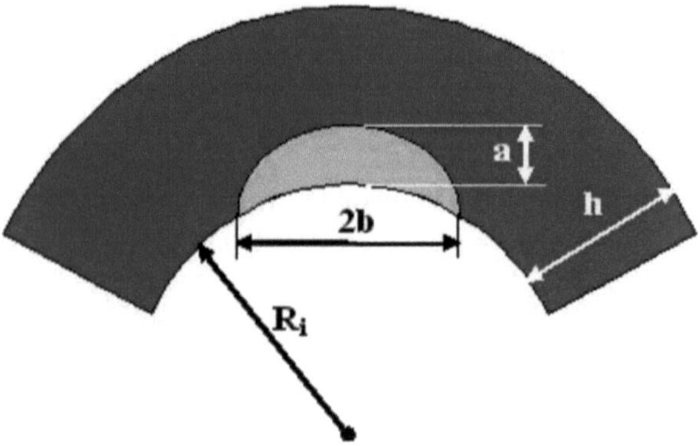

Fig. 6.23 Crack representation in a pipe

correlating the results of laboratory experiments on 2 in long specimens with the damage parameter of the specimens. In order to take into account the considerable scatter in the initiation velocity observed during the experiments, the initiation velocity is considered as a lognormally distributed random variable. The mean of $\log(v_1)$ varies linearly with D, whereas standard deviation of $\log(v_1)$ is independent of D. The mean of initiation velocity, v_1(in/year) is given by

$$\mu_{\log v_1} = F + G\log(D) \tag{6.29}$$

where F and G are constants whose values are specified in PRAISE. In this study, the following procedure is followed to govern the transition from initiation to fracture mechanics crack growth rate as described in PRAISE.

- Fracture mechanics velocity is assigned to preexisting cracks.
- Initiated cracks will grow with initiation velocity.
- At any given time, if the initiation velocity is less than fracture mechanics velocity, thereafter that particular crack grows with fracture mechanics velocity.
- If the depth of the crack is greater than 0.1 in, the growth of that crack will always be governed by fracture mechanics principles.
- The crack will not grow if the stress intensity factor of that crack is negative.

The fracture mechanics-based crack growth velocity v_2 (in/year), is calculated as follows:

$$\log(v_2) = C_{14} + C_{15}D_K \tag{6.30}$$

where D_K is stress intensity factor (K)-related damage parameter given by

$$D_K = C_{12}\log[f_2(\text{environment})] + C_{13}K \tag{6.31}$$

The constants C_{12}–C_{15} are as specified in PRAISE using experimental data. C_{14} is a random variable represented by normal distribution. Growth of existing cracks is affected by the stress field around the cracks, which is defined uniquely by the stress intensity factor. The expressions given in PRAISE have been utilized for computing the values of stress intensity factors in the depth (K_a) and length (K_b) directions.

6.9.4 Failure Criteria

The failure criteria used in this study for determining the failure probabilities of the piping component against SCC is based on net section stress and detectable leak rate (PRAISE) [28]. The part-through initial stress corrosion cracks considered can grow and become unstable part-through cracks or stable or unstable through-wall cracks. The stability of the part-through or through-wall crack is checked by comparing net section stress with the flow stress of the material. The net section stress criterion is applicable to very tough material, and the failure is due to the insufficient remaining area to support the applied loads (i.e., net section stress due to applied loads becomes greater than the flow stress of the material). The net section stress is given as

$$\sigma_{\text{net}} = \frac{\sigma_{LC} \times A_P}{A_P - A_{Cr}} \tag{6.32}$$

where σ_{LC} is the load controlled component of stress (stress due to dead weight and operating pressure), A_P the area of cross section of pipe, and A_{cr} is the area of crack given by

$$\begin{aligned} A_P &= \pi h(2R_i + h) \\ A_{cr} &= ab\left[2 + \frac{a}{R_i}\right] \end{aligned} \tag{6.33}$$

where R_i is the internal radius of pipe and h is the pipe wall thickness. A defect that grows to become a stable through-wall crack is considered to have potential for leak. It is assumed that if the leak is detectable, i.e., sufficiently large, it can lead to failure of the pipe. In order to determine if a leak is detectable, it is necessary to estimate the leak rate, which in turn requires an estimate of the crack opening area. The opening area is estimated by considering the crack to be rectangular in shape in plan with a length $2b$ and width (crack opening displacement) of δ, where

$$\delta = \frac{4\sigma b(1 - v^2)}{E} \tag{6.34}$$

where v is the Poisson's ratio, E the modulus of elasticity of pipe material, and σ is the applied stress. The leak rate is estimated using the expression

$$\frac{Q \times h^{1/2}}{2b} = \begin{cases} 0.25\,\delta^2 & \text{for} \quad \delta \leq 2\,\text{mils} \\ 0.9375\,\delta^2 - 0.875 & \text{for} \quad \delta > 2\,\text{mils} \end{cases} \tag{6.35}$$

where δ the total crack opening displacement (mils), h the pipe wall thickness (in), 2b the through-wall crack length (in), and Q is the leak rate (gal/min). If the total leak rate through all the through-wall cracks, if any, in the pipe considered, is greater than the detectable leak, the pipe is considered to have failed. In the present study, a leak rate of 0.1 gal/min is considered as the detectable leak rate, which is the regulatory requirement for leak detection system for unidentified leak [28].

6.9.5 Simulation

In this study, the stochastic evolution of cracks due to SCC in 304 austenitic stainless steel is simulated using MCS technique. The values of f_1 and f_3 are computed using the values of P_a and σ generated using MCS. Using the values of f_1, f_2 (considered as deterministic), and f_3, values of damage parameter, D, are obtained. The variation in the values of damage parameters is due to the variation in microstructural properties of the material and the operating conditions. In order to take into account the variations in the time to initiation of stress corrosion cracks, for a given value of damage parameter different values of t_I are generated using MCS. In every time step, each one of the cracks is checked for initiation. Initial crack depth and length are assigned to the initiated cracks and crack growth velocities are calculated. For all the initiated cracks in the samples generated, crack propagation velocities are calculated based on initiation and fracture mechanics considerations, as is appropriate. After each time step, the failure criterion is checked. Failure probability P_f is calculated as

$$P_f = \frac{N_f}{N} \tag{6.36}$$

where N_f is the number of failure cases and N is the total number of simulations.

6.10 Case Study on PDHR System

In the present study the failure probability of the piping system of the PDHRS is evaluated based on the above-discussed procedure. In the evaluation of failure probability time variation of loading and strength have been considered in the analysis. These are explained with step-by-step calculations as follows.

6.10.1 System Description

To carry out the fatigue analysis a case study on piping system corresponding to passive decay heat removal system (PDHRS) has been considered. The PDHRS consists of four tanks and each tank is connected to one steam generator, respectively. In each tank horizontal V-tube condenser bundles are immersed in water. The steam side of V-tubes are connected to main steam line through takeoff lines and the condensate side are connected to steam generator shell side above the tube sheet as shown in Fig. 6.24. The diameter and length of the tank is 5 and 12 m, respectively. The tank consists of 48 tubes and length of each tube is 16 m.

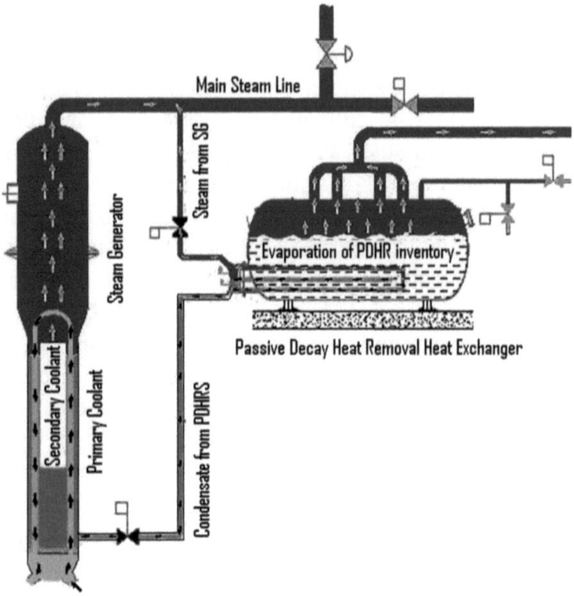

Fig. 6.24 PDHRS piping arrangement

Table 6.3 Details of piping system

S.no	Piping	Parameters	Value (m)
1	Inlet line	Outer diameter	150×10^{-3}
		Inner diameter	122.3×10^{-3}
		Thickness	30×10^{-3}
2	V-tube	Outer diameter	56.4×10^{-3}
		Inner diameter	50.8×10^{-3}
		Thickness	2.8×10^{-3}
3	Condensate line	Outer diameter	80×10^{-3}
		Inner diameter	73.2×10^{-3}
		Thickness	6×10^{-3}

The piping system has been divided into three parts viz., a piping line which carries steam from steam generators to the PDHRS tank, v-tubes inside the tank which condense the steam and a piping line which carries the condensate to the steam generator. The details of the piping system are given in Table 6.3.

Throughout the plant lifetime the system will be subjected to different loading conditions. Whenever the reactor goes from operating state to other like hot shut down, cold shut down, and full power operation the pressure and the temperatures in the system change accordingly. Eventually, this leads to the change in the stresses in the system. In the present analysis one such type of loading cycle considered is reactor changing it state from full power operation to hot shut down condition and coming back to full power operation. The temperature and pressure in

Table 6.4 Different loading cases

S.no	Piping	Parameters	Value
1	Full power	Pressure	40×10^5 (N/m^2)
		Temperature	250 °C
2	Hot shutdown	Pressure	50×10^5 (N/m^2)
		Temperature	260 °C

the system are given in Table 6.4. In this analysis v-tube piping has been analyzed for stochastic fatigue loading and stress corrosion cracking.

6.10.2 *Stochastic Fatigue Loading*

- Objective is to find the failure probability of piping system subjected to stochastic fatigue loading
- Steps involved in this analysis are as follows:
  - Determination of stresses acting on the piping system

    Dead weight stress
    Pressure stress
    Thermal stress

  - Determination of stress amplitude
  - Generation of probability distribution for stress amplitude
  - Determination of number of cycles to failure for a given stress amplitude based on S-N diagram approach
  - Evaluation of failure probability of the piping system
- **Determination of stresses acting on the piping system**: The material properties of the piping are shown in Table 6.5.
- **Dead weight stress**: This has been calculated by modeling the piping using finite element analysis and is given as

Table 6.5 Material properties

S.no	Piping	Parameters
1	Material	SS304
2	Density (ρ)	8×10^3 (kg/m^3)
3	Young's modulus (E)	1.93×10^{11} (N/m^2)
4	Ultimate tensile strength (UTS)	560×10^6 (N/m^2)
5	Yield strength (YS)	290×10^6 (N/m^2)
6	Coefficient of thermal expansion (α)	17.3×10^{-6} (/°C)
7	Thermal conductivity (K)	16.2 (W/mK)
8	Poisson's ratio (μ)	0.3

$$\sigma_{DW} = \frac{M}{Z}$$
$$Z = \frac{I}{D/2}$$

M = Maximum Bending moment
I = Moment of Inertia

- Moment of inertia for the hollow cylinder can be calculated as follows:

$$I = \frac{\pi}{64}\left(D_0^4 - D_i^4\right)$$

- The maximum bending moment obtained from the FE analysis (Fig. 6.25) is 54.1 Nm and the dead weight stress has been calculated as 8.98×10^6 N/m^2.
- **Pressure stress**: The pressure stress can be calculated by using the following equation

$$\begin{aligned}\sigma_P &= \frac{pd}{2t} \\ &= \frac{50 \times 10^5 \times 56.4 \times 10^{-3}}{2 \times 2.8 \times 10^{-3}} \\ \sigma_P &= 50.4 \times 10^6\,\text{N/m}^2 \quad \left(P = 50 \times 10^5\,\text{N/m}^2\right) \\ \sigma_P &= 40.3 \times 10^6\,\text{N/m}^2 \quad \left(P = 40 \times 10^5\,\text{N/m}^2\right)\end{aligned}$$

- **Thermal stress**: During full power operation there will not be any flow in the V-tube piping. The piping will be in equilibrium condition with the PDHR tank

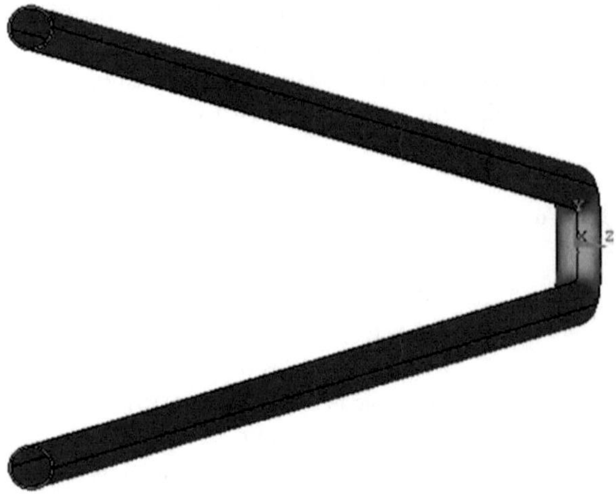

Fig. 6.25 FE model of the piping

water. However, during hot shutdown condition the steam will flow in the V-tube piping and suddenly the tubes will see the conditions of hot shut down as given in Table 6.4. Hence, the V-tube piping will experience thermal gradient stresses. In finding out the stresses caused by radial grad

- Inner surface temperature = 260 °C
- Outer surface temperature = 89 °C

ient temperature it is required to find the surface temperatures of the piping tube.

- The thermal stress acting on the piping is calculated as follows:

$$\sigma_{\text{Thermal}} = \frac{E\alpha\,\Delta T}{2(1-v)}$$
$$= \frac{1.93\times10^{11}\times17.3\times10^{-6}\times(260-89)}{2(1-0.3)}$$
$$\sigma_{\text{Thermal}} = 408\times10^{6}\,\text{n/m}^2$$

- **Determination of stress amplitude**: From the above analysis the total stresses obtained for the given two cases are as follows:
  - Case 1: 49 × 106 N/m^2
  - Case 2: 470 × 106 N/m^2
  - The stress amplitude has been calculated as 211 × 106 N/m^2
- Now the stress amplitude so obtained has been considered as a random variable and the distribution has been considered as normal distribution with the mean and standard deviation as follows:
  - μ= 211 × 106 N/m^2
  - *COV* = 0.1
  - σ=COV × μ= 21.1 × 106 N/m^2
- The lower and upper limit of the distribution has been taken as 3σ limits and are given as below
  - $\mu + 3\sigma$ = 277 × 106 N/m^2
  - $\mu - 3\sigma$ =149 × 106 N/m^2
- This means, stress amplitude can take any value in between the limits with corresponding probability value. The distribution of the stress amplitude is given in Fig. 6.26.
- Now this range has been divided into 1000 equal intervals with interval size as 1.278 × 10^5 N/m^2 (Δs). Consider a case in which the stress amplitude is 153 × 10^6 N/m^2(s) and the probability that the stress amplitude fall in the interval s and s + Δs is given as follows:

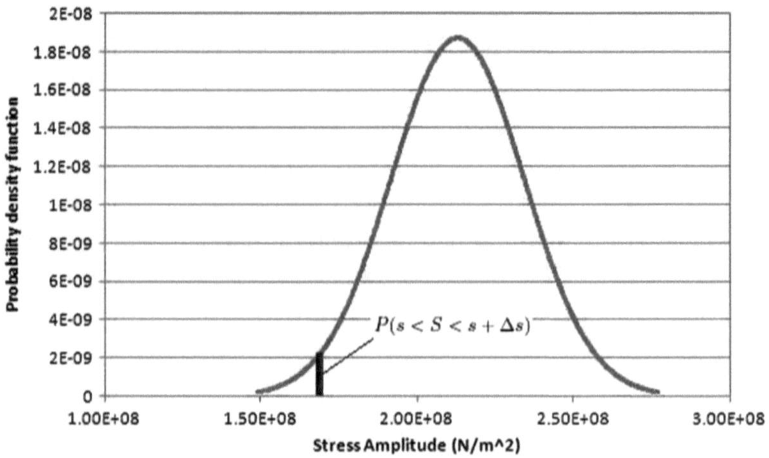

Fig. 6.26 Normal distribution for stress amplitude

$$
\begin{aligned}
s &= 153 \times 10^6 \\
\Delta s &= 1.278 \times 10^5 \\
s + \Delta s &= 153.278 \times 10^6 \\
P(s<S<s+\Delta s) &= F_S(s+\Delta s) - F_S(s) \\
F_S(s) &= \Phi\left(\frac{s-\mu}{\sigma}\right) \\
F_S(153 \times 10^6) &= 2.41447 \times 10^{-3} \\
F_S(153.1278 \times 10^6) &= 2.470146 \times 10^{-3} \\
P(s<S<s+\Delta s) &= 4.57 \times 10^{-5}
\end{aligned}
$$

- By using the S-N diagram for the given material one can obtain the number of cycles to failure. But this is also considered as a random variable and it follows lognormal distribution with median value as obtained number of cycles to failure and error factor as 10.
- For the above case, i.e., for $s = 153 \times 10^6$ N/m^2 the corresponding median number of cycles to failure is obtained as 6.06×10^6. From this information lognormal distribution parameters can be obtained as follows:

$$
\begin{aligned}
\mu_{50} &= 6.06 \times 10^6 \\
EF &= 10 \\
\sigma_{\ln N} &= \frac{\ln(EF)}{1.646} = 1.4
\end{aligned}
$$

- For the above distribution parameters the distribution can be plotted and is shown in Fig. 6.27.

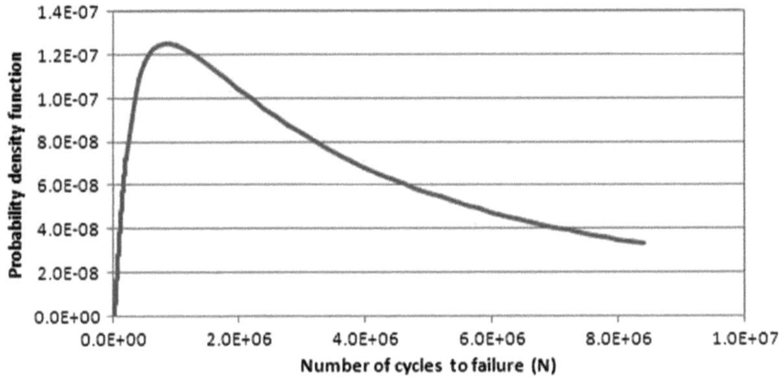

Fig. 6.27 Lognormal distribution for number of cycles to failure

- Evaluation of failure probability: From the above distribution one can obtain $P(N < n)$, which is nothing but the cumulative distribution value and it represents the probability that the component will fail for the given number of cycles.
- In the present case assume the number of cycles as 1 and the CDF value can be obtained as

$$P(N \le n) = P(N \le 1)$$
$$F_N(n) = \Phi\left(\frac{\ln(n) - \mu_{\ln N}}{\sigma_{\ln N}}\right)$$
$$F_N(1) = \Phi\left(\frac{\ln(1) - \ln(\mu_{50})}{1.4}\right)$$
$$F_N(1) = \Phi\left(\frac{\ln(1) - \ln(6.06 \times 10^6)}{1.4}\right)$$
$$F_N(1) = \Phi(-11.155)$$
$$F_N(1) = 3.37866 \times 10^{-29}$$

- Similarly, one can obtain the probability of failure value for the different number of cycles (e.g., 10, 50,100, etc.). If one considers the arrival rate of loading cycles as 6/year. Then 6 cycles represents 1 year and in 50 years of plant lifetime 300 cycles will occur. For the above case probability of failure after 50 years will be

$$F_N(300) = \Phi\left(\frac{\ln(300) - \ln(6.06 \times 10^6)}{1.4}\right)$$
$$F_N(300) = \Phi(-9.913)$$
$$F_N(300) = 7.155 \times 10^{-13}$$

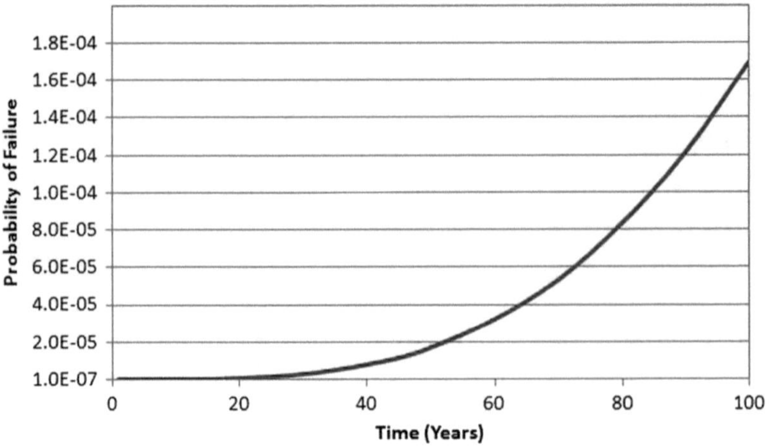

Fig. 6.28 Failure probability variation with time

- Hence, the probability of failure of the component after 50 years considering the stress amplitude as $s = 153 \times 10^6$ N/m^2 will be

$$P_f(50) = P_f(50|s) \times P(s)$$
$$P_f(50) = 7.55 \times 10^{-13} \times 4.57 \times 10^{-5}$$
$$P_f(50) = 3.27 \times 10^{-17}$$

- Similarly, the above calculations have to be repeated for different stress amplitudes and find the probability of failure of the component at different time intervals. This is shown in Fig. 6.28.

6.10.3 Stress Corrosion Cracking

- Steps involved in SCC analysis are as follows:
  - Time to initiation
  - Crack size at initiation
  - Crack propagation due to SCC
  - Failure criteria
- The inputs related to the geometry of the pipe and operating conditions are given in Table 6.6.
- The random variables with their distributions and related parameters are given in Table 6.7.

Table 6.6 Material properties

S.no	Piping	Parameters
1	Pipe wall thickness (m)	2.8×10^{3}
2	Inside diameter (m)	50.8×10^{-3}
3	Operating temperature (°C)	250.0
4	Oxygen concentration (ppm)	0.2
5	Water conductivity (μs/cm)	20.0
6	Pipe material	SS304
7	Material fracture toughness (MPa/√m)	40.0

Table 6.7 Random variables given as inputs

S.no	Variable	Distribution	Parameters	
			Mean (ksi)	COV
1	Applied stress	Normal	7.1	0.2
2	Load controlled stress	Normal	7.1	0.1
3	Flow stress	Normal	61.9	0.03

- Time to initiation: The time to initiation (t_I) is assumed to follow lognormal distribution with mean and standard deviation of log t_I as follows:

$$\mu_{\log\, t_I} = B_0 + B_1 \log(D)$$
$$\sigma_{\log\, t_I} = B_2 + B_3 \log(D)$$
$$D = f_1(\text{material}) \cdot f_2(\text{environment}) \cdot f_3(\text{loading})$$
$$f_1 = \mathrm{C}_1 (P_a)^{\mathrm{C}_2}$$
$$f_2 = \mathrm{O}_2^{\mathrm{C}_3} \cdot e^{\frac{\mathrm{C}_4}{T+273}} . \log\left(\mathrm{C}_5 \gamma^{\mathrm{C}_6}\right)$$
$$f_3 = \left(\mathrm{C}_8 \sigma^{\mathrm{C}_9}\right)^{\mathrm{C}_7}$$

- The above parameters have been estimated for the data given in Tables 6.6, 6.7, and 6.8.
- Evaluation of damage parameter D needs estimation of f_1, f_2, and f_3. f_1 is depending on P_a. P_a is assumed to follow Weibull distribution with the parameters of β = 17.3 and α = 1.05 and having the cumulative distribution function as follows:

$$F_P(p) = 1 - \mathrm{e}^{\left(-\frac{p}{\alpha}\right)^{\beta}}$$

- Since P_a is assumed as a random variable the actual value of P_a can be generated by generating random numbers. Since the value of F_P (p) lies in between 0 and 1 it can be treated as a random number. Then random variate can be generated as follows:

Table 6.8 Values of constants in the analysis

Constant	Value	Constant	Value
C_1	23	B_0	−3.1
C_2	0.51	B_1	−4.21
C_3	0.18	B_2	0.3081
C_4	−1123	B_3	0
C_5	8.7096	F_m	2.551
C_6	0.35	F_s	0.4269
C_7	0.55	G	1.3447
C_8	2.21×10^{-15}	C_{14m}	−3.16
C_9	6	C_{14s}	0.72671
C_{12}	0.8192	C_{15}	1.7935
C_{13}	0.03621		

$$P_a = e^{\left(\ln(\beta)+\frac{1}{\alpha}\ln\left[\ln\left(\frac{1}{1-F}\right)\right]\right)}$$
$$F = 0.56875\ (\text{Assume})$$
$$P_a = e^{\left(\ln(17.3)+\frac{1}{1.05}\ln\left[\ln\left(\frac{1}{1-0.56875}\right)\right]\right)}$$
$$P_a = 14.6690$$

- Now f_1 can be estimated as follows for the given information on $C_1 = 23$ and $C_2 = 0.51$:

$$f_1 = C_1(P_a)^{C_2}$$
$$f_1 = 23 \times (14.6690)^{0.51}$$
$$f_1 = 90.4881$$

- Now f_2 can be estimated for the given information on $C_3 = 0.18$, $C_4 = -1123$, $C_5 = 8.7096$, $C_6 = 0.35$, $O_2 = 0.20$, $T = 250$ °C, and $r = 20$ as follows:

$$f_2 = O_2^{C_3} \cdot e^{\frac{C_4}{T+273}} \cdot \log\left(C_5\gamma^{C_6}\right)$$
$$f_2 = 0.2^{0.18} \cdot e^{\frac{-1123}{250+273}} \cdot \log\left(8.7096 \times 20^{0.35}\right)$$
$$f_2 = 0.2809$$

- Now f_3 can be estimated for the given information on $C_7 = 0.55$, $C_8 = 2.21 \times 10^{-15}$ $C_9 = 6$, and $\sigma = 7.1$ ksi as follows:

$$f_3 = \left(C_8\sigma^{C_9}\right)^{C_7}$$
$$f_3 = \left(2.21 \times 10^{-15} \times 7.1^6\right)^{0.55}$$
$$f_3 = 5.6049 \times 10^{-6}$$

- From the above the damage parameter D is estimated as 1.4247×10^{-4}
- Now parameters of the distribution of time to initiation can be estimated for the given information on $B_0 = -3.1$, $B_1 = -4.21$, $B_2 = 0.3081$, $B_3 = 0$, and $D = 1.4247 \times 10^{-4}$ as follows:

$$\begin{aligned}
\mu_{\log t_I} &= B_0 + B_1 \log(D) \\
&= -3.1 + (-4.21) \times \log\left(1.4247 \times 10^{-4}\right) \\
\mu_{\log t_I} &= 34.1854 \\
\sigma_{\log t_I} &= B_2 + B_3 \log(D) \\
&= 0.3081 + 0 \times \log\left(1.4247 \times 10^{-4}\right) \\
\sigma_{\log t_I} &= 0.3081
\end{aligned}$$

- Since t_I is assumed as a random variable that follows lognormal distribution, the actual value of t_I can be generated as follows:

$$\begin{aligned}
F(t_I) &= \Phi\left(\frac{\ln t_I - \mu_{\log t_I}}{\sigma_{\log t_I}}\right) \\
\frac{\ln t_I - \mu_{\log t_I}}{\sigma_{\log t_I}} &= \Phi^{-1}[F(t_I)] \\
t_I &= \mathrm{e}^{\left[\Phi^{-1}[F(t_I)] \times \sigma_{\log t_I} + \mu_{\log t_I}\right]} \\
F(t_I) &= 0.0437 \text{ (Assumed)} \\
t_I &= \mathrm{e}^{\left[\Phi^{-1}[0.0437] \times 0.3081 + 34.1854\right]} \\
t_I &= \mathrm{e}^{[-1.7093 \times 0.3081 + 34.1854]} \\
t_I &= 4.1479 \times 10^{14}\ \mu\text{s} \\
t_I &= \frac{4.1479 \times 10^{14}}{3.1536 \times 10^{13}} \text{ years} \\
t_I &= 13.1528 \text{ years}
\end{aligned}$$

- From the above it can infer that the crack will initiate after 13.15 years of plant operation due to SCC.
- The crack that is so initiated will have initial size in terms of its length and depth. The crack size is treated as a random variable and is assumed to follow lognormal distribution. Crack depth is assumed as 0.001 in. Crack length follows lognormal distribution with mean and standard deviation as 0.1862 and 0.2055 in, respectively. Since these parameters are lognormal parameters, these

have to be converted to normal distribution parameters to utilize in the cumulative distribution function. This can be done by using the following expressions.

$$\mu_l = 0.1862$$
$$\sigma_l = 0.2055$$
$$\sigma_{\ln l} = \sqrt{\ln\left[1 + \left(\frac{\sigma}{\mu_l}\right)^2\right]}$$
$$\sigma_{\ln l} = 0.89254$$
$$\mu_{\ln l} = \ln(\mu_l) - \frac{(\sigma_{\ln l})^2}{2}$$
$$\mu_{\ln l} = -2.07925$$

- Once the parameters of the distribution are calculated, one can calculate the parameters based on the random number generation as follows:

$$F_L(l) = \Phi\left(\frac{\ln l - \mu_{\log L}}{\sigma_{\ln L}}\right)$$
$$\frac{\ln l - \mu_{\log L}}{\sigma_{\ln L}} = \Phi^{-1}[F_L(l)]$$
$$l = \mathrm{e}^{\left[\Phi^{-1}[F_L(l)]\times\sigma_{\ln L}+\mu_{\ln L}\right]}$$
$$F_L(l) = 0.6072\,(\text{Assumed})$$
$$l = \mathrm{e}^{\left[\Phi^{-1}[0.6072]\times 0.89254-2.07925\right]}$$
$$l = \mathrm{e}^{[0.2720\times 0.89254-2.07925]}$$
$$l = 0.1594\,\text{in}$$
$$2b = 0.1594\,\text{in}$$
$$b = 0.0797\,\text{in}$$
$$a = 0.001\,\text{in (Assumed)}$$

- The crack so initiated will grow with some velocity due to the presence of the stresses and corrosive environment. The crack growth rate can be calculated as follows for the given parameter values of $G = 0.63136$; $D = 1{:}4247 \times 10^{-4}$ and F is assumed to follow normal distribution with mean and standard deviation as -0.02266 and 0.2052, respectively.

$$\mu_{\log v_1} = F + G\log(D)$$

- Calculation of F

$$
\begin{aligned}
\mu_F &= -0.02266 \\
\sigma_F &= 0.2052 \\
\mathrm{CDF}(F) &= \Phi\left(\frac{F-\mu_F}{\sigma_F}\right) \\
\frac{F-\mu_F}{\sigma_{lF}} &= \Phi^{-1}[\mathrm{CDF}(F)] \\
F &= \Phi^{-1}[\mathrm{CDF}(F)] \times \sigma_F + \mu_F \\
\mathrm{CDF}(F) &= 0.1409\ (\text{Assumed}) \\
F &= \Phi^{-1}[0.1409] \times 0.2052 - 0.02266 \\
F &= -1.0763 \times 0.2052 - 0.02266 \\
F &= -0.2435
\end{aligned}
$$

- Calculation of v_1

$$
\begin{aligned}
G &= 0.63136 \\
D &= 1.4247 \times 10^{-4} \\
\mu_{\log v_1} &= -0.2435 + 0.63136 \log\left(1.4247 \times 10^{-4}\right) \\
\mu_{\log v_1} &= -5.8351 \\
\mathrm{COV}_{\log v_1} &= 0.2 \\
\sigma_{\log v_1} &= -5.8351 \times 0.2 \\
\sigma_{\log v_1} &= -1.1670 \\
v_1 &= \mathrm{e}^{\left[\Phi^{-1}[F(v_1)] \times \sigma_{\log v_1} + \mu_{\log v_1}\right]} \\
F(v_1) &= 0.8594\ (\text{Assumed}) \\
v_1 &= \mathrm{e}^{\left[\Phi^{-1}[0.8594] \times (-1.1670) + (-5.8351)\right]} \\
v_1 &= \mathrm{e}^{[1.0776 \times (-1.1670) + (-5.8351)]} \\
v_1 &= 8.3118 \times 10^{-4}\ \text{in/s}
\end{aligned}
$$

- Now the new crack size can be calculated based on the initiation velocity as follows:

$$
\begin{aligned}
a_{\text{new}} &= a_0 + v_1 \Delta t \\
a_0 &= 0.001\ \text{in} \\
v_1 &= 8.3118 \times 10^{-4}\ \text{in/s} \\
\Delta t &= 0.2\ \text{years} \\
a_{\text{new}} &= 0.001 + 8.3118 \times 10^{-4} \times 0.2 \\
a_{\text{new}} &= 1.1662 \times 10^{-3}\ \text{in}
\end{aligned}
$$

Also

$$b_0 = 0.0797\,\text{in}$$
$$b_{\text{new}} = 7.9866 \times 10^{-2}\,\text{in}$$

- Once the new crack sizes are available one can check for failure of the piping based on the failure criteria as follows:

$$\sigma_{\text{net}} = \frac{\sigma_{LC} \times A_P}{A_P - A_{Cr}}$$
$$A_P = \pi h(2R_i + h)$$
$$A_{cr} = ab\left[2 + \frac{a}{R_i}\right]$$
$$R_i = 1\,\text{in}$$
$$h = 0.11\,\text{in}$$
$$A_p = 7.2916 \times 10^{-1}\,\text{in}^2$$
$$A_{cr} = 1.8639 \times 10^{-4}\,\text{in}^2$$
$$\sigma_{LC} = 7.1\,\text{ksi}$$
$$\sigma_{\text{net}} = \frac{7.1 \times 7.2916 \times 10^{-1}}{7.2916 \times 10^{-1} - 1.8639 \times 10^{-4}}$$
$$\sigma_{\text{net}} = 7.1018\,\text{ksi}$$

- If the calculated net section stress exceeds the flow stress then the failure of the piping will take place. The flow stress is assumed to follow normal distribution with mean and standard deviation as 61.6 and 1.9 ksi, respectively, and is calculated as follows:

$$\mu_{\text{FlowS}} = 61.6\,\text{ksi}$$
$$\sigma_{\text{FlowS}} = 1.9\,\text{ksi}$$
$$F_{\text{FlowS}} = \Phi\left(\frac{\text{FlowS} - \mu_{\text{FlowS}}}{\sigma_{\text{FlowS}}}\right)$$
$$\frac{\text{FlowS} - \mu_{\text{FlowS}}}{\sigma_{\text{FlowS}}} = \Phi^{-1}[F_{\text{FlowS}}]$$
$$\text{FlowS} = \Phi^{-1}[F_{\text{FlowS}}] \times \sigma_{\text{FlowS}} + \mu_{\text{FlowS}}$$
$$F_{\text{FlowS}} = 0.1371\ (\text{Assumed})$$
$$\text{FlowS} = \Phi^{-1}[0.1371] \times 1.9 + 61.6$$
$$\text{FlowS} = -1.0934 \times 1.9 + 61.6$$
$$\text{FlowStress} = 59.5225\,\text{ksi}$$

- From the above it can be seen that the net section stress is well below the flow stress and the piping will not fail. Now the piping is checked for leakages based on the leak criteria for the given parameter values of $v = 0.3$ and $E = 27992.3$ ksi as follows:

$$\delta = \frac{4\sigma b(1 - v^2)}{E}$$
$$v = 0.3$$
$$E = 27992.3\,\text{ksi}$$
$$\sigma = 7.1\,\text{ksi}$$
$$\delta = \frac{4 \times 7.1 \times 7.9866 \times 10^{-2}(1 - 0.3^2)}{27992.3}$$
$$\delta = 7.3737 \times 10^{-5}\,\text{in}$$
$$\delta = 7.3737 \times 10^{-2}\,\text{mils} < 2\,\text{mils}\ \left(1\,\text{mil} = 1 \times 10^{-3}\,\text{in}\right)$$
$$\frac{Q \times h^{1/2}}{2b} = \begin{cases} 0.25\,\delta^2 & \text{for} \quad \delta \leq 2\,\text{mils} \\ 0.9375\,\delta^2 - 0.875 & \text{for} \quad \delta > 2\,\text{mils} \end{cases}$$
$$Q = \frac{0.25\,\delta^2 \times 2b}{h^{1/2}}$$
$$Q = 6.5464 \times 10^{-4}\,\text{gals/min}$$

- From the above the leak rate so obtained for the given crack size is much less than the critical value (0.1 gals/min). Hence, leak is also eliminated for the present simulation. The same procedure can be repeated for different time intervals and check for failure criteria.
- This way one can simulate as many numbers of simulations as prescribed and check for the failure criteria and can find the failure probability as explained in the previous sections.
- Finally, the failure probability obtained due to SCC is shown in Fig. 6.29. The system failure probability due to static and dynamic analysis (stochastic fatigue loading and stress corrosion cracking) as a function of time is shown in Fig. 6.30. Here the static failure probability is taken from the previous calculations as explained in Chap. 5.

6.11 Time-Dependent CDF Analysis

PSA analysis can be extended to evaluation of importance measures for the identification of critical components, which in turn assists in maintenance planning of nuclear power plant. The plant logic diagram depicting the potential failures leading to core damage is shown in Fig. 6.31. This is the traditional level 1 PSA analysis

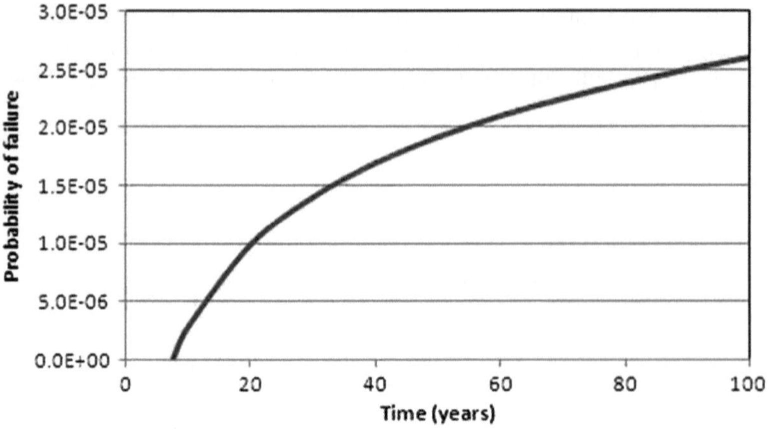

Fig. 6.29 Failure probability contribution from SCC

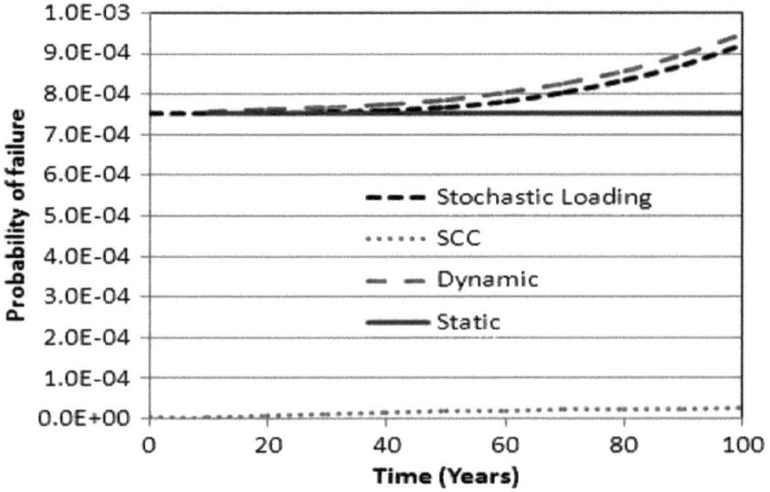

Fig. 6.30 Time-variant failure probability

which is static in nature. Where the time-dependent failure probability of the systems is not considered in calculation of CDF values. To account for the effect of random loading and aging phenomenon on the CDF evaluation one needs to carry out dynamic reliability analysis of the systems. This is further explained in the next section.

The traditional PSA analysis is only static in nature. Whereas in actual practice failure probabilities of the systems change with time due to random loading and aging phenomenon. Hence, one should consider time dependency of the accident frequency. This leads to the implementation of time-variant failure probabilities of

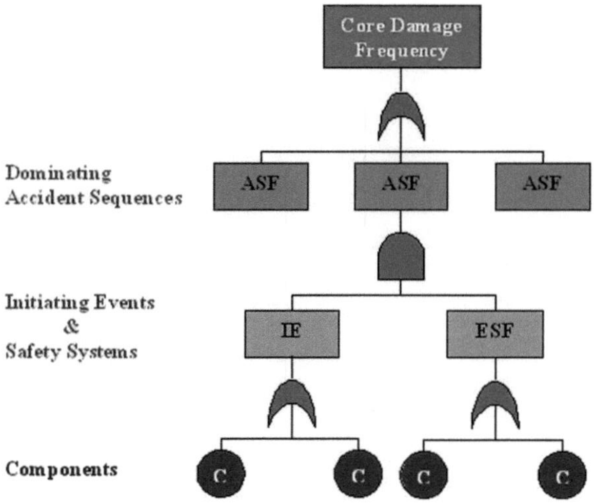

Fig. 6.31 Plant logic diagram

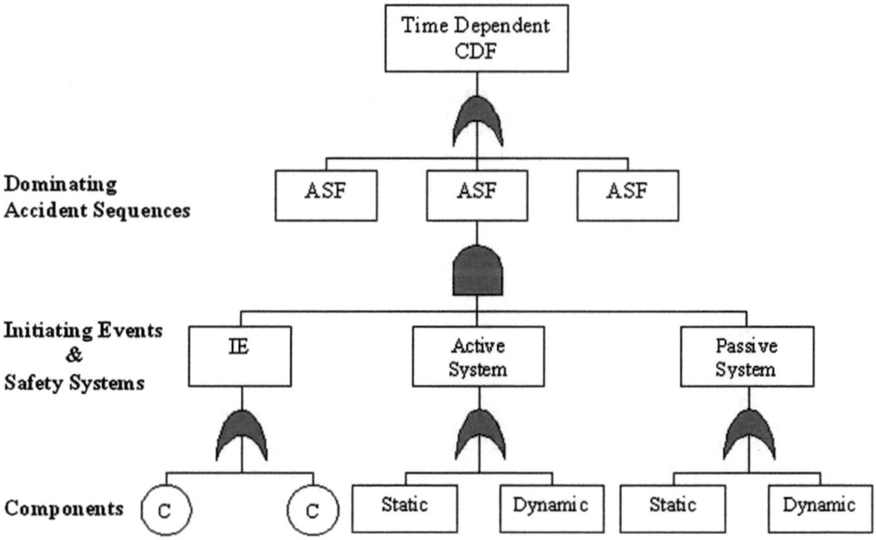

Fig. 6.32 Flow chart for time-dependent CDF Analysis

the safety systems into the PSA analysis. A flow chart showing the time-dependent CDF analysis is shown in Fig. 6.32 which is a modification of Fig. 6.31. In the present work the results obtained from the time-variant reliability analysis of

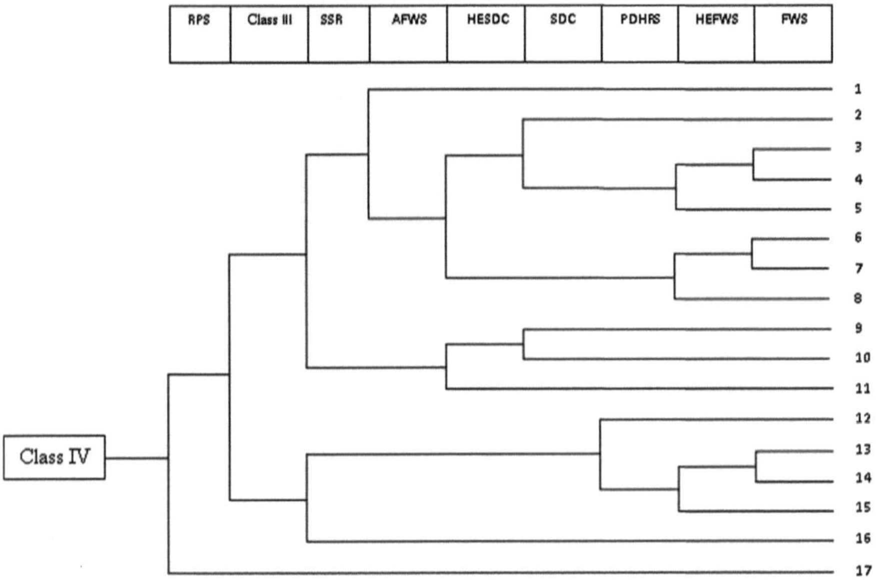

Fig. 6.33 Event tree for class IV failure

passive system have been utilized in finding out the accident frequency arising from an initiating event (Fig. 6.33) which is a function of time. The system failure probabilities are provided in Table 6.9. A graph between CDF versus time is shown in Fig. 6.34.

Table 6.9 Safety system failure probabilities

S.no	Safety system	Failure probability
1	Emergency power supply (Class III)	9.73E-04
2	Shutdown cooling system	9.70E-04
3	Fire water system (FWS)	8.02E-03
4	Auxiliary feed water system (AFWS)	2.67E-03
5	Secondary steam relief system (SSR)	3.70E-04
6	Reactor protection system (RPS)	2.03E-07
7	Human error-shutdown cooling system (HESDC)	3.40E-04
8	Human error fire water system (HEFWS)	2.50E-02

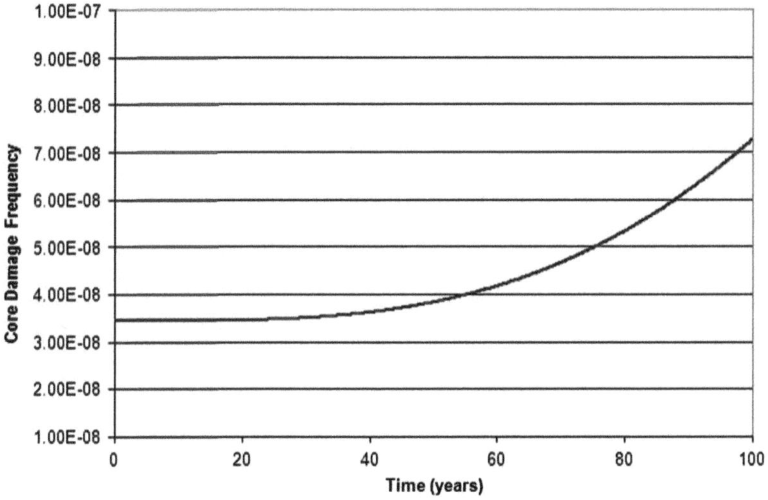

Fig. 6.34 Graph between CDF versus time

References

1. Beck AT, Melchers RE (2005) Barrier failure dominance in time variant reliability analysis. Probab Eng Mech 20:79–85
2. Madsen HO, Tvedt L (1990) Methods for time dependent reliability and sensitivity analysis. J Eng Mech 116(10):2118–2134
3. Altes J, Rackwitz R, Schulz U (1993) Time variant reliability of mechanical components. In: SMIRT-12, Elsevier Science Publishers, pp. 111–116
4. Kuschel N, Rackwitz R (2000) Optimal design under time-variant reliability constraints. Struct Saf 22:113–127
5. Torres MA, Ruiz SE (2007) Structural reliability evaluation considering capacity degradation over time. Eng Struct 29(9):2183–2192
6. Czarnecki AA, Nowak AS (2008) Time-variant reliability profiles for steel girder bridges. Struct Saf 30(1):49–64
7. Becker G, Camarinopoulos L, Kabranis D (2002) Dynamic reliability under random shocks. Reliab Eng Syst Saf 77(3):239–251
8. Streicher H, Rackwitz R (2004) Time-variant reliability-oriented structural optimization and a renewal model for life-cycle costing. Probab Eng Mech 19(1/2):171–183
9. Kopustinskas V, Augutis J, Rimkeviius S (2005) Dynamic reliability and risk assessment of the accident localization system of the lgnalina NPP RBMK-1500 reactor. Reliab Eng Syst Saf 87(1):77–87
10. Wen YK, Chen HC (1989) System reliability under time varying loads. J Eng Mech 115 (4):808–839
11. Melchers RE (1992) Load-space formulation for time-dependent structural relia-bility. J Eng Mech 108:853–870
12. Huang W, Askin RG (2004) A Generalized SSI reliability model considering stochastic loading and strength aging degradation. IEEE Trans Reliab 53(1):77–82
13. Xie L, Wang Z, Lin W (2008) System fatigue reliability modelling under stochastic cyclic load. Int J Reliab Saf 2(4):357–367

14. Zheng W, Liyang X (2008) Dynamic reliability model of components under random load. IEEE Trans Reliab 57(3):474–479
15. Newman JC Jr (1984) A crack-opening stress equation for fatigue crack growth. Int J Fract 24:131–135
16. Hari Prasad M, Rami Reddy G, Dubey PN, Srividya A, Verma AK (2013) Reliability estimation of structures under stochastic loading-A case study on nuclear piping. Nucl Eng Des 254:185–193
17. Castillo E (1988) Extreme value theory in engineering. Academic Press, Boston
18. Galambos J (1978) The asymptotic theory of extreme order statistics. Wiley, New York
19. Gumbel EJ (1958) Statistics of extremes. Columbia University Press, New York
20. Melchers RE (1999) Structural reliability analysis and prediction. Wiley, Chichester
21. Provan JW (1987) Probabilistic fracture mechanics and reliability, 1st edn. Martinus Nijhoff Publishers, The Netherlands
22. Naess A (2001) Crossing rate statistics of quadratic transformations of gaussian processes. Probab Eng Mech 16:209–217
23. Naess A, Karlsen HC (2004) Numerical calculation of the level crossing rate of second order stochastic Volterra systems. Probab Eng Mech 19:155–160
24. Sudret B (2006) Analytical derivation of the outcrossing rate in time variant reliability problems. Struct Infrastruct Eng 1–14
25. Ditlevsen O (1982) First outcrossing probability bounds. J Eng Mech 110(2):282–292
26. Rice SO (1944) Mathematical analysis of random noise. Bell Syst Tech J 23:282–332
27. Ting K (1999) The evaluation of intergranular stress corrosion cracking problems of stainless steel piping in Taiwan BWR-6 nuclear power plant. Nucl Eng Des 191(2):245254
28. NUREG/CR-5864 (1992) Theoretical and users manual for PC-PRAISE, USNRC
29. Priya C, Rao KB, Anoop MB et al (2005) Probabilistic failure analysis of austenitic nuclear pipelines against stress corrosion cracking. Proc Inst Mech Eng Part C J Mech Eng Sci 219:607–626
30. Gopika V, Verma AK, Srividya A (2005) "Risk informed studies on in-service inspection", PhD Thesis, Indian Institute of Technology, Bombay

Chapter 7
Risk Management of Nuclear and Thermal Power Plants

7.1 Introduction

According to the definition of risk, it has two components one is the likelihood of occurrence of undesirable events and the other one is the consequences of these events. If we consider a plant, there are different sources of risk associated with an event. Now consider the case of a nuclear power plant. The operation director wants to install passive systems in the place of active systems in the plant in order to increase the availability of the safety systems during power supply failures. However, this is not the requirement by the nuclear regulatory body. The director has to make a decision whether to go for these changes in the plant by considering the different sources of risks in the plant. There might be different sources of risks that are associated with this investment from different sections of the plant.

- From the safety point of view, whether these newly installed systems can demonstrate that the frequency of radio activity release will be with in the regulatory requirements or not?
- From the operations point of view, what will be the operational difficulties or benefits that will arise due to the installation and operation of the new system?
- From the financial point of view, whether the cost of the investment will be recovered during the plant life time or not?
- From the project management point of view, what is the probability that the project will be completed on time and with in the budget allocated?

One should consider all the above-mentioned risks in order to evaluate over all risk from different sources of the plant. The integration of all these risks can be done in the frame work of risk management. There are four broad categories of sources of risk that an organization can be exposed, they are namely [1]:

1. Safety related,
2. Production/Operations,

A.K. Verma et al., *Risk Management of Non-Renewable Energy Systems*,
Springer Series in Reliability Engineering, DOI 10.1007/978-3-319-16062-7_7

3. Financial/Commercial, and
4. Strategic.

Here strategic source of risk results from fundamental changes in the commercial, economic, or political environments. The above sources of risk can be integrated with a risk management model comprising of safety, operational, and financial sectors of risk embedded with in the strategic environment as shown in Fig. 7.1 [1]. These sectors interact with each other in such a way that any decision in one sector will have impact on the other sectors. The objective of risk management is to identify all the possible consequences of an event arising from one sector and find an optimal decision to minimize adverse effects in a cost-efficient manner.

There are four major steps involved in risk management in a plant for any proposed action. They are listed as follows:

- Identification of risk
- Identification of techniques or strategies to manage risks
- Implementation of the strategies
- Monitor effectiveness of solutions.

For any proposed action in the plant, it is required to identify the different sources of risk that can arise from different sectors of the plant. After successful identification of different sources of risk, one needs to assess these risks. For the assessment of the risk, well-established deterministic and probabilistic safety analysis are being used in NPPs all over the world. PSA methodology can be used to assess safety-related issues such as analyze the impact of changes in technical specifications and operational-related issues like in-service inspection of piping (focusing on where they are required more). Once the risk identification step is over, next step is to identify the techniques or strategies to manage risk. In general, the following techniques such as risk reduction, risk retention, and risk transfer are used to

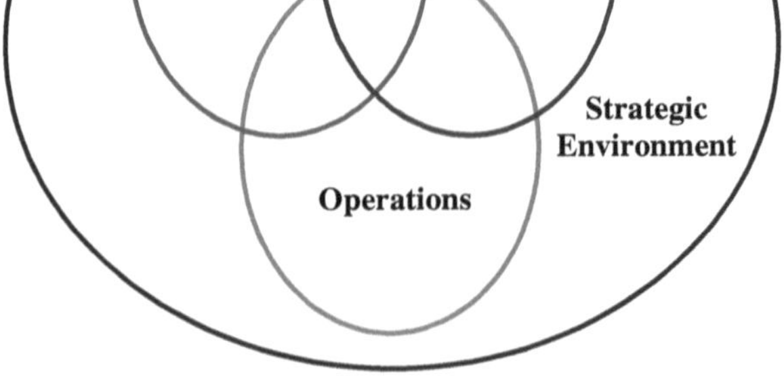

Fig. 7.1 Risk management model for a NPP operator [1]

manage the risk. Risk can be reduced either by reducing the likelihood of occurrence of events or by reducing consequences of the events if they occur. Once the proper techniques are identified, in the next steps the techniques have to be implemented and monitor the effectiveness of the solutions.

As on today, probabilistic safety assessment (PSA) has become a key tool to identify and analyze the NPP systems to address the safety issues and it can be regarded as being part of the risk management of a plant. PSA models have been successfully employed during design evaluation to assess weak links and carry out design modifications to improve the system reliability and safety. As a result of the availability of these PSA studies, one can make its use to enhance the plant safety and to operate the plants in the most efficient manner. This necessitates development of software tools like Risk Monitor, precursor analysis, etc., [2] and are explained in the following sections.

7.2 Risk Monitor

Risk Monitor is a PC-based tool [3–5], which computes the real-time safety level based on the actual status of systems and components and in addition assists plant personnel and regulatory authorities to manage day-to-day activities and can provide solutions to various regulatory decision-making issues. In Risk Monitor, risk is subjected to the change in the state of the system, which in turn depends on the state of the components and these are explained with a case study. Risk Monitoring can be defined as being the process whereby a complex technical facility is continuously monitored as regards the functioning or nonfunctioning of its different subsystems and the risk emanating from the facility is evaluated on the basis of this information.

Operation of Risk Monitor is based on PSA methods for assisting day-to-day applications. Risk Monitoring programs can assess the risk profile and are used to optimize the operation in NPPs with respect to a minimum risk level over the operating time. Risk Monitor provides safety status information for a plant and thus aids decision making about whether continued plant operation is tolerable under certain system function outages. It may also support operations and be of help in deciding on maintenance strategies allowing immediate assessment of different plant configurations.

Risk Monitor is an user friendly and can reevaluate core damage frequency (CDF) for changes in component status, test interval, initiating event frequency, etc. Plant restoration advice, when the plant is in high-risk configuration is provided in the tool. Current status of all plant equipments and equipment prioritization is also provided in the tool. What-If analysis can be done by using this tool, which provides guidance to plant operators to choose among various options.

7.2.1 Necessity of Risk Monitor

PSA models can be used to quantify risk due to changes in components status, system design, and operations consequent to changes in plant configuration. Plant configuration undergoes changes due to changes in the component status and/or operating/maintenance procedures. Some components are randomly down and/or others can be planned for test, maintenance, and repair. This results in a variation of the risk level over operating time, which is termed as risk profile, and indicates the trends which could lead to deviation from desired CDF. Thus, Risk Monitor helps in computing the risk based on the actual status of systems and components and assists plant personnel and regulatory authorities to manage day-to-day activities and can provide solutions to various regulatory decision-making issues [5]. The data flow diagram of risk monitor is shown in Fig. 7.2.

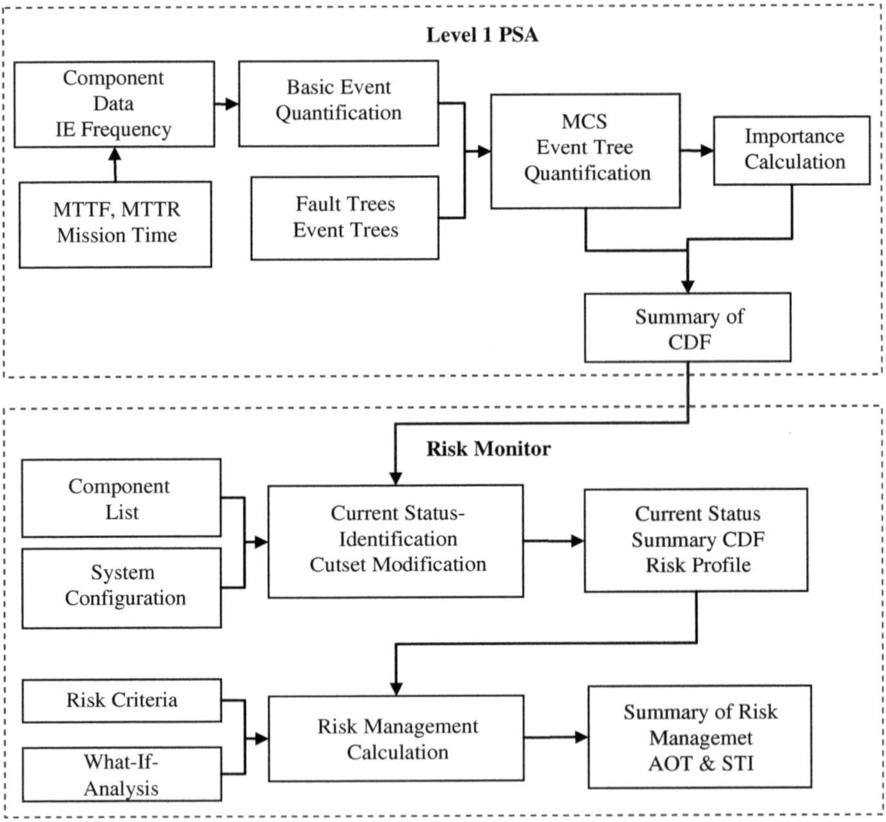

Fig. 7.2 Data flow diagram of risk monitor

7.2.2 Various Modules of Risk Monitor

In general, the Risk Monitor tool includes several modules such as

- PSA model,
- Risk profile,
- What-If Analysis, and
- Probabilistic Precursor Analysis.

These are briefly explained with a case study on typical NPP.

PSA Model The input to the risk monitor is the PSA model of a given plant [6, 7]. It generally consists of the following information about

- Initiating events
- Safety systems
- Event trees.
- Minimal cut sets of safety systems as well as core damage frequency
- Component failure data
- Common cause failure groups and common cause basic events.

List of initiating events are given in Table 7.1 and a pie chart showing the percentage contribution of IEs to CDF is shown in Fig. 7.3. Table 7.2 gives the information about safety systems along with their unavailabilities for a typical NPP.

Risk profile The Risk profile module summarizes the status of the safety systems based on the status of the components and list of components which have been taken out from the service and risk profile (CDF vs. Time). The variation of risk (in terms of CDF) along time is shown in Fig. 7.4. The different color bands in the risk profile shows different acceptance levels of risk for the plant under consideration (acceptable, not acceptable, safe, etc.). The details of the status of the different safety systems are given in Table 7.2. List of the components taken out from service and are due for restoration is given in Table 7.3.

What-If Analysis is the unique feature of the risk monitor. Before making any changes in the plant regarding the configuration of the systems, the status of the components, component maintenance strategies, inspection plans, testing intervals,

Table 7.1 Different IEs with their frequencies

S.no	IE	Frequency (/year)
1	LFEEDERBRK	2.00E-03
2	PRTBCLTB	2.20E-03
3	CLIV	1.92
4	SBLOCA	1.00E-03
5	BLINDLOCA(H)	1.00E-03
6	MCFL	2.00E-02
7	LBLOCA	2.00E-04
8	ESCFL	1.80E-03

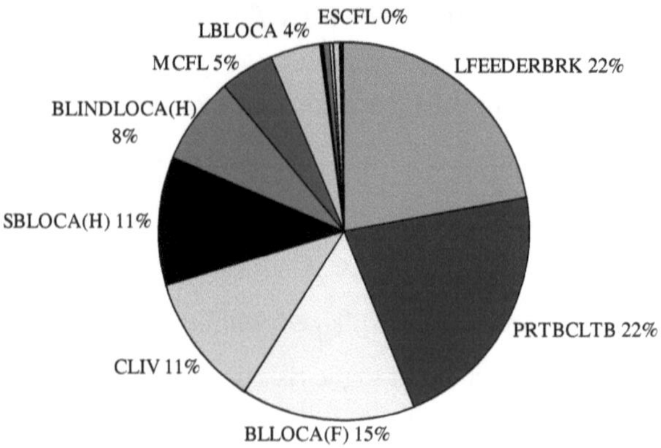

Fig. 7.3 Pie chart showing the percentage contribution of IEs to CDF

Table 7.2 Safety system unavailability and status

S.no	ID	Description	Unavailability	Status
1	ABFW	Auxiliary boiler feed water system	6.11E-03	Available
2	ALPAS	Automatic liquid poison injection system	3.16E-05	Available
3	ECCR	Emergency core cooling recirculation system	1.80E-04	Available
4	GRAB	Gravity added boron system	1.72E-05	Available
5	IRV	Instrument regulating valve	4.50E-06	Available
6	ECCI	Emergency core cooling injection system	4.88E-04	Available

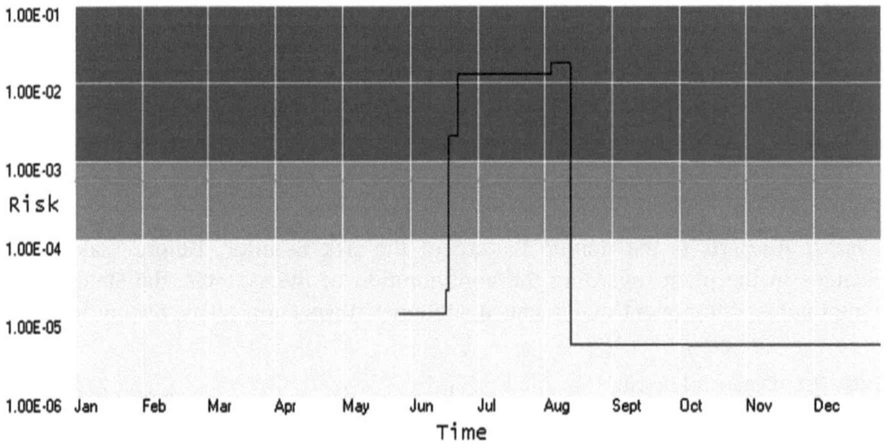

Fig. 7.4 Graphical representation of risk varying with time

etc., which may affect the technical specifications or violation of rules and regulations. With the help of What-If analysis, decision can be taken whether to go for these changes or not. For an example, if the plant personnel want to take a component for maintenance, testing, or inspection then What-If analysis suggests whether to go for this action presently or not.

Table 7.3 List of components taken out from service

S.no	Component ID	Status
1	ABFW-HX1	Out of service
2	ABFW-P3	Out of service
3	ECC-CHKV1	Out of service

In this analysis, the risk in terms of core damage frequency (CDF) or large early release frequency (LERF) is evaluated whenever there is a change in the status of the components. Depending on the risk acceptance criteria, decision can be made whether any component outage can be tolerable or not.

There are two sets of guidelines, one is based on CDF and the other one is based on the LERF [8]. These are graphically represented in Figs. 7.5 and 7.6, respectively. Both sets should be used for decision making. Table 7.4 shows the details of the component which the user wants to take it for maintenance, assume that it may be violation of technical specification. Risk Monitor helps the user whether one can perform this action based on the risk with the help of What-If analysis. The risk comparison is also shown in Table 7.4.

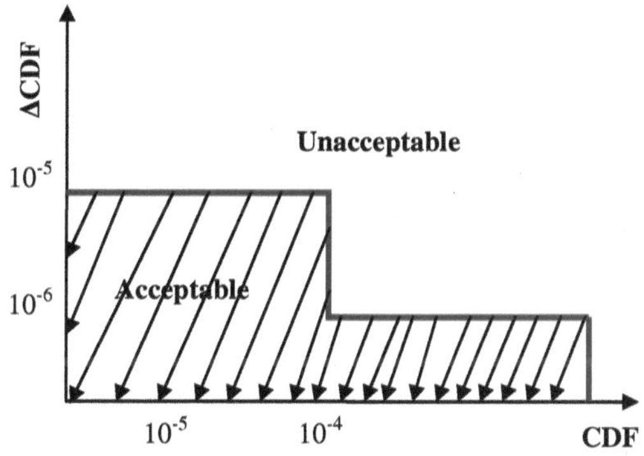

Fig. 7.5 Acceptance criteria based on core damage frequency [8]

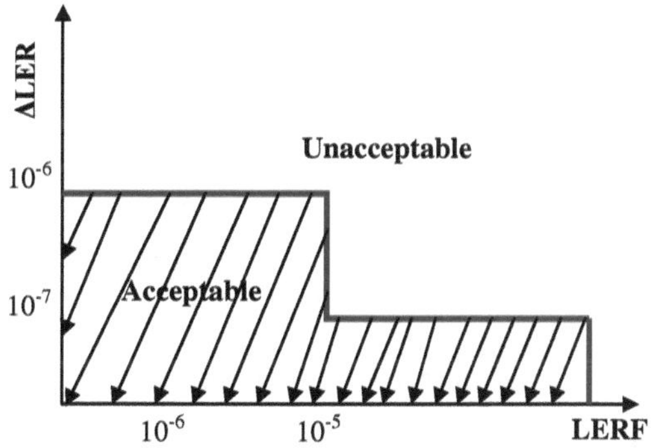

Fig. 7.6 Acceptance criteria based on large early release frequency [8]

Table 7.4 Details of the component which has been taken out from service

General		Parameter	Value
ID	KFFW-MV1	Failure rate	4.75×10^{-6}/h
Description	Motor operated valve	Test Interval	720 h
System	Fire fighting water system	Time to first test	0
Model type	Tested	Unavailability	1.708×10^{-3}
Existing CDF value			4.487×10^{-6}/year
New CDF value when this component is fully unavailable			3.078×10^{-5}/year
ΔCDF			2.629×10^{-5}/year

The reevaluated CDF when this component taken out from the service is 3.078e-5/year and the ΔCDF is 2.629×10^{-5}/year. The baseline CDF is 4.487e-6/year. If one compares these changes with Fig. 7.5 this scenario is falling in the unacceptable region, i.e., no changes are allowed. Similarly, if the component failure data parameters change, one can reevaluate the change in risk value due to the changes in these parameters and decision can be taken whether to go for these changes or not. Table 7.5 shows the different failure data parameters based on the type of component and the failure probability model. If any of the failure parameters changes accordingly, the failure probability also changes and in effect it will affect the risk value which is a function of the component failure probabilities and initiating event frequencies.

Probabilistic Precursor Analysis Significant events can be analyzed to assess their potential to develop into catastrophic accident like core damage situation or not. This is further explained in the following sections.

Table 7.5 Failure data parameters of components

Failure probability model	Parameters
Repairable components	Operating failure rate (λ_0)
	Mean time to repair (MTTR, T_R)
Non repairable components	Operating failure rate (λ_0)
	Mission time (T_M)
Periodically tested standby components	Standby failure rate (λ_S)
	Component test period (T)
	Average test duration (τ)
	Mean time to repair (MTTR, T_R)
	Frequency of scheduled maintenance (f_m)
	Mean time of scheduled maintenance (T_m)
Untested standby components	Standby failure rate (λ_S)
	Fault exposure time (T_p)
Continuously monitored standby components	Standby failure rate (λ_S)
	Mean time to repair (MTTR, T_R)

7.2.3 *Applications of Risk Monitor*

Importance Measures

Importance measures are useful in identification of critical components for the purpose of design modifications and maintenance. Two commonly used importance measures are risk achievement worth (RAW) and risk reduction worth (RRW) [9] of components in terms of CDF.

- **RAW** of a component is defined as the increase in risk, when the component is failed or removed from the service. Risk monitor RAW is modeled as follows:

$$\text{RAW} = \frac{R_1}{R} \tag{7.1}$$

where

R Baseline risk, i.e., the original risk value calculated by Risk Monitor,

R_1 Risk level when the component is removed, i.e., risk recalculated when the component unavailability is set to 1.

- **RRW** of a component is defined as the reduction in risk, when the component is fully reliable. Risk monitor RRW is modeled as follows:

$$\text{RRW} = \frac{R}{R_0} \tag{7.2}$$

where

R_0 Risk level when the component is fully available, i.e., risk recalculated when the component unavailability is set to 0.

The RRW suggests the components that can reduce the risk effectively, if modified suitably. The RAW signifies the importance of the components in achieving the present level of safety. Components having higher RAW have to be maintained immediately, in order to minimize the failure probability value. Similarly, component having higher RRW should be given attention from the design point of view, since it can enhance the reliability of the system.The details of importance measures of some of the components calculated from risk monitor are shown in the Table 7.6. These measures are calculated on the basis of unavailability of a particular system.

7.2.3.1 Allowed Outage Time (AOT)

The AOT is a maximum time during which the component can be down and its repair or maintenance can be performed. If the component is not brought back to function in the AOT, then the plant must take an appropriate action as defined in the

Table 7.6 Importance measures of various components of ABFW system

System	ABFW	
Description	Auxiliary boiler feed water system	
Unavailability	5.45E-04	
S. no	Component ID	RAW
1	ABFW-C1-FTS	50
2	ABFW-C2-FTS	48
3	ABFW-C3-FTE	42
4	ABFW-C4-FTS	35
5	ABFW-C5-FTE	25
6	ABFW-C6-FTS	10
7	ABFW-C7-FTE	8
8	ABFW-C8-FTE	7
9	ABFW-C9-RUPT	1

technical specifications (Tech-Specs). In risk monitor, the AOT is modeled (risk increased from the component outage is set to 10 % of the baseline plant risk) [10] as follows:

$$\begin{aligned} \text{ICCDP} &= (R_1 - R_0) \times \frac{\text{AOT}}{8760} \\ \text{AOT} &= \frac{\text{ICCDP}}{\Delta R} \times 8760 \end{aligned} \tag{7.3}$$

where ICCDP is the Incremental conditional core damage probability.

AOT is in Hours.

Example 7.1 Find the allowed outage time of a component for the following information

ICCDP = 1×10^{-6}
$R_1 = 3.078 \times 10^{-5}$/year
$R_0 = 5.2234 \times 10^{-6}$/year

Solution:

$$\begin{aligned} \text{AOT} &= \frac{\text{ICCDP}}{(R_1 - R_0)} \times \frac{1}{8760} \\ &= \frac{1 \times 10^{-6}}{(3.078 \times 10^{-5} - 5.2234 \times 10^{-6})} \times \frac{1}{8760} \\ &= 0.01756 \text{ years} \\ \text{AOT} &= 153.8 \text{ h} \end{aligned}$$

- From the above, it is calculated that AOT of the component is 153.8 h. Before which the component has to be put back into the operation after maintenance or inspection etc.

7.2.3.2 Surveillance Test Interval (STI)

In order to keep the plant risk under control, the individual standby components of the plant are periodically put to surveillance tests when the plant is in operation. When the periodically tested component fails between tests, its function is lost until the next test and the plant risk increases. A time period between two consecutive tests is called a STI. In risk monitor STI is modeled as follows.

$$\begin{aligned} \text{STI} &= \frac{R}{5\lambda(R_1 - R_0)} \\ &= \frac{1}{5\lambda\left(\frac{R_1}{R} - \frac{R_0}{R}\right)} \\ &= \frac{1}{5\lambda\left(\text{RAW} - \frac{1}{\text{RRW}}\right)} \end{aligned} \tag{7.4}$$

where

λ Failure rate of the component

Example 7.2 Find the surveillance rest interval time of a component for the following information

$R = 4.487 \times 10^{-6}$/year
$R_1 = 3.078 \times 10^{-5}$/year
$R_0 = 5.2234 \times 10^{-6}$/year
$\lambda = 4.75 \times 10^{-6}$/h $= 4.161 \times 10^{-2}$/year

Solution:

$$\begin{aligned} \text{STI} &= \frac{R}{5\lambda(R_1 - R_0)} \\ &= \frac{4.487 \times 10^{-6}}{5 \times 4.161 \times 10^{-2}(3.078 \times 10^{-5} - 5.2234 \times 10^{-6})} \\ \text{STI} &= 0.8439\,\text{years} = 10\,\text{months} \end{aligned}$$

- From the above it is calculated that STI of the component is 10 months

7.3 Probabilistic Precursor Analysis

Precursors are defined as the conditions, events, and sequences that precede and could lead up to accidents [11]. Based on this definition, these can be thought of loosely as "building blocks" of accidents and can include both events internal to an organization (such as equipment failures and human errors) and external events (such as earthquakes and hurricanes). Precursor analysis mainly focuses on how a particular event might have adversely developed. An event can be subjected to precursor study at two levels as mentioned below.

- One can analyze all the events and can identify whether these events are precursor to some significant events, i.e., whether an event has the potential to develop into an initiating event or can make safety systems unavailable
- Given the significant events, they can be analyzed to identify whether they have the potential to develop into core damage situation or not.

In this chapter, focus is given on second level of event analyze, i.e., the potential of a significant event to render a core damage type of situation.

7.3.1 Approaches for Precursor Analysis

There are two approaches for carrying out precursor studies namely deterministic approach and probabilistic approach. The deterministic approach uses incident reporting system of a plant to analyze the events. In this approach, all the events that occur in a plant will be recorded. Based on their severity, they will be ranked. However, all the events may not be precursor events. The International Nuclear Event Scale (INES) developed by IAEA is the popularly followed deterministic approach worldwide. INES has seven levels as shown in Fig. 7.7 [12]. In the probabilistic approach PSA based precursor analyses [13,14] is generally being used and is explained indetail in the following sections.

7.3.2 PSA-Based Precursor Analysis

In this analysis, the results of PSA model of a plant will be used to carry out this study. The PSA results mainly contain the minimal cut sets at the CDF level. With the help of MCS, once can find out the CDF value. Generally, MCS of a plant consists of initiating events and component failure probabilities. Hence, if the event is of IE type or component failure, this can be analyzed with the PSA results.

In probabilistic precursor analysis, the following measures are used to categorize the events or to know the severity of the events, namely:

- Conditional core damage probability (CCDP).
- Core damage index

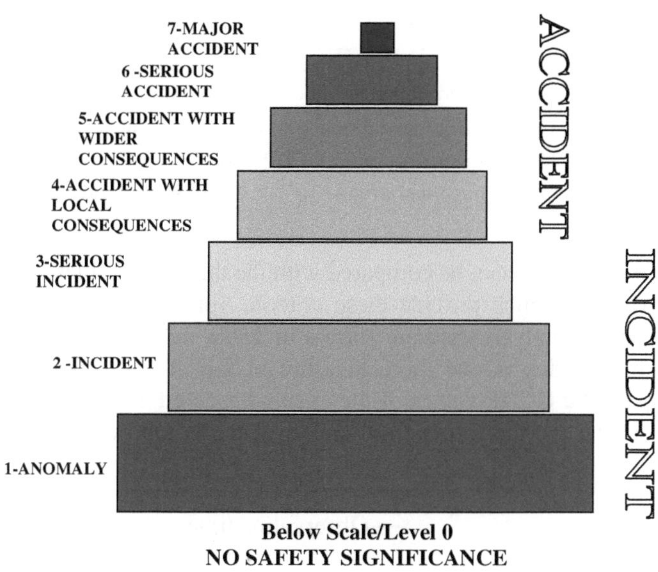

Fig. 7.7 INES levels [12]

CCDP measure can be used both for initiating event type or component failure type of events. For an initiating event type, the CCDP is the conditional probability of core damage given the event occurrence and can be calculated as follows:

$$\mathrm{CCDP_{IE}} = \frac{\mathrm{CDF_{IE}}}{\lambda_{\mathrm{IE}}} \tag{7.5}$$

where λ_{IE} is the frequency of the initiating event. If the event is of component failure type then CCDP is calculated as follows:

$$\mathrm{ICCDP} = T_{\mathrm{event}} \times \frac{(\mathrm{CDF_{event}} - \mathrm{CDF_{base}})}{A} \tag{7.6}$$

where

A	Fractional duration of power operation per year
T_{event}	Duration of the operational event (year),
$\mathrm{CDF_{event}}$	Core damage frequency during the event (1/year), and
$\mathrm{CDF_{base}}$	Baseline core damage frequency during power operation (1/year)

Here ICCDP is the increase of the core damage probability due to the event occurrence. In this case, CDF is reevaluated due to the failure of the component and change in CDF is calculated by subtracting the new CDF with the baseline CDF, which is then multiplied with the duration of the event to get CCDP value for that event. In order to know whether these events are precursor events or not, the

Table 7.7 Standards for probabilistic precursor analysis

IAEA TEC DOC-1417 [1]	Depending on its CCDP (>10^{-6}, >10^{-4}, >10^{-3}), an event is classified as a precursor, important precursor or significant precursor, respectively
STUK, Finland [15]	Three categories according to their CCDP value:
	Most important events: CCDP > 10^{-7},
	Other important events: 10^{-8} < CCDP < 10^{-7}, other events

calculated CCDP values can be compared with the threshold values. There are some standards available which provide these criteria. Some of the standards for probabilistic precursor analysis used are shown in Table 7.7.

Core damage index is one more measure in this study. This is calculated by summing up all the CCDP values of the events observed over a period of time and divide with respect to the time period and is given as follows:

$$\lambda = \sum_{j} \frac{\text{CCDP}_j}{\text{Observation time}} \tag{7.7}$$

This measure will be helpful in comparing the two plants. If the core damage index is more, the plant might be seeing more events in a given period of time or the events may be very critical from risk point of view. This is explained with a case study.

7.4 A Case Study on NPP Events

Probabilistic precursor analysis (PPA) studies were carried out on events observed in typical nuclear power plants. Incident reporting systems at NPPs systematically analyze the events and accordingly reports are issued in the form of event reports, significant event reports, etc. As a part of this study, significant event reports issued on various events that occurred during 2002–2006 were analyzed for two power plants. The criteria for classifying the precursor were adopted as per IAEA guidelines. Based on the available event information from NPP-1 and NPP-2, precursor analysis has been performed and the results are displayed in the form of graphs.

7.4.1 PPA for Plant 1

There were nine events observed during the period of 2002–2006 from NPP-1, out of which seven events were identified as IE type that are amenable for PSA analysis; whereas two events were identified as not amenable for PSA analysis (these events have not been considered in the PSA analysis, this is due to inadequacy of the existing PSA models and the PSA model needs to be updated to take care of all the events).

The results are shown graphically in Figs. 7.8 and 7.9, respectively. Figure 7.8 shows the graph between CCDP value and the events. From the graph, it can be shown that all the events are having the CCDP value well below 10^{-6}. Hence, from the analysis it is evident that no event is of "precursor" type and all are of 'no precursor" type. A graph between number of events versus year is shown in Fig. 7.9. From the graph, it is found that there is only one event of "no precursor type" in the years 2002 and 2003. There are three events in the year 2004 and two events in the year 2006. There is no event in the year 2005. All these events are of "no precursor type". Since there are no events which are of precursor type, the graphs are not shown for that category.

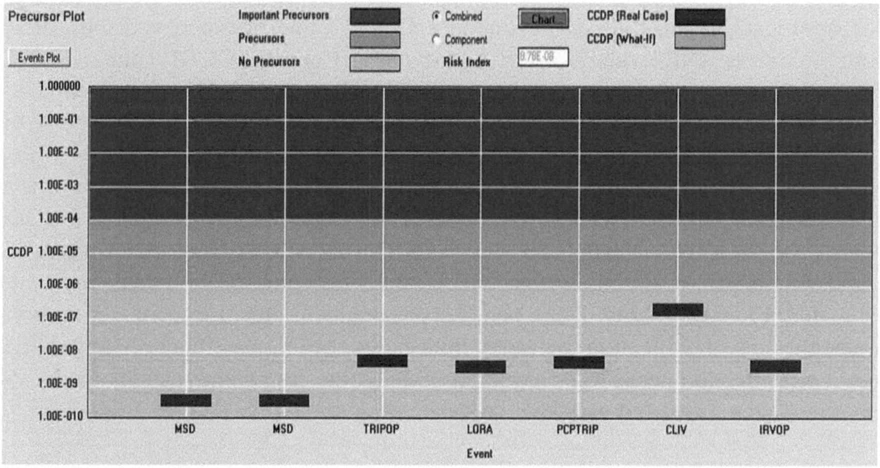

Fig. 7.8 Graph between CCDP versus event for NPP-1

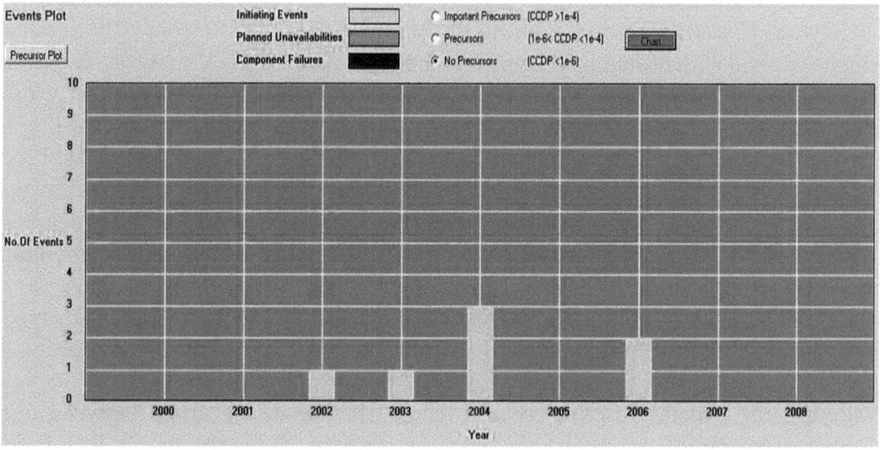

Fig. 7.9 Graph between number of events versus year for NPP-1

Table 7.8 Incident reported and their INES level along with CCDP evaluated

S.no	Event description	CCDP	INES level
1	Event 1–IE (MSD)	5.64E-10	0
2	Event 2–IE (MSD)	5.64E-10	0
3	Event 3-IE (TRIPOP)	9.00E-09	1
4	Event 4-IE (LORA)	5.91E-09	2
5	Event 5-IE (PCPTRIP)	8.24E-09	0
6	Event 6-IE (CLIV)	3.21E-07	0
7	Event 7-IE (IRVOP)	6.27E-09	1

In Table 7.8, CCDP and INES values against each event have been shown. For the first two events, the INES values have been given as "0", that means these are not significant events. The corresponding CCDP value obtained is 5.64E-10 for both the events. INES value for the third event is 1 and the CCDP value is 9.00E-09. But, for the fourth event even though the CCDP value is 5.91E-09, which is less than the previous value INES value is given as 2. Similarly, CCDP value for the sixth event is 8.24E-09, which is in between the above-mentioned two events (four and five), but the INES value is given as zero. For the eigth event INES vale is given as zero, but the CCDP value is 3.21E-07 which is greater than the CCDP value of three and four events for which INES values have been given. From the above discussion, it is evident that INES values for some of the events are under estimated. From this study, it can be observed that consistent measures for assuring uniformity in classification between PSA-based event classification and INES-based classification needs to be explored. Finally, the core damage index for NPP-1 plant has been estimated as 8.78E-08/year.

7.4.2 PPA for Plant 2

There were 11 events observed during the period of 2002–2006 from NPP-2, out of which six events identified as IE type that are amenable for PSA analysis, one event of Component failure type which is also amenable for PSA analysis. But, four events identified as not amenable for PSA analysis. The results are shown graphically in Fig. 7.10. This shows the graph between CCDP value and the events. From the graph, it can be shown that three events, whose CCDP values lie in between 10^{-4} and 10^{-6}, are of precursor type. The remaining events are of "no precursor" type (whose CCDP value lie well below 10^{-6}). There are no events of "important precursor" category.

Figure 7.11 shows the graph between total number of events of "precursor type" versus year. From the graph, it is found that there are three events of "precursor type". One event observed in the year 2002 and two events observed in the year 2006. All the events are of IE type. Figure 7.12 shows the graph between total number of events of "no precursor type" versus year. From the graph, it is found that there are four events of "no precursor type". One event observed in the year

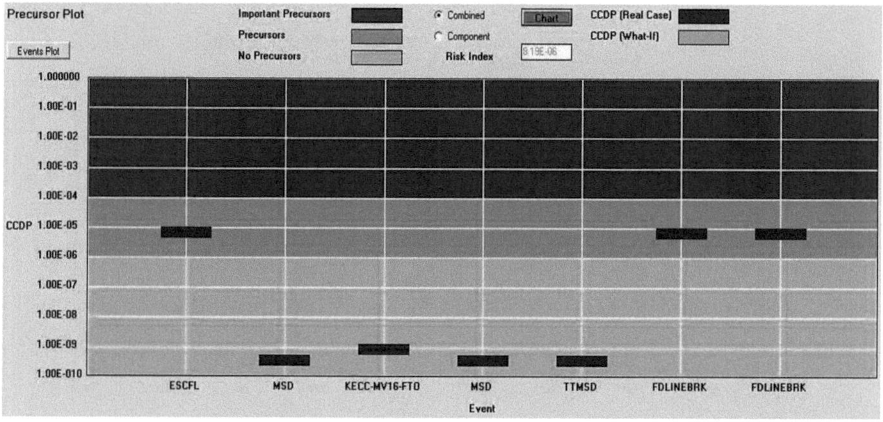

Fig. 7.10 Graph between CCDP versus event for NPP-2

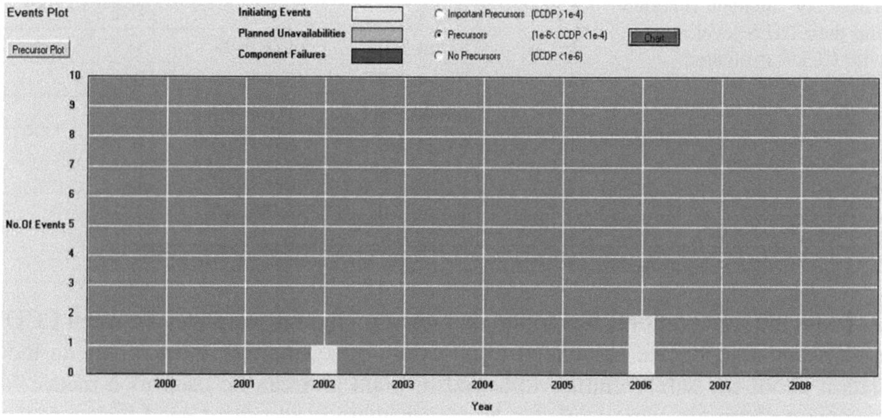

Fig. 7.11 Graph between number of events of "precursor type" versus year for plant-2

2003 which is of "component failure" type and three events observed in the year 2004 which are of IE type. Since there are no events which are of "important precursor" type, the graphs are not shown for that category.

In the results table, CCDP and INES values against each event have been shown in Table 7.9. For the first event, the INES values have been given as "1". The corresponding CCDP value obtained is 1.09E-05. CCDP value for the last event is 1.09E-05, which is similar to the 10th event, but INES value is given as zero. From the above discussion it is evident that INES values for some of the events are under estimated. The risk index for plant-2 plant has been estimated as 8.19E-06/year which is more than the plant-1. This is obvious from the results that plant-2 has seen more number of events and also there are two events of precursor type where as in the plant-1 there was no event of precursor type and the number of events observed are lesser.

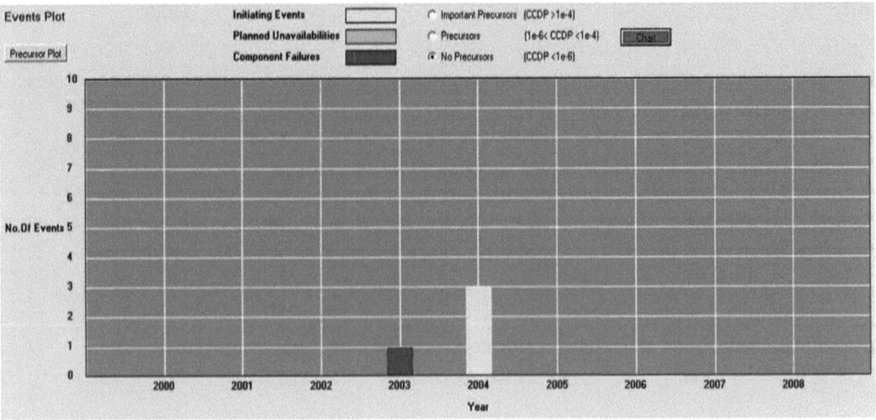

Fig. 7.12 Graph between number of events of "no precursor type" versus year

Table 7.9 Incident reported and their INES level along with CCDP evaluated

S.no	IE/SS	CCDP	INES Level
1	Event-1-IE	1.09E-05	1
2	Event-2-IE	5.64E-10	0
3	Event-3-SS	1.40E-09	0
4	Event-4-IE	5.64E-10	0
5	Event-5-IE	5.63E-10	0
6	Event-6-IE	1.09E-05	1
7	Event-7-IE	1.09E-05	0

From the observations, it is found that there is slight discrepancy between CCDP values and INES scale associated to an event. Also, the risk index gives an indication about the safety culture followed in plant and can be used as a metric for comparing between various plants. However, there is an urgent need for consensus among the values of CCDP between various standards for identifying events that fall under important precursor, precursor, and no precursor category. This is more prominent while comparing USNRC and STUK precursor programme.

7.5 Risk-Based Inspection of Thermal Power Plants

Risk-Based Inspection (RBI) procedure can be applied to different pressure equipment in a plant. With the help of RBI programme, one can identify the different equipment which have high likelihood of failure, which have high consequences of failure and which are high-risk components. With the help of this methodology, one can manage the different damage mechanisms of the equipment with the help of different inspection and maintenance plans by utilizing most cost effective inspection techniques. The RBI process is a step-by-step approach to develop, maintain and optimize the plant inspections by incorporating plant-specific knowledge, maintenance and failure history. In this section, application of RBI procedure to a typical thermal power plant is explained with example problems. As discussed in Chap. 1, a typical thermal power plant can be divided into four major systems such as:

- Boiler
- Turbine
- Generator and Electrical, and
- Common Plant

A partial list of components along with possible damage mechanisms that present in a typical thermal plant is provided in Table 7.10.

7.5.1 Calculation of Risk

According to the definition of risk, in order to quantify the risk one should know the likelihood of occurrence of an undesirable event (probability of failure or failure frequency) and the consequences if that event occurs. This can be mathematically represented as

$$\text{Risk}(t) = \text{PoF}(t) \times \text{CoF} \tag{7.8}$$

From the Eq. 7.8, it can be seen that risk is a function of time as the probability of failure of a component is always increasing function of time. There are different types of risks involved when a component fails such as safety risk which concern about the injury or fatalities of person, financial risk which takes into account the economic loss due to the failure of component and environmental risk which will consider the impact of component failure on the environment. Now we will see how the PoF and CoF are calculated for thermal power plant components. These are explained with examples in the following subsections.

7.5.1.1 PoF Calculations

Probability of failure (PoF) of a component can be calculated in different ways as listed below:

- By utilizing generic failure frequency (gff) from generic data sources. This procedure will not ensure the exact risk that is expected from the component failure due to the fact that the gff value is not plant specific and it is only a rough estimation. However, if no other information is available this can be used for preliminary calculations.
- By utilizing plant-specific failure data. The gff calculated from this information will be realistic, but it needs lot of operating experience data (large sample size). With small sample size the results will be biased.

Table 7.10 Components with damage mechanisms present in a typical thermal power plant

Component	Potential damage mechanism	Component	Potential damage mechanism
Condensate piping	FAC, F, IC, CF	Final superheater inlet header	C, F
Feedwater piping	FAC, F, IC, TF, CF	Final superheater tubes	IC, EC, F, C, SCC
Economizer inlet header	TF, F, EC, WD	Final superheater outlet header	F, C, TF
Economizer tubes	IC, E, CF	Reheater inlet header	F, C
Economizer outlet header	WD, EC	Reheater tubes	IC, EC, F, C, E
Steam Drum	TF, F, WD	Reheater outlet header	EC, F, C, TF
Furnace downcomers	WD	Main steam pipe	C, TF, F
Furnace manifolds	IC, F, CF, WD	Hot reheat pipe	C,TF,F
Furnace wall feeder pipes	IC, F, CF, WD	Cold reheat pipe	IC, CF, F, TF
Furnace wall inlet headers	IC, F, CF	HP bypass piping	F, C, TF, IC
Furnace wall tubes	IC, EC, F, TF, CF, E, OH	LP bypass piping	F, IC, TF
Furnace wall outlet headers	IC, F, CF, WD	Steam drum safety valves	EC, F, SCC, CF
Primary superheater inlet header	IC, F	Steam extraction/bled steam piping	FAC, F, IC
Primary superheater tubes	IC, EC, F, C, W, OH	LP feedwater heaters	IC, F, E
Primary superheater outlet header	IC, F,C	Deaerator	IC, FAC, CF, F, W
Superheater attemperators	IC, F, C, TF	HP feedheaters	IC, FAC, E, CF, TF, F
Secondary superheater inlet headers	F	Gland steam piping	C, F, IC, FAC
Secondary superheater tubes	IC, F, C, DMW,	Gland steam condenser	IC, FAC, F, TF
Secondary superheater outlet headers	F, C, TF	Air receivers	EC, IC, F

Legend FAC Flow accelerated corrosion, *F* Fatigue, *IC* Internal corrosion, *CF* Corrosion fatigue, *TF* Thermal fatigue, *EC* External Corrosion, *WD* Weld defect, *E* Erosion, *SCC* Stress corrosion cracking, *C* Creep, *OH* Over heating

- By combining both generic failure frequency and plant-specific failure data with Bayesian updation technique. This procedure can be utilized with small sample size of plant-specific failure data. However, larger the sample size better the estimate.

- By modifying the generic failure frequency with damage factors and management systems factor as described in API 581. With this procedure, one can utilize the plant-specific operating and design conditions of a component and can obtain better estimate of PoF.

For an example, consider the case of discharge pipe of feed water pump. The different design and operating conditions of the component are given in Table 7.11.

The general damage mechanisms that could be possible on discharge pipe are corrosion and fatigue. The failure type could be leak. Now consider the failure type as small leak happening due to fatigue. Consider the gff as 1.0×10^{-3}/year. The plant-specific failure data for this specific damage mechanism is given in Table 7.12.

From the Table 7.12, one can obtain the plant-specific failure frequency of discharge pipe as given in Eq. 7.9.

$$\lambda = \frac{R}{T_{\mathrm{OP}}} \tag{7.9}$$

where R is total number of failures = 1

T_{OP} is total operating hours of all the units considering all the components and can be calculated as follows:

$$T_{\mathrm{OP}} = 6490 \times 12 \times 2 + 7855 \times 10 \times 2 + 7650 \times 5 \times 2 + 8125 \times 10 \times 2 = 551860\,\mathrm{h}$$

$$\lambda = \frac{1}{551860} = 1.812 \times 10^{-6}/\mathrm{h} = 1.587 \times 10^{-2}/\mathrm{year}$$

By comparing the gff and the failure frequency obtained from the plant-specific failure information, it is evident that the plant-specific data is giving conservative

Table 7.11 Design and operating conditions of discharge pipe of feed water pump

Design conditions	Value	Operating conditions	Value
Design pressure	16,180 kPa	Operating pressure	14,000 kPa
Design temperature	240 °C	Operating temperature	200 °C
Nominal diameter	270 mm	Operating flow rate	125 kg/s
Design wall thickness	19 mm	Operating fluid	Water
Material	Carbon steel	State of content at ambient temperature	Gas
Operating hours	155760 h		

Table 7.12 Plant-specific failure data of discharge pipe of feed water pump

Unit	No. of components	No. of failures	Avg. Operating hours/year	Operating years
1	2	1	6490	12
2	2	0	7855	10
3	2	0	7650	5
4	2	0	8125	10

estimate than the gff obtained from the generic sources. One of the reasons could be the operating experience data may not be sufficient to estimate the actual failure frequency of the component or the gff obtained from the generic sources is less conservative as this information is obtained from different plant sources and may not be from homogeneous plants. In order to utilize both the information, one can use Bayesian updation technique to have better estimate of failure frequency. The inputs to this technique can be given as

- Consider the gff value as the median value of prior distribution with an error factor 3 and the prior distribution follows log normal distribution. Hence,

$$\lambda_{\text{median}} = 1 \times 10^{-3}/\text{year},$$
$$\text{EF} = 3$$

- Consider the evidence as 1 failure over a period of 551860 h (63 years).

As discussed in Chap. 3 by using Bayesian updation technique one can obtain the mean failure frequency as 1.67×10^{-3}/year. The prior and posterior distributions are shown in Fig. 7.13.

Now we calculate the PoF according to API-581 procedure [16]. As per this procedure, the PoF can be calculated from Eq. 7.10.

$$P_f(t) = \text{gff} \times D_f(t) \times F_{\text{MS}} \tag{7.10}$$

where

$D_f(t)$ is damage factor
F_{MS} is management systems factor

In this procedure, the generic failure frequency is adjusted with damage factor and management systems factor to account for the actual plant operating conditions,

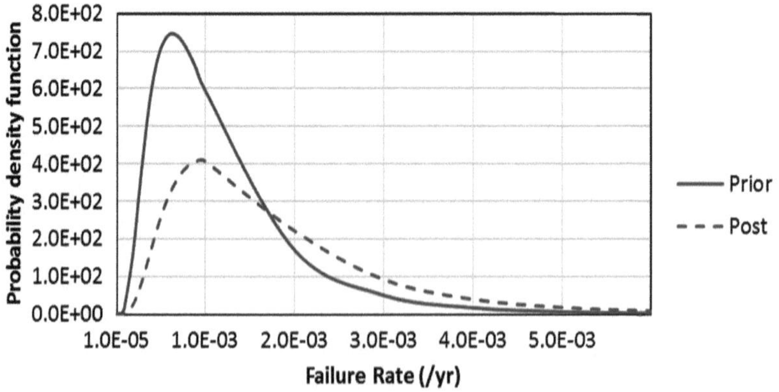

Fig. 7.13 Prior and posterior distributions of failure rate

damage mechanisms present in the component and management practices followed in the plant. The calculation of damage factor depends on the damage mechanism. Each damage mechanism will have different damage factors. They depend on component age, material type, inspection technique, design and operating conditions. The damage factor will tell how much damage is present in the component and the effectiveness of inspection technique used for quantifying that damage. Different methods have been provided in estimating damage factors for different damage mechanisms in API 581. To understand how the damage factors are calculated the reference calculations are provided here for mechanical fatigue damage mechanism. The damage factor is calculated from the Eq. 7.11.

$$\begin{aligned} D_f^{\text{MFat}} &= D_{\text{fb}}^{\text{MFat}} \times F_{\text{CA}} \times F_{\text{PC}} \times F_{\text{CP}} \times F_{\text{JB}} \times F_{\text{BD}} \\ D_{\text{fb}}^{\text{MFat}} &= \max\left[D_{\text{fb}}^{\text{PF}}, \left(D_{\text{fb}}^{\text{AS}} \times F_{\text{fb}}^{\text{AS}}\right), D_{\text{fb}}^{\text{CF}}\right] \end{aligned} \tag{7.11}$$

The factors required to calculate the damage factor in the Eq. 7.9 are explained in Table 7.13 and the values are obtained from actual plant conditions.

From the Table 7.13, the base damage factor can be calculated as follows:

$$\begin{aligned} D_{\text{fb}}^{\text{MFat}} &= \max\left[D_{\text{fb}}^{\text{PF}}, \left(D_{\text{fb}}^{\text{AS}} \times F_{\text{fb}}^{\text{AS}}\right), D_{\text{fb}}^{\text{CF}}\right] \\ &= \max[50, (50 \times 0.02), 50] \\ &= \max[50, 1, 50] \\ D_{\text{fb}}^{\text{MFat}} &= 50 \end{aligned}$$

Table 7.13 Various factors required for damage factor calculation

Factor	Description	Value	Explanation
$D_{\text{fb}}^{\text{PF}}$	Base damage factor for previous failures	50	One failure observed in the past
$D_{\text{fb}}^{\text{AS}}$	Base damage factor for shaking	50	Moderate, little vibration is observed
$F_{\text{fb}}^{\text{AS}}$	Adjustment factor for audible shaking	0.02	The vibrations are observed more than a year
$D_{\text{fb}}^{\text{CF}}$	Base damage factor for cyclic loading	50	The pipe is attached to a pump
F_{CA}	Damage factor adjustment for corrective action	2	No modifications done
F_{PC}	Damage factor adjustment for pipe complexity	1	Total pipe fittings between 6 and 10
F_{CP}	Damage factor adjustment for condition of pipe	2	Improper support
F_{JB}	Damage factor adjustment for joint type	0.2	Welded
F_{BD}	Damage factor adjustment for branch diameter	0.02	Any branch greater than 2 NPS

The final value of damage factor can be calculated as

$$\begin{aligned} D_f^{\text{MFat}} &= D_{\text{fb}}^{\text{MFat}} \times F_{\text{CA}} \times F_{\text{PC}} \times F_{\text{CP}} \times F_{\text{JB}} \times F_{\text{BD}} \\ &= 50 \times 2 \times 1 \times 2 \times 0.2 \times 0.02 \\ D_f^{\text{MFat}} &= 0.8 \end{aligned}$$

Now the PoF can be calculated from Eq. 7.9 as follows:

$$\begin{aligned} P_f(t) &= \text{gff} \times D_f(t) \times F_{\text{MS}} \\ &= 1 \times 10^{-3} \times 0.8 \times 1 \\ \text{PoF} &= 8 \times 10^{-4} \end{aligned}$$

In this calculations, management systems factor is considered as 1. As one can see from the PoF value that it is modified from the gff value considered based on the actual plant conditions.

7.5.1.2 CoF Calculations

Now consider the case of financial risk, in this case the risk can be calculated based on the Eq. 7.12

$$\text{Risk}(t) = \text{PoF}(t) \times \text{CoF}_{\text{Finanace}} \tag{7.12}$$

According to API-581, the financial consequence can be calculated from the Eq. 7.13.

$$\text{FC} = \text{FC}_{\text{cmd}} + \text{FC}_{\text{affa}} + \text{FC}_{\text{prod}} + \text{FC}_{\text{inj}} + \text{FC}_{\text{environ}} \tag{7.13}$$

The explanation of different factors and their corresponding costs due to the failure of the component for the case of small leak is provided in the Table 7.14.

In calculating the different costs associated with failure of a component, it is advised to use actual failure cost data from the plant. If the data is not available, one can use the cost calculations as described in API-581 [16]. In the present case for

Table 7.14 Various costs associated with failure of a component

Factor	Description
FC_{cmd}	Costs associated with repair of component
FC_{affa}	Costs associated with failure of adjacent component
FC_{prod}	Cost associated with production loss
FC_{inj}	Cost associated with personal injury
$\text{FC}_{\text{environ}}$	Cost associated with environmental cleanup

the small leak in the discharge pipe of the feed water pump, the cost calculations are described as follows:

- One can calculate the repair cost from the actual plant failure cost data
- When there is a leak in discharge pipe, it will not affect the adjacent components; hence there is no cost associated with adjacent component failure.
- However, for repairing the component plant may have to be shut down for 1 day. Hence, the cost associated with the production loss will be equivalent to 1 day production loss. The calculation should be based on actual capacity of the plant.
- Since the operating fluid is water above 100 °C, when water leaks it will flashes into steam at ambient temperatures. The area affected by this event will be less than 5 m. If any operating personnel is present (up to three people in this case from plant data present for a period up to 1 h/day), it can cause small injuries.
- Also there will not be any effect on environment as it is steam.

From the above discussions, the total financial cost associated with failure of discharge pipe can be calculated. Also the risk can be calculated by multiplying the PoF and CoF values.

7.5.2 Risk Matrix

Risk matrix can be defined as a matrix that is used to categorize the components into high, medium, and low importance based on probability of failure of component and consequence of its failure and this can be used as a decision matrix. Hence, the required parameters for constructing the risk matrix are PoF and CoF values. Once the PoF and CoF are known, one can plot these values on risk matrix as

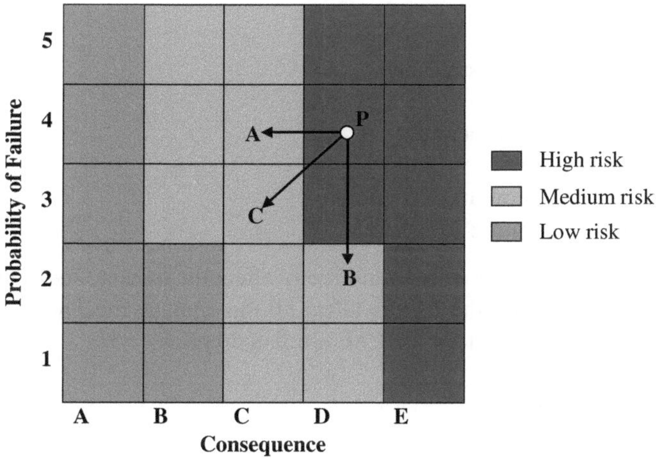

Fig. 7.14 Representation of risk matrix

shown in Fig. 7.14. The risk matrix shown in Fig. 7.14 is only an example representation. One can develop plant-specific risk matrix based on the risk targets of the plant. Based on the region in which the component falls in the risk matrix the remedies should be taken. For example, if the determined PoF and CoF lies within the green area the risk is at an acceptably low level. Here the current inspection and maintenance strategies can be continued. If it falls in red region that means the component should be looked in for more detailed evaluation of risk and the current inspection practices should be looked into and if required they should be modified. One can also think of how the risk of a component can be made low or how the component can be brought into low risk regions. Consider Fig. 7.14 in which component is falling in red region represented as point P.

Now there are three ways with which this component or Point P can be brought into low risk regions such as:

- To point A—By reducing the consequences of the component failure
- To point B—By reducing the probability of failure of the component
- To point C—By reducing both consequences and probability of failure

From the previous discussions, it is clear that both the component failure probability and consequences of component failure depends on various factors. If one can control these factors the risk can be controlled. The following are the list of controlling factors for PoF.

- Age of the component
- Total Starts per year
- Typical start-up rate/Loading rate
- Material type
- Time since last inspection
- Scope of inspection techniques
- Adequacy of inspection techniques
- Water/Steam Chemistry
- Fatigue stresses
- Local mechanical over-stressing
- Local Hot spots
- Nominal operating temperature
- Corrosion susceptibility
- Frequency of temperature excursions
- Severity of Temperature excursions

These controlling factors directly or indirectly affect the damage mechanism present in the component. Not all the factors affect all the damage mechanisms. Similarly, some of the controlling factors for CoF are listed below:

- Area affected by failure
- No. of people in area
- Time in the area
- Injury type

- Effect on environment
- Repair cost
- Shutdown cost

When these factors are controlled, the risk due to the component failure can also be controlled.

7.5.3 *Effect of Inspection Intervals on Risk*

It is clear from the Eq. 7.8 that risk is always a function of time as the probability of failure of a component increases with time (assuming that consequences do not change with time for the given scenario). It is also mentioned that the probability of failure of a component depends on several factors, one of the factors is inspection interval. If a component is inspected in regular intervals one can find out if there was any degradation present in the material or not and if the degradation rate is high one can think of replacing the component or take necessary measures to arrest the degradation. If the component is not inspected, then it is not possible to know the condition of the component and it can fail at any point of time if the degradation is presented in the component and if the rate of degradation is high. With the help of the inspection results, one can estimate the remaining life of the component also. For example, if there was corrosion present in the component with the help of thickness measurements one can estimate the corrosion rate and remaining life of the component. The effect of inspection interval on risk is shown in Fig. 7.15.

7.5.4 *Inspection Strategies Based on Risk*

Inspection strategy should be identified for each component at different inspection locations. The inspection strategy depends on the degradation mechanism present,

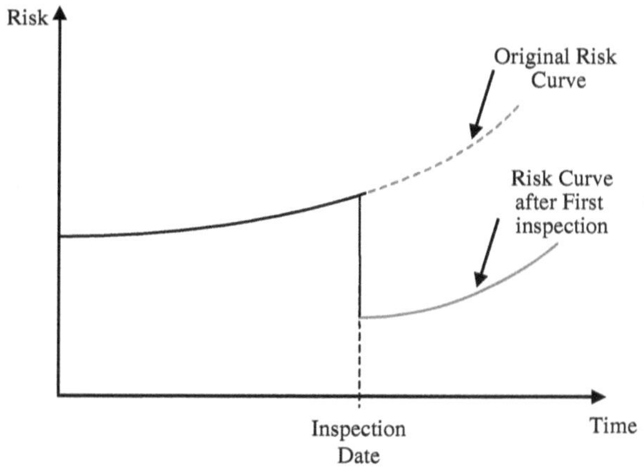

Fig. 7.15 Effect of inspection on risk curve

physical access constraints, and cost considerations. Therefore, examination methods, inspection volumes, and acceptance and evaluation criteria must be specifically designed for the damage mechanism(s) active at the inspection location. In this section, a general inspection strategy that can be adopted for components has been explained. The strategy is based on utilization of the existing inspection strategies that is followed in the plant and the results from the RBI process. The strategy is explained in the following steps.

- Collect the PoF and CoF values of all the components from the RBI results.
- Plot the PoF and CoF values on Risk Matrix that is designed for the plant
- This step will give a pictorial representation of how many components are falling in different regions of the Risk Matrix.
- Collect the present inspection strategy information of all the components which are represented on Risk Matrix.
- Superimpose the inspection strategies on the Risk Matrix. The above steps are explained with an example problem as follows.
- Consider there are 50 components for which RBI results are available.
- Table 7.15 gives the information on how the components are distributed on Risk Matrix. The same is shown in Fig. 7.16.
- If the current inspection strategy (hydro, UT etc.) as per the plant for all the components is 3 years (2GO, considering plant General Outage as 1.5 years).
- Now superimpose the current inspection strategy on risk matrix for all the components as shown in Fig. 7.17.
- From Fig. 7.17, it can be seen that the components which are in red region, orange region, and green region are having the same inspection strategies.
- According to the risk matrix the components falling in red region should be given more priority than the components in orange and green.
- If one considers that the present inspection strategy (2GO) is sufficient for the components falling in red region, then from risk point of view for the components falling in other regions the inspection interval can be increased.

Table 7.15 Distribution of components on risk matrix

S.no	Region	Components	Region	Components	Region	Components
1	5-A	2	5-B	2	5-C	2
2	4-A	1	4-B	2	5-D	1
3	3-A	1	4-C	1	5-E	1
4	3-B	2	3-C	2	4-D	2
5	2-B	1	3-D	3	4-E	1
6	2-C	1	2-D	1	3-E	1
7	1-C	2	1-D	2	2-E	2
6	2-A	3	1-E	2		
7	1-A	4				
8	1-B	3				
	Total	25	Total	15	Total	10

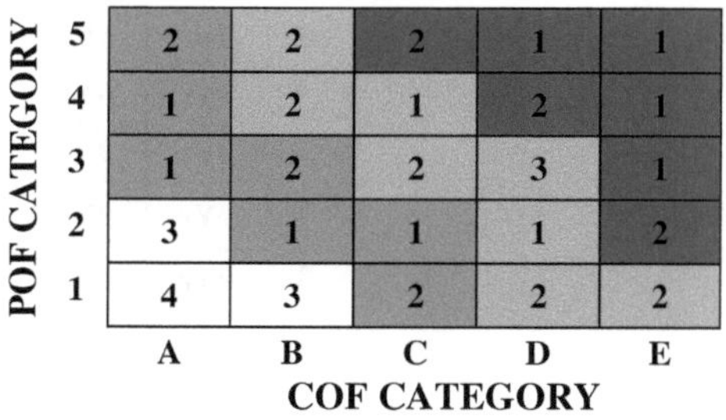

Fig. 7.16 Distribution of components on risk matrix

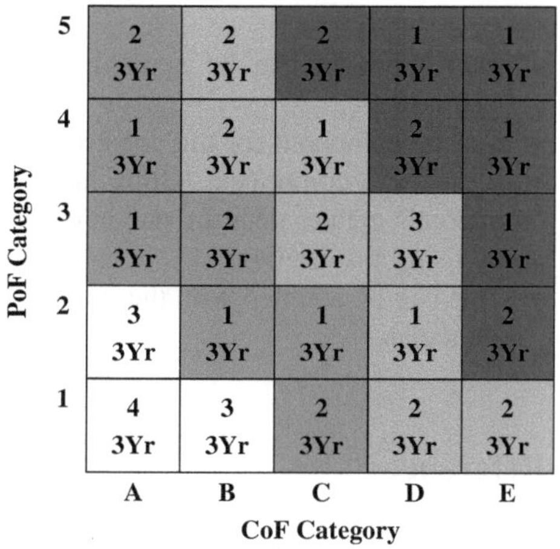

Fig. 7.17 Distribution of components on risk matrix

- In changing the inspection intervals for the components falling in orange, green, and white regions the following factors also can be taken into consideration
  - Damage mechanism presented
  - Last inspection findings (thickness)
  - General regulations
- By considering the above aspects, different inspection strategies can be proposed as given in Table 7.16.
- The inspection strategies presented in Table 7.16 are incorporated in the Risk Matrix and is shown in Fig. 7.18.

Table 7.16 Different inspection strategies proposed

Strategy	Region		
	Red	Orange	Green/white
1	2GO	3GO	4GO
2	2GO	4GO	5GO
3	2GO	4GO	6GO

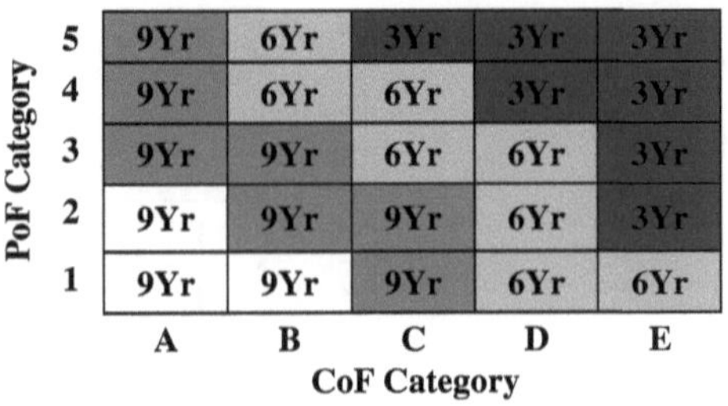

Fig. 7.18 Incorporation of new inspection strategies in risk matrix

- This approach can be applied for each damage mechanism and Inspection technique.
- Can be compared for the consistency in the strategies.
- It means, a component falling in green region may sometimes have inspection intervals greater than the one in orange and red regions. This can be eliminated by using this method.
- This can be coupled with the risk targets year wise and can further optimize the strategies.

References

1. IAEA-TECDOC-1209 (2001) Risk management: a tool for improving nuclear power plant performance, IAEA, Vienna
2. Nakai R, Kani Y (1991) A living PSA system LIPSAS for an LMFBR. In: International symposium on the use of PSA for operational safety, PSA 1991, Vienna, Austria, 3–7 June 1991
3. IAEA-TECDOC-1106 (1999) Living probabilistic safety assessment (LPSA)
4. Prasad MH, Vinod G, Saraf RK, Ghosh AK (2006) Risk monitor a tool for operational safety assessment, report BARC/2006/E/016
5. Hari Prasad M, Gopika Vinod VVS, Saraf RK, Ghosh AK (2005) Application of PSA in risk informed decision making. In: International conference on reliability, safety and hazards, Mumbai
6. Hari Prasad M, Gopika Vinod VVS, Rao S (2014) Risk management of NPPs using risk monitors. Int J Syst Assur Eng Manag. doi:10.1007/s13198-014-0295-5
7. Saraf RK, Hari Prasad M, Ghosh AK (2005) Treatment of common cause failures in risk monitor. In: International conference on reliability and safety engineering 2005, Bhubaneswar, 21–23 Dec 2005
8. Regulatory Guide-1.174 (2008) An approach for using probabilistic risk assessment in risk-informed decisions on plant-specific changes to the licensing basis. USNRC
9. IAEA-TECDOC-737 (1994) Advances in reliability analysis and probabilistic safety assessment for nuclear power reactors, IAEA, Vienna
10. Smith CL (1998) Calculating conditional core damage probabilities for nuclear power plant operations. Reliab Eng Syst Saf 59:299–307
11. IAEA-TECDOC-1417 (2004) Precursor analyses-the use of deterministic and PSA based methods in the event investigation process at nuclear power plants, IAEA, Vienna

12. http://www-ns.iaea.org/tech-areas/emergency/ines.htm
13. Kirchsteiger C (1997) Impact of accident precursors on risk estimates from accident databases. J Loss Prev Process Ind 10(3):159–167
14. De Gelder P, Hulsmans M, Gryffroy D, De Boeck B (2003) Use of the deterministic and probabilistic approach in the Belgian regulatory context. In: SMIRT-17 proceedings
15. STUK–B 92 (2007) Regulatory control of nuclear safety in Finland. Annual report
16. API Recommended Practice 581 (2008) Risk-based inspection technology, 2nd edn. API

Appendix A
Response Surface Methodology

A.1 Introduction

Response Surface Methodology (RSM) [1] is a collection of statistical and mathematical techniques useful for developing, improving, and optimizing processes. The most extensive applications of RSM are in the particular situations where several input variables potentially influence some performance measure or quality characteristic of the process. Thus performance measure or quality characteristic is called the response. The input variables are sometimes called independent variables, and they are subject to the control of the scientist or engineer.

The field of response surface methodology consists of the experimental strategy for exploring the space of the process or independent variables, empirical statistical modeling to develop an appropriate approximating relationship between the yield and the process variables, and optimization methods for finding the values of the process variables that produce desirable values of the response. In this section we will concentrate on the second strategy: statistical modeling to develop an appropriate approximating model between the response y and independent variables $\zeta_1, \zeta_2, \ldots, \zeta_k$. In general, the relationship is

$$y = f(\xi_1, \xi_2, \ldots, \xi_k) + \varepsilon$$

where the form of the true response function f is unknown and perhaps very complicated, and ϵ is a term that represents other sources of variability not accounted for in f. Usually ϵ includes effects such as measurement error on the response, background noise, the effect of other variables, and so on. Usually ϵ is treated as a statistical error, often assuming it to have a normal distribution with mean zero and variance σ^2. Then

$$E(y) = \eta = E(f(\xi_1, \xi_2, \ldots, \xi_k) + \varepsilon)$$
$$E(y) = E(f(\xi_1, \xi_2, \ldots, \xi_k)) + E(\varepsilon)$$
$$E(y) = E(f(\xi_1, \xi_2, \ldots, \xi_k))$$

A.K. Verma et al., *Risk Management of Non-Renewable Energy Systems*,
Springer Series in Reliability Engineering, DOI 10.1007/978-3-319-16062-7

The variables $\zeta_1, \zeta_2, \ldots, \zeta_k$ in the above equation are usually called the natural variables. In much RSM work it is convenient to transform the natural variables to coded variables $x_1, x_2, \ldots, x_k$, which are usually defined to be dimensionless with mean zero and the same standard deviation. In terms of the coded variables, the response function can be written as

$$y = f(x_1, x_2, \ldots, x_k)$$

Since the true form of the response function is not known one has to approximate the function and utilize in the analysis. However, successful utilization of RSM is mainly dependent on the experimenter's ability to develop a suitable approximation for f. In general, a low-order polynomial in some relatively small region of the independent variable space is appropriate. However, in many cases, either a first-order or a second-order model is used. If the experimenter is interested in only a small region of independent variable and if it contains a little curvature in the function, one can utilize the first-order models. For the case of two independent variables, the first-order model in terms of the two independent variables can be given as follows:

$$\eta = \beta_0 + \beta_1 x_1 + \beta_2 x_2$$

As can be seen from above, the equation is represented with each term containing only one input variable and there are no interaction terms, i.e., combination of different input variables. This kind of a model with no interaction between the input variables is called a main effects model, because it includes only the main effects of the two variables x_1 and x_2. If one considers the interaction between these two variables it can be added to the first-order model and the model can be represented as follows:

$$\eta = \beta_0 + \beta_1 x_1 + \beta_2 x_2 + \beta_{12} x_1 x_2$$

This is the first-order model with interaction. By addition of interaction term into the first-order model one can introduce curvature into the response function. Often, even with the introduction of curvature into the true response surface the first-order model is inadequate. In such situations one should go for the higher order terms such as second-order models. For the case of two variables, the second-order model can be represented as

$$\eta = \beta_0 + \beta_1 x_1 + \beta_2 x_2 + \beta_{11} x_1^2 + \beta_{22} x_2^2 + \beta_{12} x_1 x_2$$

In general, the first-order model with interactions can be written as follows:

$$\eta = \beta_0 + \sum_{i=1}^{n} \beta_i x_i + \sum_{i=1}^{n-1} \sum_{j=i+1}^{n} \beta_{ij} x_i x_j$$

And the second-order model can be written as follows:

$$\eta = \beta_0 + \sum_{i=1}^{n} \beta_i x_i + \sum_{i=1}^{n} \beta_{ii} x_{ii}^2 + \sum_{i=1}^{n-1} \sum_{j=i+1}^{n} \beta_{ij} x_i x_j$$

A.2 Linear Regression Model

In the practical application of RSM it is necessary to develop an approximating model for the true response surface. The underlying true response surface is typically driven by some unknown physical mechanism. The approximating model is based on observed data from the process or system and is an empirical model. Multiple regression is a collection of statistical techniques useful for building the types of empirical models required in RSM. The first-order multiple linear regression model with two independent variables can be represented as

$$y = \beta_0 + \beta_1 x_1 + \beta_2 x_2 + \varepsilon$$

The independent variables are often called predictor variables or regressors. The term linear is used because the above equation is a linear function of the unknown parameters β_0, β_1, β_2. In general, the response variable y may be related to k regressor variables. The model can be represented as

$$y = \beta_0 + \beta_1 x_1 + \beta_2 x_2 + \cdots + \beta_k x_k + \varepsilon$$

and is called a multiple linear regression model with k regressor variables. The parameters β_j, $j = 0, 1, \ldots, k$ are called the regression coefficients. Models that are more complex in appearance than the above equation may often still be analyzed by multiple linear regression techniques. For example, consider adding an interaction term to the first-order model in two variables

$$y = \beta_0 + \beta_1 x_1 + \beta_2 x_2 + \beta_{12} x_1 x_2 + \varepsilon$$

As another example, consider the second-order response surface model in two variables

$$y = \beta_0 + \beta_1 x_1 + \beta_2 x_2 + \beta_{11} x_1^2 + \beta_{22} x_2^2 + \beta_{12} x_1 x_2 + \varepsilon$$

In general, any regression model that is linear in the parameters (the β values) is a linear regression model, regardless of the shape of the response surface that it generates.

A.3 Estimation of the Parameters in Linear Regression Models

The method of least squares is typically used to estimate the regression coefficients in a multiple linear regression model. Suppose that n observations on the response variable are available, say $y_1, y_2, \ldots, y_n$. Along with each observed response y_i, we will have an observation on each regressor variable, let x_i denote the ith observation. Now consider a simple linear regression model relating y to x which can be given as follows:

$$y = \beta_0 + \beta_1 x + \varepsilon$$

For each of the observations we can write the above equation as follows:

$$y = \beta_0 + \beta_1 x_i + \varepsilon_i$$

The objective is to minimize the sum of the squares of the error term. Let this sum be denoted by L as follows:

$$L = \sum_{i=1}^{n} \varepsilon_i^2$$

L will also be equal to the following equation:

$$L = \sum_{i=1}^{n} (y_i - \beta_0 + \beta_1 x_i)^2$$

We need to find out the values of β_i such that L is minimized. Thus

$$\frac{\partial L}{\partial \beta_0} = -2 \sum_{i=1}^{n} (y_i - \beta_0 + \beta_1 x_i) = 0$$

$$\frac{\partial L}{\partial \beta_1} = -2 \sum_{i=1}^{n} (y_i - \beta_0 + \beta_1 x_i) x_i = 0$$

Simplifying the above equations we get

$$n\beta_0 + \beta_1 \sum_{i=1}^{n} x_i = \sum_{i=1}^{n} y_i$$

$$\beta_0 \sum_{i=1}^{n} x_i + \beta_1 \sum_{i=1}^{n} x_i^2 = \sum_{i=1}^{n} y_i x_i$$

Now we have two equations to solve for the two unknowns β_0 and β_1. The solution can be written as follows:

$$\beta_0 = \frac{\sum_{i=1}^{n} y_i - \beta_1 \sum_{i=1}^{n} x_i}{n}$$

$$\beta_0 = \frac{\sum_{i=1}^{n} y_i - \beta_1 \sum_{i=1}^{n} x_i}{n}$$

and

$$\beta_1 = \frac{\sum_{i=1}^{n} y_i x_i - \frac{\sum_{i=1}^{n} y_i \sum_{i=1}^{n} x_i}{n}}{\left[\sum_{i=1}^{n} x_i^2 - \frac{\left(\sum_{i=1}^{n} x_i\right)^2}{n}\right]}$$

A.4 Multiple Linear Regression

In scalar notation, the first-order multiple regression model can be given by the following equation:

$$y = \beta_0 + \beta_1 x_1 + \beta_2 x_2 + \cdots + \beta_k x_k + \varepsilon$$

The above equation can be represented in matrix form as follows and it is more compact to use.

$$\begin{Bmatrix} y_1 \\ y_2 \\ y_3 \\ \cdot \\ \cdot \\ \cdot \\ y_n \end{Bmatrix} = \begin{bmatrix} 1 & x_{11} & x_{21} & \cdots & x_{k1} \\ 1 & x_{12} & x_{22} & \cdots & x_{k2} \\ 1 & x_{13} & x_{23} & \cdots & x_{k3} \\ \cdot & \cdot & \cdot & \cdots & \cdot \\ \cdot & \cdot & \cdot & \cdots & \cdot \\ \cdot & \cdot & \cdot & \cdots & \cdot \\ 1 & x_{1n} & x_{2n} & \cdots & x_{kn} \end{bmatrix} \begin{Bmatrix} \beta_0 \\ \beta_1 \\ \beta_2 \\ \cdot \\ \cdot \\ \cdot \\ \beta_k \end{Bmatrix} + \begin{bmatrix} \varepsilon_1 \\ \varepsilon_2 \\ \varepsilon_3 \\ \cdot \\ \cdot \\ \cdot \\ \varepsilon_n \end{bmatrix}$$

It can also be represented as

$$y = X\beta + \varepsilon$$

The least square estimator of β can be given as follows:

$$\beta = (X'X)^{-1}X'y$$

Now consider the second-order multiple linear regression model without interactions as follows:

$$Y = f(x_i; a_0, a_1, a_2 \ldots)$$

$$\begin{aligned} Y = {} & a_0 + a_1x_1 + a_2x_2 + a_3x_3 + a_4x_4 + a_5x_5 + a_6x_6 \\ & + b_1x_1^2 + b_2x_2^2 + b_3x_3^2 + b_4x_4^2 + b_5x_5^2 + b_6x_6^2 + \varepsilon \end{aligned}$$

The above equation can be represented in a matrix form as follows:

$$\begin{Bmatrix} y_1 \\ y_2 \\ y_3 \\ \cdot \\ \cdot \\ \cdot \\ y_n \end{Bmatrix} = \begin{bmatrix} 1 & x_{11} & x_{21} & \cdot & \cdot & x_{61} & x_{11}^2 & x_{21}^2 & \cdot & \cdot & \cdot & x_{61}^2 \\ 1 & x_{12} & x_{22} & \cdot & \cdot & x_{62} & x_{12}^2 & x_{22}^2 & \cdot & \cdot & \cdot & x_{62}^2 \\ 1 & x_{13} & x_{23} & \cdot & \cdot & x_{63} & x_{13}^2 & x_{23}^2 & \cdot & \cdot & \cdot & x_{63}^2 \\ \cdot & \cdot & \cdot & \cdot & \cdot & \cdot & \cdot & \cdot & \cdot & \cdot & \cdot & \cdot \\ \cdot & \cdot & \cdot & \cdot & \cdot & \cdot & \cdot & \cdot & \cdot & \cdot & \cdot & \cdot \\ \cdot & \cdot & \cdot & \cdot & \cdot & \cdot & \cdot & \cdot & \cdot & \cdot & \cdot & \cdot \\ 1 & x_{1n} & x_{2n} & \cdot & \cdot & x_{6n} & x_{1n}^2 & x_{24}^2 & \cdot & \cdot & \cdot & x_{6n}^2 \end{bmatrix} \begin{Bmatrix} \beta_0 \\ \beta_1 \\ \beta_2 \\ \cdot \\ \cdot \\ \cdot \\ \beta_k \end{Bmatrix} + \begin{bmatrix} \varepsilon_1 \\ \varepsilon_2 \\ \varepsilon_3 \\ \cdot \\ \cdot \\ \cdot \\ \varepsilon_n \end{bmatrix}$$

It can also be represented as

$$y = x\beta + \varepsilon$$

The procedure of estimation of the parameters is explained as follows:

$$L = \sum_{i=1}^{n} \varepsilon^2$$

$$L = \sum_{i=1}^{n} (Y_i - f_i)^2 \qquad (R \text{ is the residual})$$

$$L = \sum_{i=1}^{n} \left[y_i - \begin{pmatrix} a_0 + a_1x_1 + a_2x_2 + a_3x_3 + a_4x_4 + a_5x_5 + a_6x_6 \\ +b_1x_1^2 + b_2x_2^2 + b_3x_3^2 + b_4x_4^2 + b_5x_5^2 + b_6x_6^2 \end{pmatrix} \right]^2$$

$$\frac{\partial L}{\partial a_k} = -2\sum_{i=1}^{n} \left[y_i - \begin{pmatrix} a_0 + a_1x_1 + a_2x_2 + a_3x_3 + a_4x_4 + a_5x_5 + a_6x_6 \\ +b_1x_1^2 + b_2x_2^2 + b_3x_3^2 + b_4x_4^2 + b_5x_5^2 + b_6x_6^2 \end{pmatrix} \right] x^k = 0$$

$$\frac{\partial L}{\partial a_0} = 2\sum_{i=1}^{n} \left[y_i - \begin{pmatrix} a_0 + a_1x_1 + a_2x_2 + a_3x_3 + a_4x_4 + a_5x_5 + a_6x_6 \\ +b_1x_1^2 + b_2x_2^2 + b_3x_3^2 + b_4x_4^2 + b_5x_5^2 + b_6x_6^2 \end{pmatrix} \right] (-1) = 0$$

$$\sum_{i=1}^{n} y_i - \sum_{i=1}^{n} a_0 - \sum_{i=1}^{n} a_1x_{1i} - \ldots \sum_{i=1}^{n} b_6x_{6ii}^2 = 0$$

$$a_0n + a_1\sum_{i=1}^{n} x_{1i} + \cdots + b_6\sum_{i=1}^{n} x_{6ii}^2 = \sum_{i=1}^{n} y_i$$

$$\begin{bmatrix} n & \sum_{i=1}^{n} x_{1i} & \cdots & \sum_{i=1}^{n} x_{6ii}^2 \end{bmatrix} \begin{Bmatrix} a_0 \\ a_1 \\ \vdots \\ b_6 \end{Bmatrix} = \left\{ \sum_{i=1}^{n} y_i \right\}$$

Similarly, one can obtain the remaining equations for different coefficients. After simplification one can write the above equations in the form of matrix as shown below:

$$\begin{bmatrix} n & \sum_{i=1}^{n} x_{1i} & \cdots & \sum_{i=1}^{n} x_{6i}^2 \\ \sum_{i=1}^{n} x_{1i} & \sum_{i=1}^{n} x_{1i}^2 & \cdots & \sum_{i=1}^{n} x_{1i}x_{6i}^2 \\ \vdots & \vdots & \vdots & \vdots \\ \sum_{i=1}^{n} x_{6i}^2 & \sum_{i=1}^{n} x_{1i}x_{6i}^2 & \cdots & \sum_{i=1}^{n} \left(x_{6i}^2\right)^2 \end{bmatrix} \begin{bmatrix} a_0 \\ a_1 \\ \vdots \\ b_6 \end{bmatrix} = \begin{bmatrix} \sum_{i=1}^{n} y_i \\ \sum_{i=1}^{n} x_i y_i \\ \vdots \\ \sum_{i=1}^{n} x_{6i}^2 y_i \end{bmatrix}$$

The above matrix can be generalized and can be written for any function as follows:

$$\begin{bmatrix} n & \sum_{i=1}^{n} x_i & \cdots & \sum_{i=1}^{n} x_i^k \\ \sum_{i=1}^{n} x_i & \sum_{i=1}^{n} x_i^2 & \cdots & \sum_{i=1}^{n} x_i^{k+1} \\ \vdots & \vdots & \vdots & \vdots \\ \sum_{i=1}^{n} x_i^k & \sum_{i=1}^{n} x_i^{k+1} & \cdots & \sum_{i=1}^{n} x_i^{2k} \end{bmatrix} \begin{bmatrix} a_0 \\ a_1 \\ \vdots \\ a_k \end{bmatrix} = \begin{bmatrix} \sum_{i=1}^{n} y_i \\ \sum_{i=1}^{n} x_i y_i \\ \vdots \\ \sum_{i=1}^{n} x_i^k y_i \end{bmatrix}$$

By solving the above matrix one can obtain the regression coefficients as follows:

$$XA = Y$$
$$A = X^{-1}Y$$

If we compare the above equation with the original equation

$$y = x\beta + \varepsilon$$

Then

$$X = x'x$$
$$Y = x'y$$
$$A = \beta$$

Hence this equation also represents the same as follows:

$$\beta = (x'x)^{-1}x'\,y$$

Appendix B
Simulation Techniques

B.1 Introduction

In any structural reliability analysis the concept of limit state is used to define the failure of the structure. A limit state is a boundary between desired and undesired performance of a structure. This boundary is often represented mathematically by a limit state function or performance function. The limit state can be linear or nonlinear functions of the basic variables. Several techniques have been evolved for solving the structural reliability problems, namely simulation techniques, variance reduction techniques, first-order reliability method (FORM), and second-order reliability method (SORM). Structural reliability methods can be applied to time invariant and time variant problems. In time invariant problems random variables are independent of time whereas in the later case one should deal with the time-dependent random variables and the related failure probabilities. In this appendix simulation techniques are briefly explained in the following sections.

B.2 Methodology

The basic idea behind simulation is, as the name implies, to numerically simulate a given phenomenon and then observe the number of times a given event of interest occurs. The basic concept behind simulation is relatively simple, but the procedure can become computationally intensive. The Monte Carlo method is a special technique that can be used to generate the results numerically without actually doing any physical testing. The Monte Carlo simulation can be explained with a limit state function consisting load and resistance as the random variables and is given as follows:

$$Y = R - S \tag{B.1}$$

A.K. Verma et al., *Risk Management of Non-Renewable Energy Systems*,
Springer Series in Reliability Engineering, DOI 10.1007/978-3-319-16062-7

The probability of failure is the probability that $R < S$ as shown in Eq. 5.1.

$$P_f = P(R \leq S) \tag{B.2}$$

The basic procedure is as follows:

1. Generate random numbers (u_i, in our case it is u_R, u_S) for the basic random variables between 0 and 1.
2. Generate a random variate for the random variables using the corresponding probability distributions from the following formula:
   $x_i = F_X^{-1}(u_i)$
   where F_X^{-1} is the inverse of F_X
   F_X is the cumulative distribution function (CDF) of the random variable x.
3. Calculate the value of the limit state function (Y)
4. Store the calculated value of Y
5. Repeat steps 1–4 until a sufficient number of Y values have been generated.
6. Estimate the probability of failure as follows:

$$P_f = \frac{\text{Number of times that } Y < 0}{\text{Total Number of simulated values}}$$

It is important to recognize that this estimate of probability is only an estimate. However, the estimate improves as the number of simulations increases. Let N be the total number of simulations and n be the number of times that a particular criterion was achieved. Then the estimated probability can be calculated as the ratio

$$P = \frac{n}{N} \tag{B.3}$$

Let P_{true} be the theoretically correct probability. It can be shown that the expected value, variance, and coefficient of variation of the estimated probability P as follows:

$$\begin{aligned} E[P] &= P_{\text{true}} \\ \sigma_P^2 &= \frac{1}{N}[P_{\text{true}}(1 - P_{\text{true}})] \\ V_P &= \sqrt{\frac{(1 - P_{\text{true}})}{N(P_{\text{true}})}} \end{aligned} \tag{B.4}$$

From the above equation we can observe that the uncertainty in the estimate of the probability decreases as the total number of simulation, N, increases. These relationships are helpful in determining how many simulations are required to estimate a probability and limit the uncertainty in the estimate. Generation of random variates from different distributions is explained below.

B.3 Exponential Distribution

- Probability density function

$$f_X(x) = \lambda e^{-\lambda x} \qquad X \geq 0$$

- Cumulative distribution function

$$F_X(x) = \int_0^x f_X(x)\,dx$$

$$F_X(x) = 1 - e^{-\lambda x}$$

$$e^{-\lambda x} = 1 - F_X(x)$$

$$\text{Let} \qquad \text{Rnd} = \text{Random Number}$$

$$\text{Rnd} = 1 - F_X(x)$$

- Range of $F_X(x)$ is 0–1 and range of $(1-F_X(x))$ will also be 0–1. So it can be replaced with a random number.

$$e^{-\lambda x} = \text{Rnd}$$

$$\ln\left(e^{-\lambda x}\right) = \ln(\text{Rnd})$$

$$-\lambda x = \ln(\text{Rnd})$$

$$x = -\frac{1}{\lambda}\ln(\text{Rnd})$$

Example B.1 Assume x represents the failure time of a component and λ represents failure rate of the component. Generate random variate for the following information:

$$\lambda = 1 \times 10^{-2}/\text{year}$$

$$\text{Rnd} = 0.8756$$

Solution:

$$x = -\frac{1}{\lambda}\ln(\text{Rnd})$$

$$x = -\frac{1}{1 \times 10^{-2}}\ln(0.8756)$$

$$x = -\frac{1}{1 \times 10^{-2}}(-0.13285)$$

$$x = 13.2846\ \text{year}$$

- From the above it is calculated that the component can fail after 13.28 years if it is kept under testing or in the operation.

B.4 Uniform Distribution

- Probability density function

$$f_X(x) = \frac{1}{b-a} \qquad a \leq X \leq b$$

- Cumulative distribution function

$$F_X(x) = \int_0^x f_X(x)\,\mathrm{d}x$$

$$F_X(x) = \frac{x-a}{b-a}$$

$$x = a + (b-a)F_X(x)$$

$$\text{Let} \quad \text{Rnd} = \text{Random Number}$$

$$\text{Rnd} = F_X(x)$$

$$x = a + (b-a) \times \text{Rnd}$$

Example B.2 Assume x represents the strength of the material and it varies between 150 and 300 MPa. Then generate random variate for the following information.

$$a = 150\,\text{MPa}$$
$$b = 300\,\text{MPa}$$
$$\text{Rnd} = 0.2364$$

Solution:

$$x = a + (b-a) \times \text{Rnd}$$
$$x = 150 + (300 - 150) \times 0.2364$$
$$x = 150 + 35.46$$
$$x = 185.46\,\text{MPa}$$

- From the above it is calculated that the component strength can take a value of 185.46 MPa.

Normal Distribution

- Probability density function

$$f_X(x) = \frac{1}{\sigma\sqrt{2\pi}} e^{-\frac{1}{2}\left(\frac{x-\mu}{\sigma}\right)^2} \qquad -\infty \le X \le \infty$$

- Cumulative distribution function

$$F_X(x) = \int_{-\infty}^{x} f_X(x)\,dx$$

$$F_X(x) = \Phi\left(\frac{x-\mu_X}{\sigma_X}\right)$$

$$\frac{x-\mu_X}{\sigma_X} = \Phi^{-1}[F_X(x)]$$

$$x = \Phi^{-1}[F_X(x)] \times \sigma_X + \mu_X$$

$$\text{Let} \quad \text{Rnd} = \text{Random Number}$$

$$\text{Rnd} = F_X(x)$$

$$x = \Phi^{-1}[\text{Rnd}] \times \sigma_X + \mu_X$$

Example B.3 Assume x represents the strength of the material and it follows normal distribution. Generate random variate for the following information.

$$\mu_X = 200\,\text{MPa}$$

$$\sigma_X = 10\,\text{MPa}$$

$$\text{Rnd} = 0.2578$$

Solution:

$$x = \Phi^{-1}[\text{Rnd}] \times \sigma_X + \mu_X$$

$$x = \Phi^{-1}[0.2578] \times 10 + 200$$

$$\Phi^{-1}[0.2578] = -0.65014$$

$$x = (-0.65014) \times 10 + 200$$

$$x = 193.4986\,\text{MPa}$$

- From the above it is calculated that the component strength can take a value of 193.4986 MPa.

B.5 Lognormal Distribution

- Probability density function

$$f_X(x) = \frac{1}{\sigma_{\ln X}\sqrt{2\pi}x} e^{-\frac{1}{2}\left(\frac{\ln x - \mu_{\ln X}}{\sigma_{\ln X}}\right)^2} \qquad X \geq 0$$

- Cumulative distribution function

$$F_X(x) = \int_0^x f_X(x)\,dx$$

$$F_X(x) = \Phi\left(\frac{\ln x - \mu_{\ln X}}{\sigma_{\ln X}}\right)$$

$$\frac{\ln x - \mu_{\ln X}}{\sigma_{\ln X}} = \Phi^{-1}[F_X(x)]$$

$$\ln x = \Phi^{-1}[F_X(x)] \times \sigma_{\ln X} + \mu_{\ln X}$$

$$x = e^{\left[\Phi^{-1}(F_X(x)) \times \sigma_{\ln X} + \mu_{\ln X}\right]}$$

$$\text{Let} \quad \text{Rnd} = \text{Random Number}$$

$$\text{Rnd} = F_X(x)$$

$$x = e^{\left[\Phi^{-1}(\text{Rnd}) \times \sigma_{\ln X} + \mu_{\ln X}\right]}$$

Example B.4 Assume x represents the strength of the material and it follows log normal distribution. Generate random variate for the following information.

$$\mu_X = 150\,\text{MPa}$$

$$\sigma_X = 5\,\text{MPa}$$

$$\text{Rnd} = 0.6543$$

Solution:

- In case of lognormal distribution the distribution parameters in the pdf are $\mu_{\ln X}$, $\sigma_{\ln X}$
- These are the parameters of normal distribution (ln X)
- Now one has to convert the lognormal distribution parameters (μ_{lX}, σ_X) to normal distribution parameters to use in the calculations.

- The conversion are given as follows:

$$\mu_{\ln X} = \ln(\mu_X) - \frac{1}{2}\ln\left[\left(\frac{\sigma_X}{\mu_X}\right)^2 + 1\right]$$

$$\sigma_{\ln X} = \sqrt{\ln\left[\left(\frac{\sigma_X}{\mu_X}\right)^2 + 1\right]}$$

$$\mu_{\ln X} = \ln(150) - \frac{1}{2}\ln\left[\left(\frac{5}{150}\right)^2 + 1\right]$$

$$\mu_{\ln X} = 5.01008$$

$$\sigma_{\ln X} = \sqrt{\ln\left[\left(\frac{5}{150}\right)^2 + 1\right]}$$

$$\sigma_{\ln X} = 0.033324$$

$$x = \mathrm{e}^{\left[\Phi^{-1}(\mathrm{Rnd}) \times \sigma_{\ln X} + \mu_{\ln X}\right]}$$

Let $z = \Phi^{-1}(\mathrm{Rnd}) \times \sigma_{\ln X} + \mu_{\ln X}$

$$x = \mathrm{e}^{[z]}$$

$$z = \Phi^{-1}(0.6543) \times 0.033324 + 5.01008$$

$$z = 0.3970 \times 0.033324 + 5.01008$$

$$z = 5.023308$$

$$x = \mathrm{e}^{[z]} = \mathrm{e}^{5.023308} = 151.913\,\mathrm{MPa}$$

- From the above it is calculated that the component strength can take a value of 151.913 MPa.

B.6 Weibull Distribution

- Probability density function

$$f(x) = \frac{\beta}{\alpha}\left(\frac{x}{\alpha}\right)^{\beta-1}\mathrm{e}^{\left[-\left(\frac{x}{\alpha}\right)^{\beta}\right]} \quad X \geq 0$$

- Cumulative distribution function

$$F_X(x) = \int_0^x f_X(x)\,\mathrm{d}x$$

$$F_X(x) = 1 - e^{-\left(\frac{x}{\alpha}\right)^\beta}$$

$$e^{-\left(\frac{x}{\alpha}\right)^\beta} = 1 - F_X(x)$$

$$e^{\left(\frac{x}{\alpha}\right)^\beta} = \frac{1}{1 - F_X(x)}$$

$$\ln\left[e^{\left(\frac{x}{\alpha}\right)^\beta}\right] = \ln\left[\frac{1}{1 - F_X(x)}\right]$$

Let Rnd = Random Number

Rnd $= F_X(x)$

$$\left(\frac{x}{\alpha}\right)^\beta = \ln\left[\frac{1}{\text{Rnd}}\right]$$

$$\ln\left[\left(\frac{x}{\alpha}\right)^\beta\right] = \ln\left[\ln\left(\frac{1}{\text{Rnd}}\right)\right]$$

$$\beta\ln\left[\left(\frac{x}{\alpha}\right)\right] = \ln\left[\ln\left(\frac{1}{\text{Rnd}}\right)\right]$$

$$\ln\left[\left(\frac{x}{\alpha}\right)\right] = \frac{1}{\beta}\ln\left[\ln\left(\frac{1}{\text{Rnd}}\right)\right]$$

$$\text{Let} \quad z = \frac{1}{\beta}\ln\left[\ln\left(\frac{1}{\text{Rnd}}\right)\right]$$

$$\left(\frac{x}{\alpha}\right) = e^z$$

$$x = \alpha\, e^z$$

Example B.5 Assume x represents the time to failure of a component and it follows Weibull distribution. Generate random variate for the following information.

$$\alpha = 2$$
$$\beta = 2.5$$
$$\text{Rnd} = 0.4573$$

Solution:

$$x = \alpha e^{z}$$

where $$z = \frac{1}{\beta}\ln\left[\ln\left(\frac{1}{\text{Rnd}}\right)\right]$$

$$z = \frac{1}{2.5}\ln\left[\ln\left(\frac{1}{0.4573}\right)\right]$$

$$z = -0.09815$$

$$x = \alpha \mathrm{e}^{z}$$

$$x = 2 \times \mathrm{e}^{-0.09815}$$

$$x = 1.81303 \text{ years}$$

- From the above it is calculated that the component can fail after 1.81 years.

Appendix C
Fuzzy Set Theory

C.1 Introduction

The performance of engineering structures and systems is directly affected by uncertainties associated with loads, materials, geometry, boundary conditions, etc. Accounting for these uncertainties has been accomplished through the use of probabilistic design methods. The probabilistic analysis has become an efficient tool for quantification and modeling of uncertainty. The most common approach used to represent uncertainty regarding a quantity, either epistemic or aleatory, is to use probability distributions. However, the application of probabilistic analysis will become difficult in certain situations, especially when one has limited information about the parameters. This is because the probabilistic analysis needs information about the probability density function and its basic parameters (mean, standard deviation, etc.) of the parameter under consideration. However, a large amount of data is required for obtaining the above-mentioned information but it is often not practically possible. In most cases only the maximum and minimum values for uncertain parameter can be obtained and therefore, uncertain parameter can be expressed only with interval between minimum and maximum. Under this condition, one can utilize the concept of fuzzy set theory in which uncertain parameter can be expressed as a fuzzy number.

Fuzzy set theory has been used to model systems that are hard to define precisely. Fuzzy set theory is suited to make decisions in complex systems when the context of the problems is often unclear. It has been commonly used for imprecise information in a nonprobabilistic sense and allows integration of information of various parameters into the modeling and evaluation process. As a methodology, fuzzy set theory incorporates imprecision and subjectivity into the model formulation and solution process. The fuzzy set theory was introduced by Zadeh [2] to deal with problems in which the absence of sharply defined criteria is involved. It has been considered as a modeling language to approximate situations in which fuzzy phenomena and criteria exist. Using the term principle of incompatibility, Dr. Zadeh states "As the complexity of a system increases, our ability to make precise and yet significant statements about its behavior diminishes until a threshold is reached beyond which precision and significance (or relevance) become almost mutually exclusive characteristics" [3].

A.K. Verma et al., *Risk Management of Non-Renewable Energy Systems*,
Springer Series in Reliability Engineering, DOI 10.1007/978-3-319-16062-7

Fuzzy set theory may be regarded as a generalization of classical set theory. In fuzzy approach, the input parameters are treated as a fuzzy number and the variability is characterized by the membership function which can be obtained based on the available information. In classical set theory the membership function of a set is 1 within the boundaries of the set and is 0 outside. A fuzzy set is defined in terms of a membership function which maps the domain of interest. The shape of the curves shows the membership function for each set. The membership function of each fuzzy set can be a triangular, trapezoidal (See Appendix), normal function, etc., and is treated as a possibility distribution. The membership function _*A*(*x*) describes the membership of the elements *x* of the fuzzy set *A* that assigns a value between 0 and 1 that represents the degree of membership of an element *x* to set *A*. The concept of membership function is explained in Fig. C.1 for a trapezoidal fuzzy membership function and the membership function is given as follows:

$$\mu_A(x) = \begin{cases} 0 & x < a \\ \frac{x-a}{b-a} & a \le x \le b \\ 1 & b \le x \le c \\ \frac{d-x}{d-c} & c \le x \le d \\ 0 & x > d \end{cases}$$

The **support** of a fuzzy set *A* is the crisp set that contains all the elements that have nonzero membership grades in *A*.
The **core** of a fuzzy set *A* is the crisp set that contains all the elements that have the membership grades of one in *A*.
The **boundary** is the crisp set that contains all the elements that have the membership grades in between 0 and 1 in *A*.

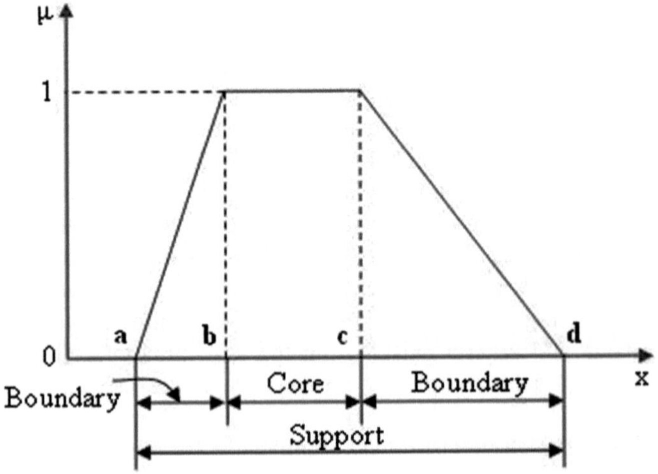

Fig. C.1 Some characteristics of a trapezoidal fuzzy membership function

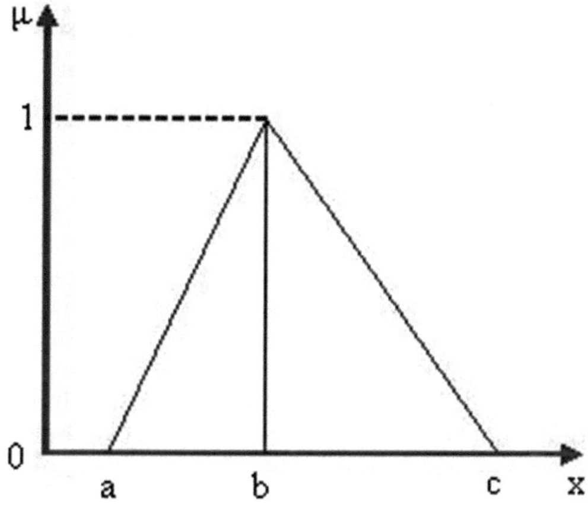

Fig. C.2 Triangular fuzzy membership function

Similarly, the triangular membership function is shown in Fig. C.2 and the membership function is given as follows:

$$\mu_A(x) = \begin{cases} 0 & x < a \\ \frac{x-a}{b-a} & a \leq x \leq b \\ 1 & x = b \\ \frac{c-x}{c-b} & b \leq x \leq c \\ 0 & x > c \end{cases}$$

If the available information is probability distribution, transformation from probability to possibility methods can be used.

C.2 Probability to Possibility Transformation

Consider a random variable *X*, which follows normal distribution with mean and standard deviation as μ and σ respectively. The density function of the variable is given as

$$f_X(x) = \frac{1}{\sigma\sqrt{2\pi}} e^{-\frac{1}{2}\left(\frac{x-\mu}{\sigma}\right)^2} \quad -\infty < X < +\infty$$

Cumulative distribution function of *X* can be given as:

$$F_X(x) = \int_{-\infty}^{x} f_X(x)\,\mathrm{d}x = \Phi\left(\frac{x-\mu}{\sigma}\right)$$

In the fuzzy membership function the maximum membership value is 1. For normal distribution f_{max} will occur at mean value. Hence, the fuzzy membership function for the normal distribution can be given as follows:

$$f_{max} = \frac{1}{\sigma\sqrt{2\pi}}$$

$$M_X(x) = \frac{f}{f_{max}} = e^{-\frac{1}{2}\left(\frac{x-\mu}{\sigma}\right)^2} \quad -\infty < X < +\infty$$

In general, for any distribution one can divide the range of the parameter into very small intervals and one can find the probability of occurrence of the parameter in each interval as follows:

$$P(x_i \le X \le x_i + \Delta x) = F(x_i + \Delta x) - F(x_i) = \Delta F_i$$

Find the ΔF_{max} from the above information. Now membership function for each interval can be written as follows:

$$M_i = \frac{\Delta F_i}{\Delta F_{max}}$$

$$M_i = \frac{1}{\Delta F_{max}} \Delta F_i = Cp_i \quad i = 1, 2, \ldots, n$$

In the above expression C is a constant depends on the distribution. The above expression can be used to transform possibility to probability distribution as follows:

$$M_i = \frac{\Delta F_i}{\Delta F_{max}}$$

$$p_i = \frac{1}{C} M_i \quad i = 1, 2, \ldots, n$$

where

$$C = \sum_{i=1}^{n} M_i \quad \sum_{i=1}^{n} p_i = 1$$

Appendix D
Stochastic Process Theory

D.1 Introduction

A stochastic process can be viewed as a family of random variables. A collection of time functions for a stochastic process is typically called an ensemble. Hence, Ensemble of all the possible time functions (time histories) that might result from the experiment is known as a random process or stochastic process. In stochastic process it is common practice to use braces to denote a set or collection of items. So one can write $\{l(t)\}$ for a stochastic process that gives a random variable $l(t)$ for any particular value of t. The parameter t may be called the index parameter for the process, and the set of possible t values is then the index set. The basic idea is that for every possible t value there is a random variable $l(t)$.

Figure D.1 illustrates the idea of a statistical sample, or ensemble, from a stochastic process, using the notation ${}^{i}l(t)$ for the ith sample time history observed for the process. It shows five time histories observed in separate, independent observations of the particular $\{l(t)\}$ process. A section across this ensemble at any particular time gives a statistical sample for the random variable corresponding to that t value.

D.2 Probability Density Function

One of the objectives of the analysis of a stochastic process is to determine the probability of finding extreme or peak values, or of determining the percentage of time that a random variable exceeds a given value. Figure D.2 shows a sample of a random signal with respect to time defined over duration T.

A.K. Verma et al., *Risk Management of Non-Renewable Energy Systems*,
Springer Series in Reliability Engineering, DOI 10.1007/978-3-319-16062-7

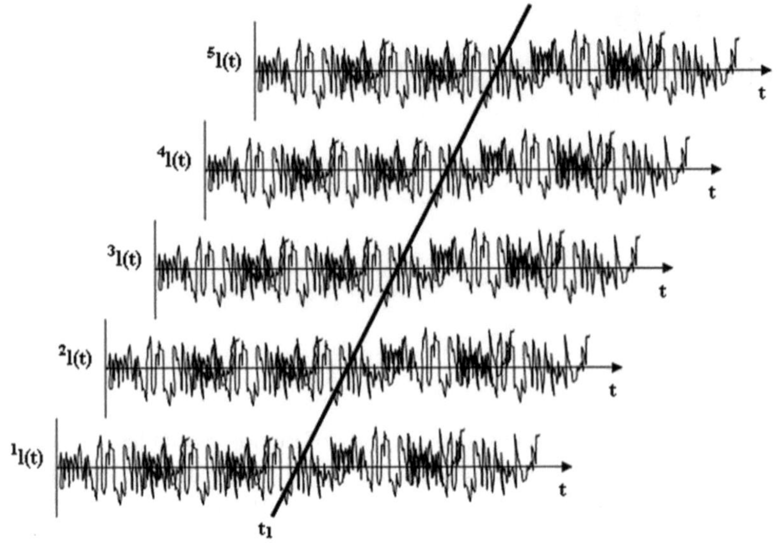

Fig. D.1 Ensemble of time histories of $l(t)$

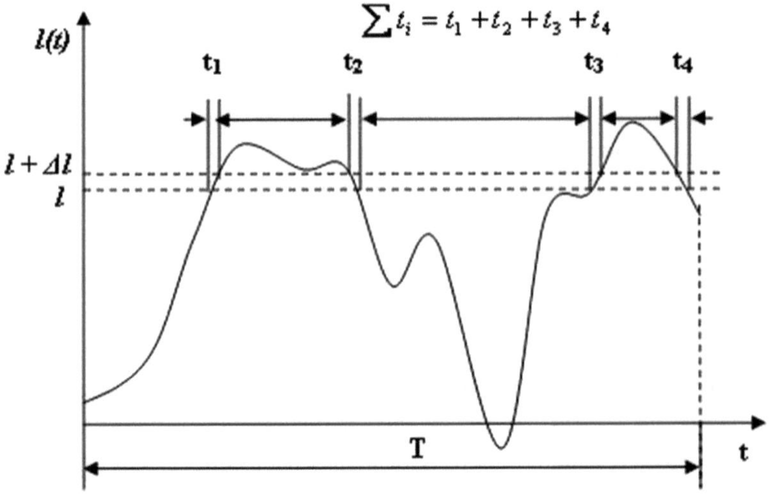

Fig. D.2 Sample of random signal

The probability that this function $l(t)$ is in the interval l, $l + \Delta l$ is equal to the percentage of time during which it has values in this interval. This probability can be expressed mathematically as follows:

$$P[l < l(t) < l + \Delta l] = \frac{\sum_i t_i}{T}$$

If this interval Δl is small, a density function probability $p(l)$ is defined as

$$P[l < l(t) < l + \Delta l] = p(l)\Delta l$$
$$p(l) = \frac{1}{T}\frac{\sum_i t_i}{\Delta l}$$

To precisely define $p(l)$, it is necessary to consider very small intervals Δl and of very long duration T, so that mathematically, the probability density function can be defined as

$$P[l < l(t) < l + \Delta l] = p(l)\Delta l$$
$$p(l) = \lim_{\Delta l \to 0}\left[\lim_{T \to \infty}\left(\frac{1}{T}\frac{\sum_i t_i}{\Delta l}\right)\right]$$

D.2.1 Distribution Function

The probability that the signal is inside the limits $a < l(t) < b$ is obtained by integration of $p(l)$ and is given as follows:

$$P[a < l(t) < b] = \int_a^b p(l)\,\mathrm{d}l$$

Since the probability that $l(t)$ within the limits $-\infty$, $+\infty$ is equal to 1, it follows that

$$\int_{-\infty}^{\infty} p(l)\,\mathrm{d}l = 1$$

and the probability that l exceeds a given level L is simply

$$P[L \leq l(t)] = 1 - \int_{l(t)}^{\infty} p(l)\,\mathrm{d}l$$

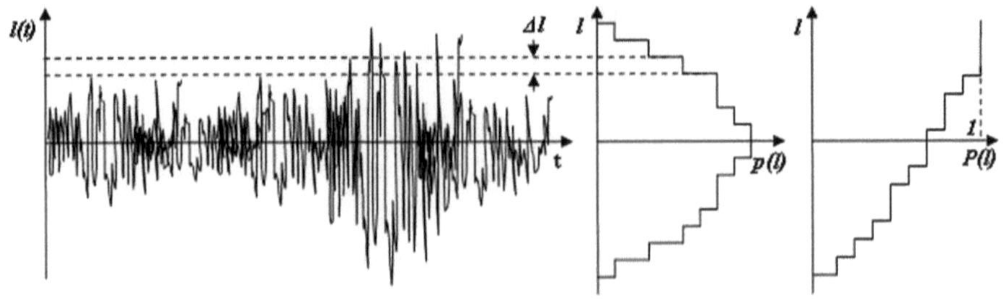

Fig. D.3 Distribution of instantaneous values of the signal

Figure D.3 shows how one passes from the signal $l(t)$ to the probability density and the distribution function.

D.2.2 Gaussian Random Process

A Gaussian random process $l(t)$ is one such that the ensemble of the instantaneous values of $l(t)$ obeys a law of the form:

$$p[l(t)] = \frac{1}{\sigma\sqrt{2\pi}} \mathrm{e}^{-\frac{1}{2}\left(\frac{l(t)-\mu}{\sigma}\right)^2}$$

where μ and σ are constants.

A Gaussian process is fully determined by knowledge of the mean value μ and of the standard deviation σ.

D.3 Ensemble Averages: Through the Process

D.3.1 'n' Order Average

Let us consider N recordings of a random phenomenon varying with time ${}^{i}l(t)$ [i€(1, N)] for t varying from 0 to T (Fig. D.4). The ensemble of the curves ${}^{i}l(t)$ constitutes the process $\{{}^{i}l(t)\}$. If one has (N) records of the phenomenon, one can calculate, for a given t_1 the mean:

$$\bar{l}(t) = \frac{{}^{1}l(t_1) + {}^{2}l(t_1) + \cdots + {}^{n}l(t_1)}{N}$$

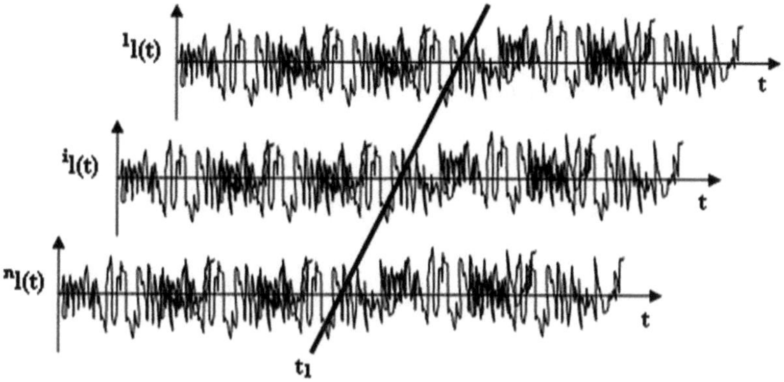

Fig. D.4 Through the process study

If the values ${}^{i}l(t)$ belong to an infinite discrete ensemble, the moment of order n is defined as

$$E[l^n(t_1)] = \lim_{N \to \infty} \sum_{i=1}^{N} \frac{{}^{1}l^n(t_1)}{N}$$

By considering the ensemble of the samples at the moment t_1, the statistical nature of $l(t_1)$ can be specified by its probability density:

$$p[l(t_1)] = \lim_{\Delta t \to 0} \frac{P[l \leq l(t_1) \leq l + \Delta l]}{\Delta l}$$

and by the moments of the distribution:

$$E[l^n(t_1)] = \int_{-\infty}^{\infty} l^n(t_1)\, p[l(t_1)]\, \mathrm{d}l(t_1)$$

if the density $p[l(t_1)]$ exists and is continuous. The moment of order 1 is the mean or expected value. The moment of order 2 is the quadratic mean.

D.3.2 Central Moments

The central moment of order n (with regard to the mean) in the case of a discrete ensemble can be written as

$$E\{[l^n(t_1) - m]^n\} = \lim_{N\to\infty} \sum_{i=1}^{N} \frac{[^i l(t_1) - m]^n}{N}$$

In the case of continuous density function it can be written as follows:

$$E\{[l(t_1) - m]^n\} = \int_{-\infty}^{\infty} [l(t_1) - m]^n \, p[l(t_1)] \, dl(t_1)$$

The variance is the central moment of order 2.

D.3.3 Autocorrelation Function

Given a random process $^i l(t)$, the autocorrelation function is the function defined, in the discrete case, by:

$$R(t_1, t_1 + \tau) = \lim_{N\to\infty} \frac{1}{N} \sum_i {}^i l(t_1) {}^i l(t_1 + \tau)$$

$$R(t_1, t_1 + \tau) = E[l(t_1), l(t_1 + \tau)]$$

and, for a continuous process, by:

$$R(\tau) = \int_{-\infty}^{\infty} l(t_1) l(t_1 + \tau) \, p[l(t_1)] \, \mathrm{d}l(t_1)$$

D.3.4 Cross-Correlation Function

Given the two processes $\{l(t)\}$ and $\{u(t)\}$ (for example, the excitation and the response of a mechanical system), the cross-correlation function can be given as:

$$R_{lu}(t_1, t_1 + \tau) = E[l(t_1), u(t_1 + \tau)]$$

The correlation is a number measuring the degree of resemblance or similarity between two functions of the same parameter (time generally).

D.3.5 Autocovariance

Autocovariance is given by

$$C(t_1, t_1 + \tau) = E[l(t_1) - \bar{l}(t_1)][l(t_1 + \tau) - \bar{l}(t_1 + \tau)]$$
$$C(t_1, t_1 + \tau) = R(t_1, t_1 + \tau) - \bar{l}(t_1)\bar{l}(t_1 + \tau)$$
$$C(t_1, t_1 + \tau) = R(t_1, t_1 + \tau) \quad \text{(If the mean values are zero)}$$

D.3.6 Covariance

One defines the covariance as the quantity:

$$C_{lu} = E[l(t_1) - \bar{l}(t_1)][u(t_1 + \tau) - \bar{u}(t_1 + \tau)]$$

D.4 Stationarity

A phenomenon is strictly stationary if every moment of all orders and all the correlations are invariable with time t_1. If only one recording of the phenomenon $l(t)$ is available, one defines sometimes the autostationarity of the signal by studying the stationarity with n samples taken at various moments of the recording, by regarding them as samples obtained independently during n measurements (Fig. D.5). For a stationary process, the autocorrelation function can be written as

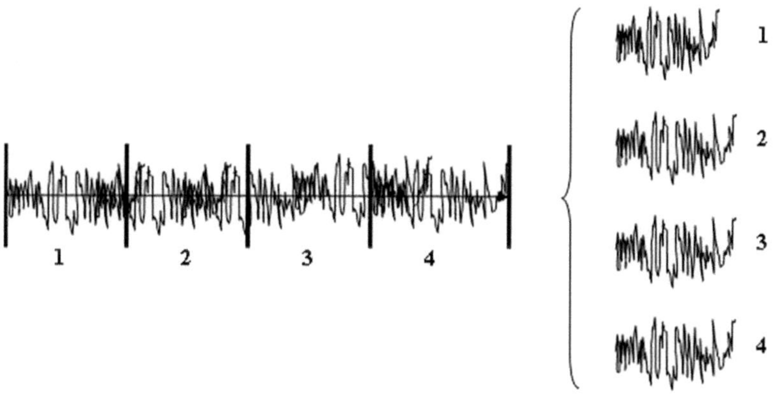

Fig. D.5 Study of autostationarity

$$R(\tau) = E[l(0)l(\tau)]$$

$$R(\tau) = \lim_{N\to\infty} \frac{1}{N}\sum_{i=1}^{N} {}^{i}l(0)^{i}l(\tau)$$

$$R(\tau) = E[l(0)l(0)] = E\left[l^2(t)\right]$$

Similarly, the above discussed terms can be defined along the process (only one sample is available). This is called as temporal averages.

D.5 Ergodicity

A process is known as ergodic if all the temporal averages exist and have the same value as the corresponding ensemble averages calculated at an arbitrary given moment. A ergodic process is thus necessarily stationary. In practice, one simply calculates the temporal averages by making the assumption that the process is stationary and ergodic.

References

1. Myers RH, Montgomery DC (2002) Response surface methodology: process and product optimization using designed experiments, 2nd edn. Wiley, USA
2. Zadeh LA (1965) Fuzzy sets, Inf Control 8(3):338–353
3. Zadeh LA (1973) Outline of a new approach to the analysis of complex systems and decision processes, IEEE Trans Syst Man Cybern 3(1)

MIX
Papier aus verantwortungsvollen Quellen
Paper from responsible sources
FSC® C105338

If you have any concerns about our products,
you can contact us on
ProductSafety@springernature.com

In case Publisher is established outside the EU,
the EU authorized representative is:
Springer Nature Customer Service Center GmbH
Europaplatz 3, 69115 Heidelberg, Germany

Printed by Libri Plureos GmbH
in Hamburg, Germany